CATALOGUE

OF

FOSSIL MAMMALIA.

PART V.

Richard Lydekker

Catalogue of the Fossil Mammalia in the British Museum Natural History

Part V

Containing the Group Tillodontia
The Orders Sirenia, Cetacea, Edentata,
Marsupialia, Monotremata
and Supplement

Elibron Classics
www.elibron.com

Elibron Classics series.

ISBN 1-4021-6985-X (paperback)
ISBN 1-4021-1933-X (hardcover)

This Elibron Classics Replica Edition is an unabridged facsimile of the edition published in 1887 by Taylor and Francis, London.

CATALOGUE

OF THE

FOSSIL MAMMALIA

IN THE

BRITISH MUSEUM,

(NATURAL HISTORY)

CROMWELL ROAD, S.W.

PART V.

CONTAINING

THE GROUP TILLODONTIA, THE ORDERS SIRENIA, CETACEA, EDENTATA, MARSUPIALIA, MONOTREMATA, AND SUPPLEMENT.

BY

RICHARD LYDEKKER, B.A., F.G.S., ETC.

LONDON:
PRINTED BY ORDER OF THE TRUSTEES.
1887.

PRINTED BY TAYLOR AND FRANCIS,
RED LION COURT, FLEET STREET.

PREFACE.

It is no small satisfaction to be able to announce the completion of the fifth and final part of this Catalogue, of which the first was published less than three years since.

Of its utility to the student of Mammalogy there can be no question. If it contained a bare list of specimens with their localities and references to the works in which they are described, it would be of much value; but when, as is often the case, both distinctive characters and descriptions are given in addition, it becomes a still more important work of reference, which must be indispensable in every Biological library.

The Collection, under the excellent skill of Mr. William Davies, assisted by Mr. Arthur Smith Woodward, has been revised, arranged, and named, so as to correspond in order, as nearly as the exigencies of exhibition-space will permit, with the several suborders and families proposed in this Catalogue, references being everywhere affixed to figured and described specimens.

To these Assistants and to Mr. Lydekker my best thanks are due for the admirable manner in which they have cooperated to complete the cataloguing and arrangement of this important section of the Palæontological Collection.

HENRY WOODWARD.

British Museum
(Natural History),
Department of Geology,
August 1, 1887.

INTRODUCTION.

WITH the present part the List of Fossil Mammalia now represented in the collection of the Museum is completed [1]. The Supplement issued herewith comprises all specimens that have been acquired since the date of publication of the parts in which they should properly have been recorded, as well as some of which the affinity or serial position had not been determined at the time when such parts were written, and also certain others which after the transference of the collection from Bloomsbury had been deposited in drawers out of their proper serial position, and thus escaped notice. The specimens are recorded in about 9820 entries, but many of such entries include more than a single specimen. The named species (omitting one or two which have been subsequently found to be synonyms) are 719 in number, and are arranged under 301 generic and 100 family headings—106 out of the total being regarded as indistinguishable from existing forms. Allowing for the above-mentioned cases of synonymy (which are noticed in the sequel), it may be affirmed that among the better-known forms the list of genera and species does not err on the side of redundancy, the writer having endeavoured as far as possible in this respect to follow the lead of the Director of the Museum in not recognizing such as are founded upon trivial characters [2]. In imperfectly known forms, however, this rule cannot be enforced, and it may eventually be found that among the smaller forms, and in groups like the Tertiary

[1] The specimen belonging to the so-called *Antilope torticornis*, Aymard, from the Pliocene of Auvergne, noticed by Rütimeyer in his "Rinder der Tertiär-Epoche," Abh. schw. pal. Ges., pp. 84–85, as being in the Museum, cannot be identified with any specimen in the collection.

[2] See Preface to 'List of Cetacea in the British Museum' (1885).

Phocidæ and *Trichechidæ* and the Physeteroid Cetacea, reductions in these respects will be necessary[1]. There is of course the danger that the principle in question may be carried to an extreme, and it will perhaps prove that in some instances the remains of more than a single species have been included under one name; this, however, the writer regards as a less serious error than its reverse.

It may be observed that this work, as its title implies, is merely a Catalogue of the species and genera represented in the Collection of the Museum, and consequently that it does not embrace all the known Fossil Mammalia. I am indeed well aware how valuable a work of the latter nature would be if it could be made reliable and complete; but when I consider the number of names that have been relegated to the rank of synonyms from the examination of the specimens in the Museum, and the great uncertainty still existing as to the generic and specific determination of an immense number of the forms unrepresented in the Collection, I am convinced that the time has not yet arrived when such a work could be profitably undertaken. This is indeed well illustrated at the present time in the case of the Mammals of the infra-Pampean formation[2] of South America, whose history is now undergoing a phase very like that experienced in the case of those of North America some few years ago. Thus, in the Bulletins of the Academy of Sciences of Cordova, Signor Ameghino proposes a host of new generic and specific names for the Mammals from this formation; but as these memoirs are without illustrations, it is often totally impossible to form any conclusion as to the validity of these generic and specific terms; of which Dr. Burmeister, of Buenos Ayres, considers a large proportion are not entitled to stand[3].

In respect of classification I have endeavoured throughout not to form any entirely new scheme, and have in the main followed the

[1] If I had been describing the remains of these forms *de novo* I certainly should not have made the number of genera and species that have been founded; but as the specimens at present available are not sufficient to say decisively that some of these forms are equivalents of others, the only course has been to refer the specimens to the genera and species to which they appear to belong.

[2] The so-called *Eutemnodus* mentioned in pt. i. p. 21, note 5, and the *Macrauchenia* in pt. iii. p. 16, are from this formation, and are therefore older than the Pleistocene Pampean. Dr. Burmeister refers the infra-Pampean to the Pliocene, while Signor Ameghino, on the other hand, regards it as representing the European Oligocene.

[3] Another difficulty occurs with regard to existing species, all of which probably date back to the Pleistocene, although only a moiety have been actually recorded as fossils.

general plan adopted by Prof. Flower[1], although I have found it necessary to introduce some modifications; and in cases where good working classifications of particular groups have been prepared by specialists, such have frequently been followed. Thus in the Selenodont Artiodactyla the classification of Prof. Rütimeyer[2], of Basle, has been adopted, while for the Perissodactyla Prof. Cope[3] is the authority. A new scheme for the Artiodactyla and Perissodactyla has recently been proposed by Dr. Schlosser[4] of Munich, to which the reader's attention may be directed. This scheme adopts a phylogenetic system of families, and therefore differs very widely from the systems generally employed, which run, so to speak, transversely to such phylogenetic stems[5]. I have not seen the necessity for adopting the order Bunotheria of Prof. Cope, since it appears to me that their general primitive nature is the only bond of union between the very diverse forms which are included in it; and in this view I have the support both of Prof. Flower and Dr. Schlosser. Neither have I retained Professor Marsh's orders Pantotheria and Allotheria for the Mesozoic Mammalia; the majority of which appear so nearly related to existing Marsupials that it has been a question whether some of them should not be included in the modern families. In his latest memoir Prof. Marsh[6] comes to the conclusion that the *Allotheria*, which comprehend the Diprotodont genera, should be regarded as at least a suborder of Marsupialia; while the *Pantotheria*, or Polyprotodont genera, are considered as being probably placental[7].

[1] In the 9th edition of the 'Encyclopædia Britannica,' in the Catalogue of the Mammalia in the Museum of the Royal College of Surgeons, and in a paper published in the 'Proceedings of the Zoological Society' for 1883.

[2] Verh. nat. Ges. Basel, vol. vii. art. 2, pp. 29–60 (1883). In that memoir the writer did not propose family names for the various groups, and it has been necessary to accord as well as possible with such grouping.

[3] Proc. Amer. Phil. Soc. vol. xix. pp. 377–403 (1881). Amendments proposed by the author of this memoir in later papers have been adopted.

[4] Morphol. Jahrb. vol. xii. pp. 1–136 (1886).

[5] See Cope, Amer. Nat. vol. xx. p. 720 (1886).

[6] Amer. Journ. ser. 3, vol. xxxiii. p. 345 (1887).

[7] The characters given by Prof. Marsh (*op. cit.*) as distinctive of the Pantotheria do not differentiate them from the Marsupialia, and are not all applicable to many of the forms. Thus in some examples of *Perameles* and *Chœropus* the canine has either a double or a distinctly grooved root (*infrà*, p. 255); while many of the English Mesozoic genera show a most distinct inflection of the angle of the mandible. The dental succession of *Triconodon* (*infrà*, p. 257) appears both to Mr. O. Thomas and myself to be absolutely conclusive evidence of its marsupial affinities.

The necessity of adapting the sequence of the different orders to the requirements of the parts into which the work is divided has entailed some departure from an ideal scheme, and the following table of the orders and suborders arranged in the manner which appears to best represent their mutual relationships is accordingly appended:—

Subclass I. EUTHERIA.

Order I. Primates.
Suborder 1. Anthropoidea.
2. Lemuroidea.

Order II. Chiroptera.
Suborder 1. Megachiroptera.
2. Microchiroptera.

Order III. Insectivora.
Suborder 1. Dermoptera.
2. Insectivora Vera.

Order IV. Carnivora.
Suborder 1. Carnivora Primigenia (or Creodonta).
2. Carnivora Vera.
3. Carnivora Pinnipedia.

Order V. Rodentia.
Suborder 1. Simplicidentata.
2. Duplicidentata.

Group *Tillodontia*.

Order VI. Ungulata.
Suborder 1. Proboscidea.
2. Amblypoda.
3. Hyracoidea.
4. Condylarthra.
5. Toxodontia.
6. Perissodactyla.
7. Artiodactyla.

Order VII. Sirenia.

Order VIII. Cetacea.
Suborder 1. Mystacoceti.
2. Archæoceti.
3. Odontoceti.

Order IX. Edentata.

Subclass II. METATHERIA.

Order X. Marsupialia.

Suborder 1. Diprotodontia.

2. Polyprotodontia.

Subclass III. PROTOTHERIA.

Order XI. Monotremata.

At the commencement of the work I had no intention of making it a descriptive one, and proposed to confine myself to giving notes on such specimens, species, or groups as appeared to require them. As the work proceeded, however, it seemed advisable, in the case of many groups, to notice some of the more important distinctive characters available in the case of fossils; and in the later parts, where I have found it necessary to make some considerable emendations, both in specific and generic characters, the work has become to a considerable extent descriptive, although it has not been thought necessary to make it entirely so.

In a work of this nature, it is unfortunately almost impossible to avoid certain clerical and other slight errors; but since I have had occasion in the course of another work to verify the dental formulæ and descriptive paragraphs, I believe all such errors have been detected.

In a science like Palæontology it is inevitable that before a work like the present can be completed, some of its earlier portions require revision; and I accordingly take this opportunity of directing attention to a few instances where either amendments of nomenclature or of individual determinations are advisable, or where it is desirable to state the grounds on which such determinations rest. Instances are also mentioned where specimens have been figured since the earlier parts were published.

Among the Primates the type specimens of *Semnopithecus palæindicus* (pt. i. p. 2) and *Cynocephalus subhimalayanus* (p. 4) have been figured in ser. 10, vol. iv. pl. i. of the 'Palæontologia Indica;' while a specific name has been assigned to the second *Cynocephalus* noticed on the latter page, under which it is entered in the Supplement. Among the Insectivora, two erroneous identifications made on the authority of Dr. Fraas[1] require correction, while attention should also be directed to some observations of Dr. Filhol which had escaped notice. In the first place, *Erinaceus arvernensis* of Blainville (pt. i. pp. 18, 19), which is the type of *Amphechinus* of Aymard, and is identified by Dr. Fraas with *Parasorex socialis*, is

[1] Fauna von Steinheim, p. 4.

regarded by Dr. Filhol[1] as belonging to the *Erinaceidæ*. In describing *Palæoerinaceus edwardsi* from St. Gérand-le-Puy, the latter writer also states[2] that the mandibles entered in pt. i. p. 17 under the name of *Erinaceus arvernensis*, Gervais (non Blainv.), "sont absolument semblables à ceux que j'ai pu étudier provenant de Saint Gérand;" although in the next paragraph he proceeds to observe that "je ne crois pas qu'on puisse essayer de confondre le fossile trouvé à Saint Gérand avec ceux découverts dans les terrains de Cournon et des Chouffours, qui constituent des horizons géologiques différents." Now since the types of *E. arvernensis*, Gervais, are from Cournon, it is very difficult to reconcile these two statements; but if they be really the same as *Palæoerinaceus edwardsi*, they must be referred to that genus, although the earlier specific name should be retained. *Palæoerinaceus*, it may be observed, differs from *Erinaceus* by the greater width of the palate, which is without vacuities. In the second place, the identification by Dr. Fraas in the passage cited of *Plesiosorex soricinoides* (*Erinaceus soricinoides*, Blainville) with *Parasorex socialis*, Meyer, is erroneous, and the specimens entered in pt. i. p. 19 under the former name are accordingly re-entered in the Supplement under the latter.

In the Carnivora, a mandible of *Machærodus* from the Norfolk Forest-bed, of which a cast is entered in the Supplement, has afforded strong reasons for doubting the correctness of the identification by Kaup and Gaudry of the Eppelsheim and Pikermi form with the Val d'Arno *M. cultridens*. *Hyæna antiqua* (pt. i. p. 87) is in all probability a synonym of *H. striata* (*vide* Supplement, p. 315). The remarkable humerus and tibia (Nos. M. 1710–1) from the Quercy Phosphorites, entered on p. 148 of the same part, and provisionally referred (after Dr. Filhol) to a Canoid, have been found to agree so closely with the corresponding bones of the Condylarthra[3], as to suggest the probability of their belonging to that suborder. The only known Condylarthrous genus from those deposits is *Hyracodontherium*[4], which has been shown by Dr. Schlosser[5] to be nearly allied to *Meniscotherium*[6]; but the specimens in question appear too large for the one known species of the European genus. In the Rodentia it should have been

[1] Ann. Sci. Géol. vol. x. art. 3, pp. 20–21.

[2] *Op. cit.* p. 21.

[3] See Cope, Amer. Nat. vol. xviii. p. 799, fig. 7, and p. 904, fig. 27 (1884).

[4] See Filhol, Ann. Sci. Géol. vol. viii. pl. xiii. figs. 283–4 (1877).

[5] Morphol. Jahrb. vol. xii. p. 22 (1886).

[6] Schlosser (*op. cit.*) refers these genera to the Perissodactyla, a view which is not accepted by Cope (Amer. Nat. vol. xx. p. 721 [1886]).

mentioned that *Castoroides* (pt. i. p. 221) is regarded by Coues and Allen[1] as forming the type of a distinct family, which is placed in the Hystricomorpha. Alston's identification of *Arvicola ambiguus* (pt. i. p. 231) with *Myodes* (or *Cuniculus*) *torquatus* had escaped notice, and the specimens are therefore re-entered under the latter name in the Supplement. Among the Artiodactyla the type crania of the Siwalik Antelopes referred in pt. ii. to *Strepsiceros*, *Hippotragus*, *Cobus*, and *Alcelaphus*, have been figured in the 'Palæontologia Indica,' ser. 10, vol. iv. pls. ii.–iv. (1886); and *Alcelaphus bakeri* (pt. ii. p. 56) has been found to be identical with *A. palæindicus*[2]. *Metriotherium*, incidentally mentioned in part ii. p. 60, more probably belongs to the *Anoplotheriidæ*. *Cænotherium quinquedentatum* (pt. ii. p. 175) is regarded by Dr. Schlosser[3] as identical with *C. courtoisi*, Gervais, which was incorrectly referred by Kowalevsky[4] to *Hyopotamus*. The specimen provisionally referred in the second part (p. 182) to *Xiphodon gelyensis* is made the type of a new species in the Supplement, since Dr. Schlosser[5] has shown that the so-called *X. gelyensis* does not belong to *Xiphodon*. The same writer suggests[6] that *Diplopus* (pt. ii. p. 216) is identical with *Dichodon*, but the bones of the former appear decidedly too large for the one known species of that genus. In the genus *Merycopotamus*, since the publication of pt. ii., the following specimens, viz. *M. dissimilis* (p. 209), Nos. 18441, 18442, and *M. nanus* (p. 211), Nos. 16551, 18407, and 153409, have been figured by the writer in the 'Palæontologia Indica,' ser. 10, vol. iv. pls. v., vi. In the Perissodactyla, the North-American forms originally referred by Prof. Cope to *Pachynolophus* (pt. iii. p. 13) have been made the types of the genus *Heptodon*, Cope[7]. In the Toxodontia it has been mentioned by Messrs. Gervais and Ameghino[8] that the cranium of *Toxodon*, No. 49197 (pt. iii. p. 168), in having the first pair of incisors larger than the second, differs from the type specimen of *T. platensis*, and agrees with that known as *T. burmeisteri*, Giebel[9]; and the same feature is also shown in No. 19948. The mandible

[1] "Monogr. North American Rodentia," Rep. U.S. Geol. Surv. Terrs. p. 419 (1877).

[2] Palæontologia Indica, *op. cit.* p. 14.

[3] *Op. cit.* p. 54.

[4] Phil. Trans. 1873, p. 25.

[5] *Op. cit.* p. 297.

[6] *Ibid.* p. 298.

[7] See "Vertebrata of the Tertiaries of the West," Rep. U.S. Geol. Surv. Terrs. vol. iii., Book i. p. 653 (1884).

[8] Mammifères Fossiles de l'Amérique Méridionale, p. 83 (1880).

[9] Zeitschr. ges. Naturwis. vol. xxviii. p. 134 (1866).

No. 19949 agrees, however, in the character of its incisors with the one referred by Burmeister[1] to *T. platensis*, but in having the canine separated by a diastema from $\overline{i.\ 3}$ resembles that of *T. burmeisteri*; and accordingly Sir R. Owen's reference of all the specimens to the type species has been followed. It may also be observed that whereas Dr. Burmeister describes *T. burmeisteri* as considerably larger than *T. platensis*, No. 49197 is much smaller than the type of the latter. The comparison of a considerable series of specimens is therefore required to determine whether the variation in the form of the incisors of *T. burmeisteri* is a character of more than individual value.

It had escaped notice that the molar of *Rhinoceros antiquitatis* figured in part iii. p. 93, as well as that of *Mastodon angustidens*, No. M. 2900 (pt. iv. p. 35), are figured by Grew in his 'Catalogue of the Rarities of Gresham College,' pl. xix. (1761); and it thus appears that both specimens were probably presented to the Museum by the Council of the Royal Society. The mandible of *Hipparion gracile*, No. 248 (pt. iii. p. 53), is figured by Kaup in the 'Nova Acta Ac. Cæs. Leop.-Car.' vol. xvii. pl. xii. B. fig. 3. The distribution of the Indian Pliocene and Pleistocene Elephants to the eastward has been more fully worked out by Prof. K. Martin[2] since the publication of pt. iv. In this memoir *E. bombifrons*, *E. clifti*, *E. namadicus*, and *E. hysudricus* are recorded definitely from Java; while the name *E. (Stegodon) trigonocephalus* is applied to two crania, from the same island, with a dentition of the type of that of *E. insignis*, and it is suggested that the Javan specimens noticed in pt. iv. p. 90 may belong to the same form rather than to *E. insignis*. The Indian *Sus hysudricus* is also recorded from the same region.

Whenever practicable the references to the authorities for the generic and specific names have been verified after they were in type, and I have much pleasure in expressing my debt to Mr. B. B. Woodward for his aid in this respect. I am also indebted to Mr. O. Thomas for much valuable assistance and advice in regard to the classification and nomenclature of the Marsupialia, and I may observe that it was only after full consultation with him that the generic terms *Triacanthodon*, *Peraspalax*, and probably *Peralestes* were relegated to the rank of synonyms. Mr. W. Davies has again

[1] An. Mus. Buenos Aires, vol. pls. ix.–xi. *T. Oweni*, Burmeister, is apparently identical with the typical *T. platensis*.

[2] "Beitr. Geol. Ost-Asiens u. Austral," Samml. geol. Reichs-Mus. Leiden, vol. iv. pt. 2 (1887).

been good enough to undertake the laborious task of checking the entries of each specimen, and has also, as before, given me the advantage of his great personal knowledge of the history of individual specimens. Thanks are due to Mr. Medlicott, late Director of the Geological Survey of India, to Prof. E. D. Cope, to Prof. H. A. Nicholson and Messrs. Blackwood, to the Councils of the Geological, Zoological, and Palæontographical Societies, as well as to the authors of various memoirs, for permission to use woodcuts from their publications.

The *Beckles Collection*, mentioned for the first time in this part, was purchased in 1876 from Mr. S. H. Beckles of Hastings.

RICHARD LYDEKKER.

The Lodge, Harpenden,
Hertfordshire.
June 10, 1887.

POSTSCRIPT.

Since the above was sent to press the abstract of a memoir by Dr. Osborn, of Princeton, on Mesozoic Mammalia (Proc. Ac. Nat. Sci. Philad. June 21, 1887), has reached me. In this memoir the genus *Amphitylus* is founded on the undetermined specimen, No. M. 2299, p. 272, of the text, and *Athrodon* upon the unnamed maxillæ catalogued on p. 291; the latter being made the type of a new family, which is regarded as allied to the *Phascolomyidæ*. The maxilla of *Bolodon* is figured, and the mandible No. 36822 mentioned on p. 274 also figured and definitely referred to *Amphitherium*. An emendation on the classification of Prof. Marsh is proposed. I may observe that I cannot consider it proved that the maxillæ on which *Athrodon* is founded do not belong to one of the genera of which only the mandible is described, although I am not prepared to say which; and I may add that the resemblance of the molars of *Spalacotherium* and other forms to those of *Chrysochloris* is analogous to that existing between the corresponding teeth of *Perameles* and *Tupaia*, and does not therefore appear to afford grounds for the reference of the former to the Insectivora. The name *Diplocynodon*, Marsh, is preoccupied by Pomel for a genus of *Crocodilidæ*.

In addition to the above, the following memoirs would have been noticed in the text had copies of them reached this country in time.

These comprise a memoir on Cetacea from the Belgian Crag, by Prof. Van Beneden, in the 'Ann. Mus. R. Hist. Nat. Belg.' vol. xiii. pt. 5 (1886), containing figures and descriptions of *Amphicetus*, *Heterocetus*, *Mesocetus*, *Idiocetus*, and *Isocetus*; one by Dr. W. B. Scott, "On some New and Little-known Creodonts," in the 'Journ. Ac. Nat. Sci. Philad.' vol. ix. pt. 2 (1886); and a third, by Dr. Burmeister, forming a continuation of the 'Atlas de la Description Physique de la République Argentine,'—Osteologie der Gravigraden, pt. i. *Scelidotherium* and *Mylodon* (Buenos Ayres, 1886)[1]. In the latter (p. 103) the MS. name *Scelidotherium magnum* is revived for *S. tarijense* (*infrà*, p. 102); while the new species *Mylodon oweni*[2] (p. 111) may be identical with *M. lettsomi* (*infrà*, p. 108). The cranium represented in pl. xiv. figs. 2, 3, as *Scelidotherium leptocephalum* appears to belong to *S. bravardi* (*infrà*, p. 96), and shows (as I thought to be the case) that the length of the nasals is less than half that of the entire cranium.

July 26, 1887. R. L.

[1] This is the portion of the Atlas alluded to by myself in the 'Proc. Zool. Soc.' 1886, p. 491; its publication having apparently been long delayed.

[2] The names applied to these forms by Bravard can hardly be reckoned as more than MS. ones.

SYSTEMATIC INDEX.

[Existing species are denoted by an asterisk.]

Family MACROPODIDÆ (*continued*).

SUPPLEMENT.

Families and species mentioned in the earlier parts are marked †.

LIST OF WOODCUTS.

ABBREVIATIONS OF SERIALS QUOTED IN PART V.,

AND NOT GIVEN IN THE LISTS IN THE PRECEDING PARTS.

[Where not otherwise stated, the works are in 8vo.]

Abh. mittelrhein. geol. Vereins.—Abhandlungen der mittelrheinische geologischen Vereins. *Darmstadt.*

Act. Ac. R. Sci. Bordeaux.—Actes de l'Académie Royale (Nationale) des Sciences, Belles-Lettres et Arts de Bordeaux. *Bordeaux.*

An. Mus. Buenos Aires.—Vol. III. bears the title quoted in part iii.; the earlier are entitled Anales do Museo Publico.

Atti R. Ac. Linc.—Atti della Reale Accademia dei Lincei. 4to. *Rome.*

Bol. Ac. Nac. Cordova.—Boletin de la Academia Nacional de Ciencias de Cordova (Cordoba). *Buenos Ayres.*

Bull. Ac. Imp. St. Pétersbourg.—Bulletin de l'Académie Impériale des Sciences de St. Pétersbourg. 4to. *St. Petersbourg.*

Ext. Proc.-Verb. Ac. Montpellier.—Académie des Sciences et Lettres de Montpellier: Extraits des Procès-Verbaux des Séances. *Paris.*

Giorn. I. R. Ist. Lombardo.—Giornale dell' I. R. Istituto Lombardo di Scienze, Lettere ed Arti. *Milan.*

London Geological Journal.—The London Geological Journal: Charlesworth. *London,* 1846–47.

Lond. and Edin. Philos. Mag.—The London and Edinburgh Philosophical Magazine and Journal of Science. *London.*

Mém. Ac. Montpellier.—Mémoires de l'Académie des Sciences et Lettres de Montpellier. *Montpellier.*

Mém. A. Sci. Dijon.—Mémoires de l'Académie des Sciences, etc., de Dijon.—Partie des Sciences. *Dijon.*

Mem. Ac. Sci. Ist. Bologna.—Memorie dell' Accademia delle Scienze dell' Istituto di Bologna. 4to. *Bologna.*

Mem. R. Ist. Veneto.—Memorie del Reale Istituto Veneto di Scienze, etc. 4to. *Venice.*

Mem. Wern. Soc.—Memoirs of the Wernerian Natural History Society. *Edinburgh.*

Morphol. Jahrb.—Morphologisches Jahrbuch, ein Zeitschrift für Anatomie und Entwickelungsgeschichte. *Leipsic.*

Overs. K. Danske Vid. Selsk. Forhandl.—Oversigt over det K. Videnskabernes Selskabs Forhandlinger. *Copenhagen.*

Proc. Linn. Soc. N. S. Wales.—Proceedings of the Linnean Society of New South Wales. *Sydney.*

Sitz. k. Ac. Wiss. Wien.—Sitzungsberichte der kaiserlichen Akademie der Wissenschaften; mathematisch-naturwissenschaftliche Classe. *Vienna.*

Sitz. k. preuss. Ak. Wiss.—Sitzungsberichte der königlich-preussischen Akademie der Wissenschaften zu Berlin. *Berlin.*

Trans. Amer. Phil. Soc.—Transactions of the American Philosophical Society, etc. 4to. *Philadelphia.*

Verh. nat. Ges. Basel.—Verhandlungen der naturforschenden Gesellschaft in Basel. *Basle.*

Vid. Medd. Nat. Foren. Kjöbenhavn.—Videnskabelige Meddelelser fra den Naturhistoriske Forening i Kjöbenhavn. *Copenhagen.*

CORRIGENDA ET ADDENDA.

Part I., p. 159. If *Melursus* be included in *Ursus* the incisive formula will be $\frac{(2-3)}{3}$.

201. In dentition of *Phocidæ*, *for* I. $\frac{3}{2}$ *read* I. $\frac{(2-3)}{(1-2)}$.

209. " " *Sciurus*, *for* Pm. $\frac{1}{1}$ *read* Pm. $\frac{(1-2)}{1}$.

" II., p. 79. In description of fig. 7, for *Cervus* read *Rangifer*.

215, line 16 from bottom, *for* four *read* five.

249, note 5, *after* only *insert* other.

259. The first premolar is absent in *Potamochœrus*.

" III., p. 15, lines 6 and 8 from bottom, *for* $\underline{\text{pm. 3}}$, $\underline{\text{pm. 4}}$, *read* $\overline{\text{pm. 3}}$, $\overline{\text{pm. 4}}$.

45, line 14 from top, and 49, line 9 from top, *for* ecto- *read* entocuneiform; and in former, *for* separate *read* united.

161, line 12 from bottom, *after* $\overline{\text{m. 3}}$ *insert* frequently.

172, line 15 from top, *for* ect- *read* entepicondylar.

" IV., p. 15, line 19 from top, *for* M. $\frac{3.3.3}{3.3.3}$ *read* M. $\frac{3.3.4}{3.3.4}$.

" V., p. 186. The arch of No. M. 5 has been found and fitted to the specimen.

CATALOGUE

OF

FOSSIL MAMMALIA.

PART V.

Ordinal Position Uncertain.

Group *TILLODONTIA.*

The systematic position of the forms included in this group (e. g. *Anchippodus*, *Psittacotherium*[1], *Calamodon*, &c) must be regarded as unsettled. Marsh considers them as combining the characters of the Ungulata, Rodentia, and Carnivora; they are classed by Cope[2] in his Order Bunotheria; while Schlosser[3] regards some of them as approximating to the Edentata.

Family ANCHIPPODONTIDÆ.

Genus **ANCHIPPODUS,** Leidy[4].

Syn. *Trogosus*, Leidy[5].

This genus appears closely allied to, if not identical with, *Tillotherium*, Marsh[6] (woodcut, fig. 1), in which the dentition, according to Marsh[7], is I. $\frac{2}{2}$, C. $\frac{1}{1}$, Pm. $\frac{3}{2}$, M. $\frac{3}{3}$. The lower cheek-teeth resemble

[1] Cope, Amer. Nat. vol. xvi. p. 157 (1882).

[2] Vertebrata of the Tertiaries of the West (Rep. U: S. Geol. Surv. Terr. vol. iii.), Book i. p. 194 (1884).

[3] Morphol. Jahrb. vol. xii. p. 288 (1886).

[4] Proc. Ac. Nat. Sci. Philad. 1868, p. 232.

[5] *Ibid.* 1871, p. 115.

[6] Amer. Journ. ser. 3, vol. v. p. 485 (1873).

[7] *Op. cit.* vol. ix. p. 221 (1875). Cope (Vertebrata of the Tertiaries, &c., p. 195), who regards the outermost cutting-tooth as an incisor, states that it is absent in *Tillotherium* and present in *Anchippodus*, and that in the former there are seven, and in the latter six lower cheek-teeth.

those of *Palæotherium*, $\overline{\text{m. 3}}$ having a large third lobe; the inner incisors are very small, and the second pair large and rodent-like; there is a small diastema. The skeleton approaches that of the primitive Carnivora.

Fig. 1.

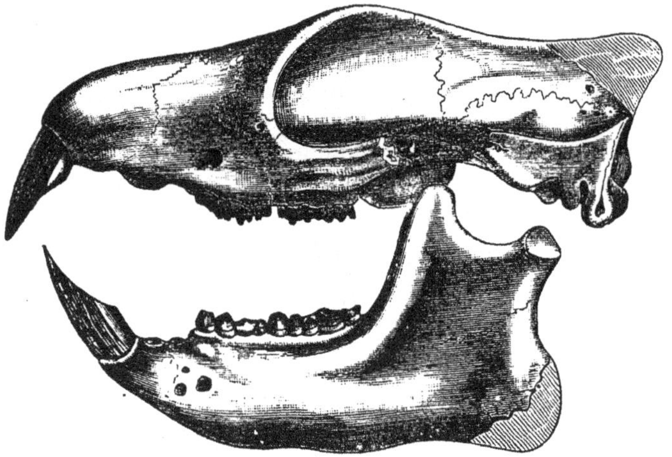

Tillotherium fodiens, Marsh.—The skull; from the Eocene of North America. $\frac{1}{4}$. (After Marsh.)

Anchippodus riparius, Leidy[1].

Syn. *Trogosus castoridens*, Leidy[2].
Palæosyops minor, Marsh[3].
Anchippodus minor, Marsh[4].

This is the type species; the writer follows Leidy in regarding the form to which the three names quoted as synonyms have been applied as specifically the same.

Hab. North America.

47503. Cast of a lower true molar. The original was obtained from the Eocene of Monmouth County, New Jersey, U.S.A., and is the type of the genus; it is described and figured

[1] Proc. Ac. Nat. Sci. Philad. 1868, p. 232.
[2] *Ibid.* 1871, p. 115.
[3] Amer. Journ. ser. 3, vol. ii. p. 36 (1871).
[4] *Ibid.* vol. v. p. 485 (1873).

by Leidy in the Journ. Ac. Nat. Sci. Philad. ser. 2, vol. vii. p. 403, pl. xxx. figs. 45, 46.

Presented by Professor J. Leidy, 1876.

47501. Cast of the greater portion of the mandible. The original was obtained from the Bridger (Lower) Eocene of Wyoming, U.S.A., and is described and figured by Leidy in the "Extinct Vertebrate Fauna of the Western Territories" (Rep. U.S. Geol. Surv. Terr.), p. 71, pl. v. figs. 1–3 (1873), under the name of *Trogosus castoridens*, of which it is the type. *Presented by Professor J. Leidy*, 1876.

Family PLATYCHŒROPIDÆ.

The type genus appears to the writer[1] to be closely allied to, if not identical with, the North-American *Esthonyx*. The position of the latter has been much discussed. It was at one time placed by

Fig. 2.

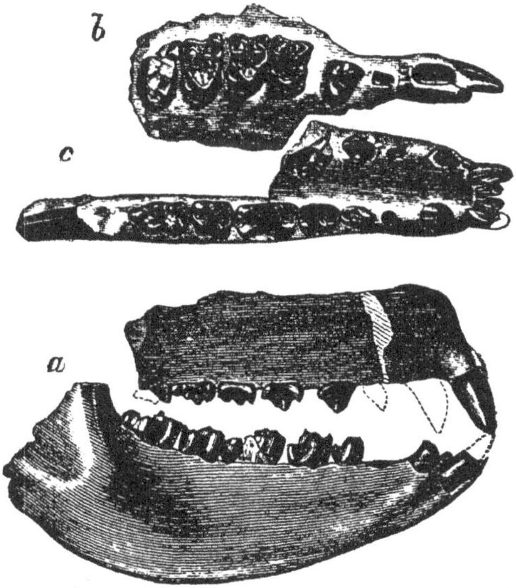

Esthonyx burmeisteri, Cope. –The palate and mandible; from the Wasatch Eocene of Wyoming, U.S.A. ¾. *a*, lateral view of part of cranium and mandible; *b*, oral view of right half of palate; *c*, oral view of mandible. (From the 'Amer. Nat.')

---

[1] Geol. Mag. dec. 3, vol. ii. p. 360; see also p. 526 of the same.

Cope[1] in the Insectivora, as an annectant form between *Erinaceus* and *Tillotherium*, but was subsequently[2] referred to the same family as *Deltatherium*[3]. The latter view is not accepted by Schlosser[4], who regards the genus as allied to the Tillodontia. In *Esthonyx* the dentition (woodcut, fig. 2), is I. $\frac{2}{3}$, C. $\frac{1}{1}$, Pm. $\frac{3}{3}$, M. $\frac{3}{3}$. The larger pair of incisors are scalpriform, but do not grow from persistent pulps; the cheek-teeth are very similar to those of *Anchippodus*.

It will be obvious that the systematic position of *Platychœrops* entirely depends on its apparent relationship to *Esthonyx*[5].

Genus **PLATYCHŒROPS,** Charlesworth[6].

Syn. *Miolophus*, Owen[7].

Platychœrops richardsoni, Charlesworth[8].

Syn. *Miolophus planiceps*, Owen[9].

Hab. Europe (England).

M. 3566. Cast of the imperfect palate, showing m. 2 and m. 3 on both sides, the right pm. 4, and the alveoli of pm. 3 and m. 1 of the two sides. The original, which is the type of the genus and species, is preserved in the Museum at York, and is described and figured by Owen, *op. cit.* pl. x. fig. 1, as *Miolophus planiceps*; it was obtained from the London Clay of Herne Bay, Kent.

Made in the Museum.

[1] Vertebrata of the Tertiary Formations of the West (Rep. U.S. Geol. Surv. Terr. vol. iii.), Book i. p. 197 (1884).

[2] Amer. Nat. vol. xviii. p. 480 (1884).

[3] *Vide infrà*, Supplement.

[4] Morphol. Jahrb. vol. xii. p. 288 (1886).

[5] Since the above was written, the author has received a letter from Mr. H. F. Osborn, of Princeton, New Jersey, who during a recent visit to England examined the cast of *Platychœrops*, in which he states his opinion, after an examination of the American specimens of *Esthonyx*, that the latter is not generically separable from the former.

[6] Rep. Brit. Assoc. for 1854—Trans. of Sections, p. 80 (1855).

[7] Geol. Mag. dec. 1, vol. ii. p. 339 (1865).

[8] *Loc. cit.*

[9] *Loc. cit.*

Order SIRENIA.

Family HALITHERIIDÆ.

This family agrees with the *Halicoridæ* in the presence of upper incisors, but differs in having the crowns of the cheek-teeth covered with enamel.

Genus **HALITHERIUM,** Kaup[1].

Syn. *Pugmeodon*, Kaup[2].
Halianassa, Meyer[3].
Chirotherium, Bruno[4].
Metaxytherium, Christol[5].
Felsinotherium, Capellini[6].

The writer follows Gaudry ('Les Enchaînements, &c.—Mamm. Tert.,' p. 35) in including *Felsinotherium* in the present genus. There are a pair of tusk-like upper incisors (smaller than in *Halicore*) and either five or six cheek-teeth in each jaw; small nasals are present, at least in some species; the premaxillæ and the mandibular symphysis are much deflected, and the angle of the mandible has a descending plate; the innominate is more developed than in existing genera, and there is a styliform femur. It has been observed that in *H. schinzi* some of the permanent anterior cheek-teeth were in some instances preceded by milk-molars; the hinder cheek-teeth resemble those of *Hippopotamus*, presenting, when partially worn, a trefoil on one column of each ridge. Germs of inferior incisors were developed. The synonymy of the species is so involved that some of the following names must be regarded as provisional.

Halitherium forestii (Capellini[7]).

Syn. *Felsinotherium forestii*, Capellini[8].

This species is distinguished by its very large size and the great extent of the downward flexure of the premaxillæ and of the two

[1] Neues Jahrb. 1838, p. 319.
[2] *Loc. cit.* [3] *Ibid.* p. 667.
[4] Mem. Ac. R. Sci. Turin, ser. 2, vol. i. p. 160 (1839).
[5] Ann. Sci. Nat. sér. 2, vol. xv. p. 307 (1841).
[6] Atti Soc. Ital. Sci. Nat. vol. viii. p. 281 (1865).
[7] Mem. Ac. Sci. Ist. Bologna, ser. 3, vol. i. p. 617 (1871).—*Felsinotherium.*
[8] *Loc. cit.*

extremities of the mandible; there are five cheek-teeth; and the posterior border of the palate is deeply notched. This species may be identical with the so-called *Chirotherium brocchii*, Blainv., in which case the latter name should be adopted.

Hab. Italy.

48428. Cast of the type cranium and mandible. The originals were obtained from the Lower Pliocene (Plaisancien) of Felsino, Bologna, and are figured by Capellini, *op. cit.* pls. i.–iv.; a reduced figure being given by Gaudry in 'Les Enchaînements, &c.—Mammifères Tertiaires,' p. 35, fig. 21. There are five cheek-teeth on the right side of the cranium. *Purchased*, 1877.

48432. Cast of the left scapula; same history. Figured, *op. cit.* pl. vii. *Purchased*, 1877.

48429. Cast of the atlas vertebra; same history. Figured, *op. cit.* pl. vi. figs. 1, 2. *Purchased*, 1877.

48430. Cast of the axis vertebra; same history. Figured, *op. cit.* pl. vi. figs. 3, 4. *Purchased*, 1877.

48431. Cast of a lumbar vertebra; same history. Figured, *op. cit.* pl. vi. figs. 8, 9. *Purchased*, 1877.

Halitherium fossile (Blainville [1]).

Syn. *Manatus fossilis*, Blainville [2].
Metaxytherium cordieri, Christol [3].
Halianassa cordieri, Meyer [4].

The length of the space occupied by the last three upper molars is 0,071; and the species is thus larger than *H. schinzi*; there is no third lobe to the last lower molar (fig. 3).

Hab. Europe.

M. 2408. Two upper and lower hinder molars, perhaps belonging to this species; from the Middle Miocene of Chazé-Henry (Maine-et-Loire), France. These teeth agree very closely

[1] Ostéographie—Genus *Manatus*, p. 81 (1844).—*Manatus*.
[2] *Loc. cit.* The name is said to be taken from Cuvier.
[3] Ann. Sci. Nat. sér. 2, vol. xv. p. 307 (1841).
[4] In Bronn's 'Index Palæontologicus,' p. 562 (1848).

with those of the Italian Pliocene form figured in Blainville's 'Ostéographie,' Genus *Manatus*, pl. ix., under the name of *Chirotherium brocchii*. *Purchased*, 1885.

Fig. 3.

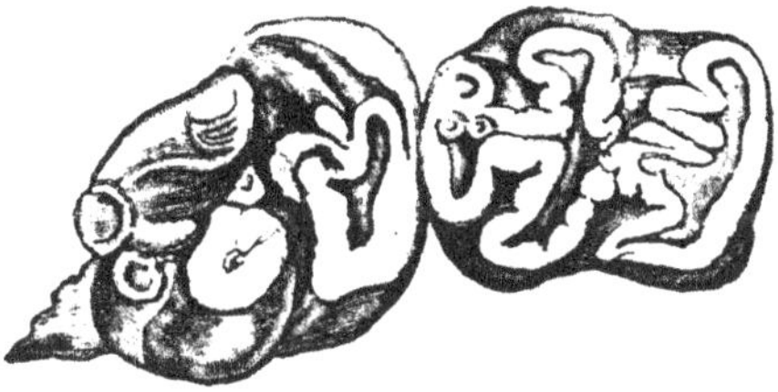

Halitherium fossile.—The penultimate and last right lower molars; from the Middle Miocene of Angers (Maine-et-Loire). ½. (After De Blainville.)

Halitherium canhami, Flower [1].

This imperfectly known species appears allied to *H. schinzi*, but the cranium is larger and more massive, with differently shaped nasals and proportionately larger teeth.

Hab. England.

46055. Cast of the imperfect and water-worn cranium. The original (which is the type) is from the bone-bed at the base of the Red Crag at Foxhall, near Woodbridge, Suffolk, and is preserved in the Ipswich Museum; it is described and figured by Flower, *op. cit.* p. 1, pl. i.
Presented by the Council of the Royal College of Surgeons, 1874.

Halitherium schinzi, Kaup [2].

Syn. *Pugmeodon schinzi*, Kaup [3].
?*Halitherium cuvieri*, Kaup [4].
Manatus guettardi, Blainville [5].
Halitherium guettardi, Gervais [6].

This species is about equal in size to *Manatus australis*; distinct nasals are present, and the number of the cheek-teeth is six. The

[1] Quart. Journ. Geol. Soc. vol. xxx. p. 1 (1874).
[2] Neues Jahrb. 1838, p. 319.—*Pugmeodon*. [3] *Loc. cit.*
[4] *Ibid.* 1840, p. 675.
[5] Ostéographie, Genus *Manatus*, p. 108 (1844).
[6] Zool. et Pal. Franç. 1st ed. p. 144 (1848–52).

Fig. 4.

Halitherium schinzi.—Skeleton; from the Lower Miocene (Middle Oligocene) of Flonheim, Hessen-Darmstadt.

Original specimen preserved in the Museum at Darmstadt. Length 7 feet 8 inches.

(From the 'Quart. Journ. Geol. Soc.' vol. xli. p. 406, 1885.)

posterior border of the palate is not deeply notched; the length of the space occupied by the last three upper molars, according to Kaup, is 0,061, but it is longer in No. 19957; there is a third lobe to the last lower molar.

Hab. Europe.

M. 1981. Cast of the nearly complete skeleton (woodcut, fig. 4). The original was obtained from the Lower Miocene (Middle Oligocene) of Flonheim, near Alzey, Hessen-Darmstadt, and is preserved in the Museum at Darmstadt. It is described and figured by Lepsius in the Abh. mittelrhein. geol. Vereins, vol. i. pt. 1 (1881). The last molar is not protruded. The woodcut is reproduced from a paper by H. Woodward in the 'Quart. Journ. Geol. Soc.' vol. xli. p. 466. *Purchased*, 1884.

41095. Cast of the hinder portion of the cranium. The original is from Darmstadt. *Purchased*, 1868.

19957. The imperfect cranium; from Flonheim. Only the last three cheek-teeth remain, the antepenultimate being much worn; the length of the three teeth is 0,068. *Purchased*, 1846.

44149. The supraoccipital and parietals; from Flonheim. *Purchased*, 1873.

19957 a. The imperfect mandible, showing the last three cheek-teeth in a much-worn condition; from Flonheim. *Purchased*, 1846.

36766. The mandible, wanting the hinder portion of the left ramus; from Flonheim. The last molar is *in alveolo*, while the fifth tooth from the hinder end of the series is quite unworn, and appears to be a true *dent de remplacement*. *Purchased*, 1859.

27887. The penultimate right lower molar, in a half-worn condition; from Flonheim. The homologous and following teeth of *H. fossile* are represented in woodcut, fig. 3. *Presented by John Brown, Esq.*, 1852.

44153. The right scapula; from Flonheim. *Purchased*, 1873.

41093. Cast of the imperfect right scapula. The original is from the Lower Miocene (Middle Oligocene) of Ufhofen, Hessen-Darmstadt. *Purchased*, 1868.

43978. The imperfect right scapula; from Darmstadt. *Purchased*, 1872.

19957 b. Two imperfect specimens of the scapula; from Darmstadt. *Purchased*, 1846.

41094. Cast of the left humerus. The original is from Darmstadt. *Purchased*, 1868.

43979. The left humerus; from Darmstadt. *Purchased*, 1872.

44155. Two specimens of the humerus, wanting the distal extremity; from Darmstadt. *Purchased*, 1873.

44156. The distal extremity of the right humerus; from Darmstadt. *Purchased*, 1873.

19957. The right radius and ulna; from Darmstadt. *Purchased*, 1846.

48993. Cast of an ulna, wanting the olecranon. The original is from Flonheim. *Presented by Prof. von Koch*, 1878.

19957 c. The left innominate; from Flonheim. *Purchased*, 1846.

44154. The imperfect left innominate; from Flonheim. *Purchased*, 1873.

48993 a. Casts of two specimens of the left innominate and the femur. The originals are from Flonheim. *Presented by Prof. von Koch*, 1878.

29648 &c. Twenty-five ribs; from Flonheim. *Purchased.*

43977. The presternum; from Darmstadt. *Purchased*, 1872.

48993 b. Six casts of portions of the sternum. The originals are from Flonheim. *Presented by Prof. von Koch*, 1878.

1762. Cast of the atlas vertebra. The original is from Flonheim, and is figured in De Blainville's 'Ostéographie,' Genus *Manatus*, pl. viii. *Purchased.*

19957 d. Two cervical vertebræ; from Flonheim. *Purchased*, 1846.

1762 a. Cast of the arch of the second dorsal vertebra. The original is from Flonheim. *Purchased.*

27887 b. The arch of the second dorsal vertebra; from Flonheim. *Presented by John Brown, Esq.*, 1852.

19957 c. An early dorsal vertebra; from Flonheim. *Purchased*, 1846.

27887 c. An early dorsal vertebra; from Flonheim. *Presented by John Brown, Esq.*, 1852.

19957 j. Three middle dorsal vertebræ; from Flonheim. *Purchased*, 1846.

44150–1. Two small specimens of late dorsal vertebræ; from Flonheim. *Purchased*, 1873.

19957 d. Three associated late dorsal vertebræ; from Flonheim. *Purchased*, 1846.

19957 k. Two late dorsal vertebræ; from Flonheim. *Purchased*, 1846.

19957 b. The imperfect first lumbar vertebra; from Flonheim. *Purchased*, 1846.

M. 3540. A lumbo-sacral vertebra; from Flonheim. *Purchased.*

M. 423. A lumbo-sacral vertebra; from Darmstadt. *Enniskillen Collection. Purchased*, 1882.

44152. The last sacral and first two caudal vertebræ; from Flonheim. *Purchased*, 1873.

19957 e. Three early caudal vertebræ; from Flonheim. *Purchased*, 1846.

43975. Three caudal vertebræ; from Flonheim. *Purchased.*

19957 f. Two early caudal vertebræ; from Flonheim. *Purchased*, 1846.

19957 g. Two late caudal vertebræ; from Flonheim. *Purchased*, 1846.

M. 3539. Two imperfect early caudal vertebræ; from Flonheim. *Purchased.*

M. 423 a. Two caudal vertebræ; from Darmstadt. *Enniskillen Collection. Purchased*, 1882.

19957 h. An imperfect caudal vertebra; from Flonheim. *Purchased*, 1846.

1762 b. Casts of six caudal vertebræ. The originals are from Flonheim. *Purchased.*

19957 i. A chevron bone; from Flonheim. *Purchased*, 1846.

M. 423 b. A chevron bone; from Darmstadt. *Enniskillen Collection. Purchased*, 1882.

The following specimens are not specifically determined.

40859. Fragment of a left maxilla, showing the last two molars; from (*Fig.*) the Middle or Lower Miocene of Gozo, near Malta. This specimen is described and figured by Leith-Adams in the Quart. Journ. Geol. Soc. vol. xxxv. p. 526, pl. xxv. figs. 3, 3 *a*, and provisionally referred to *H. schinzi*. The teeth (which are of small size) do not, however, agree with those of any examples of that species in the Museum.

Presented by Admiral Spratt, C.B., 1867.

M. 1938. An unworn last left lower molar, with a large third lobe; locality unknown.

Presented by Prof. Sir R. Owen, K.C.B., 1884.

M. 2410. Two unworn lower molars; locality unknown.

Purchased, 1885.

35578. Fragment of a rib; from the Miocene of Malta.

Presented by Sir W. Reid, 1860.

Halitherium (?) sp.

Hab. N. America.

41846 a. The centrum of a lumbar vertebra, and portions of two ribs; from the Phosphate beds (Eocene) of the Ashley river, South Carolina, U.S.A. These specimens were associated with *Zeuglodon*.

Presented by Cowlan Gravely, Esq., 1869.

Genus **PRORASTOMUS**, Owen[1].

Dentition:—I. $\frac{3}{3}$, C. $\frac{1}{1}$, Pm. and M. $\frac{8}{8}$[2] (?). This genus appears to be a generalized form allied to *Halitherium*, but having a fuller dentition and lacking the distinct deflection of the premaxillæ and the mandibular symphysis. The true molars are of a Dinotherioid type, and the upper incisors are not tusk-like.

Prorastomus sirenoides, Owen[3].

This is the type and only known species, and is of smaller size than *Halitherium schinzi*.

Hab. West Indies.

[1] Quart. Journ. Geol. Soc. vol. xi. p. 541 (1855).
[2] This is the number given by Owen, but it is doubtful if it is correct.
[3] *Loc. cit.*

44897. The skull; from Tertiary beds in the island of Jamaica.
(*Fig.*) This specimen is the type, and is figured by Owen in the 'Quart. Journ. Geol. Soc.' vol. xi. pl. xv. figs. 1–6, and vol. xxxi. pls. xxviii., xxix.
Presented by Prof. Sir R. Owen, K.C.B., 1874.

44897 a. Fragment of the atlas vertebra, associated with the preceding specimen.
Presented by Prof. Sir R. Owen, K.C.B., 1874.

M. 3565. Shaft of a humerus; from a Tertiary deposit in the island of Sombrero, near St. Kitts.
Presented by the Hon. J. Price.

Genus **EOTHERIUM,** Owen [1].

This genus, which is provisionally referred to the present family, is only known by the type species.

Eotherium ægyptiacum, Owen [2].

Syn. (?) *Manatus coulombi*, Filhol [3].

This species is founded on the cast of the brain-cavity, which resembles that of *Halitherium* and is narrower than in *Manatus*; but it appears highly probable that the lower cheek-teeth from the same locality, described under the name of *Manatus coulombi*, are specifically identical. These teeth are practically indistinguishable from those of existing species of *Manatus*; but the number is unknown.

Hab. Egypt.

46722. Natural cast of the brain-cavity; from the Eocene (Nummulitic) of Mokattam, near Cairo. This specimen is the type,
(*Fig.*) and is described and figured by Owen in the 'Quart. Journ. Geol. Soc.' vol. xxxi. p. 100, pl. iii. figs. 1–4.
Presented by Prof. Sir R. Owen, K.C.B., 1875.

46722 a. Fragment of a rib; from Mokattam. Noticed by Owen, *op. cit.* *Presented by Prof. Sir R. Owen, K.C.B.*, 1875.

[1] Quart. Journ. Geol. Soc. vol. xxxi. p. 100 (1875).
[2] *Loc. cit.*
[3] Bull. Soc. Philom. Paris, sér. 7, vol. ii. p. 124 (1878).

Fig. 5.

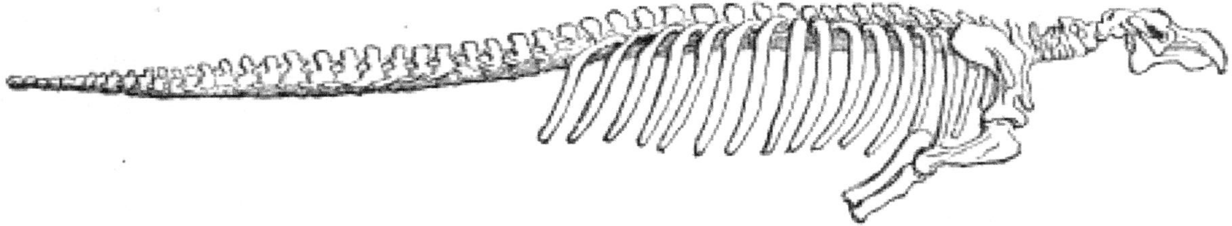

Rhytina gigas (Zimm.).—Outline sketch of nearly entire skeleton; from the peat of Behring Island, North Pacific Ocean.

Length of original specimen 19 feet 6 inches.

(From the 'Quart. Journ. Geol. Soc.' vol. xli. [1885].)

Family RHYTINIDÆ.

In this family (represented only by one species) there are no functional teeth, their function being discharged by horny oral plates. The premaxillæ are moderately deflected, and distinct nasals wanting.

Genus **RHYTINA,** Illiger[1].

Rhytina gigas (Zimmermann[2]).

Syn. *Manati gigas*, Zimmermann[3].
Trichechus manatus, var. *borealis*, Gmelin[4].
Trichechus borealis, Shaw[5].
Rytina stelleri, Desmarest[6].

Hab. North Pacific.

M. 2489. The nearly complete skeleton[7] (woodcut, fig. 5); from (*Fig.*) the peat of Behring Island. Described and figured by H. Woodward in the Quart. Journ. Geol. Soc. vol. xli. p. 459. *Purchased*, 1885.

M. 2489 a. The petrotympanic and auditory ossicles of the last specimen. *Purchased*, 1885.

[1] Prodr. Syst. Mamm. et Avium, p. 141 (1811).—*Rytina.*
[2] Geogr. Geschichte, vol. ii. p. 426 (1780).—*Manati.*
[3] *Loc. cit.*
[4] Syst. Nat. vol. i. p. 61 (1788).
[5] General Zoology, vol. i. p. 240 (1800).
[6] Nouv. Dict. d'Hist. Nat. vol. xxix. p. 574 (1819).
[7] All the bones do not belong to the same individual, and some missing bones have been replaced by plaster models.

Order CETACEA.

Suborder *MYSTACOCETI.*

Functional teeth are wanting and the palate is provided with baleen; the cranium is symmetrical; the nasals form a roof to the narial passages; and the tympanic is involuted and anchylosed to the periotic.

Family BALÆNIDÆ.

This family includes all the members of the Suborder.

A.—Balænine Section.

At least usually, some or all of the cervical vertebræ are anchylosed together. The tympanic is deep and angular, its inflation comparatively slight, the involucrum[1] not fig-shaped, and frequently with no well-marked depression at the anterior extremity of the superior border of the inner surface for the eustachian canal.

Genus **BALÆNA,** Linn.[2]

Including *Balænula*, Van Beneden[3].
Balænotus, Van Beneden[4].

Fig. 6.

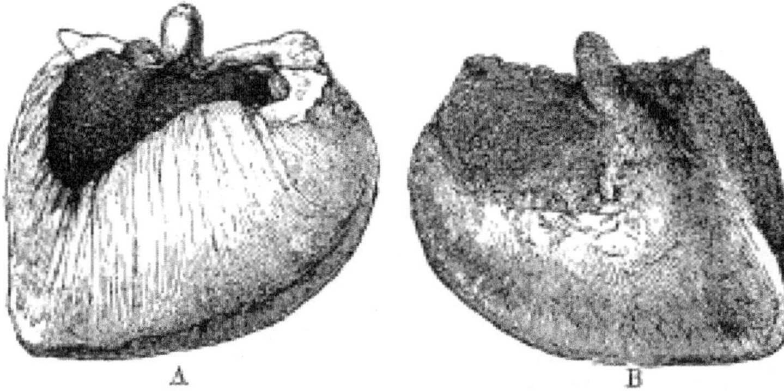

Balæna mysticetus.—The right tympanic of an immature individual, from the inner (A) and outer (B) aspects; recent. ½. (From the 'Proc. Zool. Soc.')

[1] This term may be applied to the reflected superior portion of the inner wall of the bone.

[2] Syst. Nat. ed. 12, vol. i. p. 105 (1766).

[3] Bull. Ac. R. Belg. sér. 2, vol. xxxiv. p. 11 (1872).

[4] *Ibid.* p. 13.

In existing species the whole of the cervical vertebræ are anchylosed together; but in the fossil forms which have been separated under the above-mentioned names this anchylosis is incomplete; the centra of all the vertebræ are short. In the tympanic (fig. 6) the superior border of the inner surface is usually nearly straight; but in one fossil species it may be sinuous.

Balæna mysticetus, Linn.[1]

Hab. Arctic Seas.

38141–2. Two imperfect scapulæ; from a depth of 8 feet below the surface at Deptford. In their large coracoid and acromion these specimens agree with the scapula of the present species; they are, however, in a comparatively fresh condition, and it is highly probable that they were brought to the locality where they were found by human agency.
Presented by the Commissioners of the Metropolitan Board of Works, 1864.

40641. The conjoint cervical vertebræ, in an imperfect condition; from the banks of the Thames at Limehouse. This specimen (which may not be a true fossil) agrees with the figure given by Van Beneden and Gervais in the 'Ostéographie des Cétacés,' pls. iv., v. figs. 5, 6, showing the characteristic contour of the condylar cups of the atlas, the comparatively wide interval between them, and the large size of the nutrient foramen. The remark appended to the last specimen also applies in the present instance.
Purchased, 1867.

Balæna affinis, Owen[2].

Syn. *Balænodon affinis*, Owen[3].

The type tympanic indicates a species closely allied to *B. mysticetus*, and other specimens show that the fossil was fully equal in size to the living form. The tympanic is characterized by its elongated form, the comparative straightness of the inferior border, the height of the inner wall at the eustachian part of the aperture, the produced antero-inferior angle, the well-marked keel on the inferior surface (more developed in the young than in the adult), the

[1] Syst. Nat. ed. 12, vol. i. p. 105 (1766).
[2] Proc. Geol. Soc. vol. iv. p. 283 (1843).
[3] Brit. Foss. Mamm. & Birds, table facing p. xlvi (1846).

approximate parallelism of the free border of the inner wall with the inferior border, the descent of the flattened anterior surface to the inferior angle, and the comparatively slight development of the involucrum itself. Specimens of this type of tympanic preserved in the Brussels Museum (Nos. 114, 1637, 1303) are referred to *B. primigenia*, an identification which, if correct, would necessitate the inclusion of the latter in the present species.

Hab. Europe (England and Belgium).

M. 3512. Cast of the imperfect left tympanic. The original (which is apparently lost) was obtained from the Red Crag of Felixstowe, Suffolk, and is the type of the species; it is described and figured by Owen in the 'British Fossil Mammals and Birds,' p. 530, fig. 221. Its length is 0,121, and it corresponds very closely with a specimen in the Brussels Museum (No. 1303), which measures 0,120.
No history.

46682. The imperfect left tympanic; from Woodbridge. The involucrum is more thickened than in the last specimen.
Purchased, 1875.

39011. The imperfect left tympanic; from the Red Crag of Felixstowe. *Bowerbank Collection. Purchased*, 1865.

39013. The imperfect right tympanic, in a rolled condition; from Felixstowe. *Bowerbank Collection. Purchased*, 1865.

46681. (*Fig.*) The imperfect and partially-rolled left tympanic; from Woodbridge. This specimen (woodcut, fig. 7) is larger than any of the preceding examples, and agrees with a tympanic in the Brussels Museum (No. 114) referred to *B. primigenia*, which has a length of 0,145; it has been noticed by the writer in the Quart. Journ. Geol. Soc. vol. xliii. p. 8. *Purchased*, 1875.

39010. The imperfect left tympanic, of rather smaller size; from Felixstowe. *Bowerbank Collection. Purchased*, 1865.

The two following specimens may belong either to this species or to B. primigenia.

48978. The imperfect left periotic; from Woodbridge.
Purchased, 1878.

23262. The mperfect left periotic ; from Woodbridge.
Purchased, 1852.

25587. A smaller periotic, perhaps belonging to *B. insignis*; from Woodbridge. *Presented by* — *Laws, Esq.*, 1857.

Fig. 7.

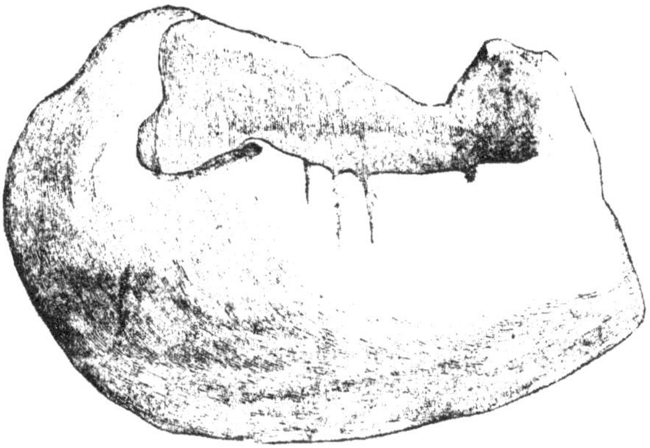

Balæna affinis.—The left tympanic; from the Red Crag of Woodbridge. ½.

Balæna biscayensis, Gray [1].

This species, which may be identical with *B. australis*, Desmoulins [2], is provisionally recorded by Newton from the Norfolk Forest-bed in the Quart. Journ. Geol. Soc. vol. xlii. p. 321.

Hab. North Atlantic.

39361. The conjoint cervical vertebræ in a slightly imperfect condition : from the bed of the Thames at Wapping. This specimen differs from the corresponding vertebral mass of *B. mysticetus*, No. 40641, and agrees with the figure of the same part in the present species given by Van Beneden and Gervais in their 'Ostéographie des Cétacés,' pl. vii. figs. 8–11, the characteristic features being the contour of the condylar cups of the atlas, the narrow interval between them, the small nutrient foramen, and the nearly horizontal direction of the inferior border of the lateral arch in the second and third vertebræ. *Purchased*, 1861.

[1] Proc. Zool. Soc. 1864, p. 200.

[2] See Flower, List of Cetacea in Brit. Mus. p. 2 (1885).

The following specimens are provisionally referred to this species.

8116. A dorsal vertebra; from silt at Lewes, Sussex.
Mantell Collection. Purchased, 1836.

33463. The centrum of a dorsal vertebra; dredged off the eastern coast. *Layton Collection. Purchased*, 1858.

39224. An early caudal vertebra; dredged from the Thames at Wandsworth. *Purchased*, 1865.

M. 3541. The centrum of a later caudal vertebra, wanting the epiphyses; probably dredged from the sea. *No history*.

33460. The centrum of a late caudal vertebra; dredged off the eastern coast. *Layton Collection. Purchased*, 1858.

46282. The left humerus; dredged from the North Sea.
Owles Collection. Purchased, 1874.

Balæna primigenia, Van Beneden[1].

This species is of large size and has all the cervical vertebræ anchylosed together. The tympanics provisionally included under this specific name present considerable variations, but, as they all seem to pass into one another, these variations cannot apparently be considered of specific value. Generally the tympanic is characterized by the convex inferior border, the absence of a produced antero-inferior angle, and the comparative lowness of the eustachian portion of the aperture. This type of tympanic indicates a species allied to the existing Whales of the southern and temperate seas (*B. biscayensis* and *B. australis*); but some specimens appear to show a relationship to *Neobalæna*.

Hab. Europe (Belgium and England).

Variety A.

In this, which may be called the type variety, the inner wall of the tympanic is very high, its superior border oblique, the flattening of the anterior surface extending nearly or quite down to the inferior border, the involucrum considerably thickened, and the inferior border somewhat angulated.

M. 1248. Cast of the right tympanic. The original was obtained from the Antwerp Crag, and is preserved in the Brussels

[1] Bull. Ac. R. Belg. sér. 2, vol. xxxiv. p. 9 (1872).

Museum; it is figured by Van Beneden in the Ann. Mus. R. Hist. Nat. Belg. vol. iv. pt. 2, pl. xix. figs. 1–4.
Purchased, 1883.

M. 1249. Cast of the right tympanic. The history of the original is the same as that of the last specimen; and it is figured by Van Beneden, *op. cit.* figs. 9–12. *Purchased*, 1883.

46686. The nearly perfect left tympanic of an immature individual; from the Red Crag of Woodbridge, Suffolk. This specimen has not been rolled, and its owner evidently lived during the deposition of the Red Crag. *Purchased*, 1875.

M. 3509 a. Cast of the slightly imperfect right tympanic. The original (woodcut, fig. 8), which has not been rolled, was

Fig. 8.

Balæna primigenia.—The right tympanic; from the Red Crag of Suffolk. ½. (From the 'Quart. Journ. Geol. Soc.')

obtained from the Red Crag of Suffolk, and is in the possession of W. Colchester, Esq., of Ipswich. The obliquity of the inner wall is less marked than in No. M. 1248, and agrees in that respect with No. M. 1249; the specimen is described and figured by the writer in the 'Quart. Journ. Geol. Soc.' vol. xliii. p. 9, woodcut fig. 1, the figure being reproduced here.
Made in the Museum, 1886.

Variety B.

The obliquity of the superior border of the inner wall of the tympanic is extremely developed, so that the structure of this bone foreshadows that of *Neobalæna*; the involucrum is thickened and the inferior border angulated.

M. 3510. Cast of the imperfect right tympanic. The original is from the Red Crag of Woodbridge, Suffolk, and is preserved in the Museum of Practical Geology, Jermyn Street; it is described and figured by the present writer in the 'Quart. Journ. Geol. Soc.' vol. xliii. p. 9, pl. ii. figs. 1, 1 *a*. A specimen in the Brussels Museum (No. 1248) presents somewhat similar characters.

Made in the Museum, 1886.

Variety C.

The tympanic resembles in many respects that of the type variety; but is much compressed laterally, has no thickening of the involucrum, and the flattening of the anterior surface does not extend much below the superior border of the inner wall; the inferior border is not angulated.

46685. The slightly imperfect and partially rolled right tympanic; from the Red Crag of Woodbridge. Noticed by the writer in the 'Quart. Journ. Geol. Soc.' vol. xliii. p. 10.

Purchased, 1875.

Variety D.

The inner wall of the tympanic is low, its superior border nearly parallel with the long axis of the bone, the flattening of the anterior surface not extending below that border, and the inferior border of the bone not angulated.

43399. (*Fig.*) The imperfect right tympanic; from the Red Crag of Woodbridge. This specimen, which has not suffered from rolling, is described and figured by the present writer in the 'Quart. Journ. Geol. Soc.' vol. xliii. p. 10, pl. ii. figs. 2, 2 *a*. *Purchased*, 1875.

M. 3509. Cast of a rather smaller and somewhat rolled imperfect right tympanic. The original was obtained from Woodbridge, and is preserved in the Museum of Practical Geology, Jermyn Street; it agrees very closely in character with the preceding specimen, and is noticed by the writer, *loc. cit.* *Made in the Museum*, 1886.

28257. An imperfect left tympanic, provisionally referred to this variety; from Woodbridge. *Purchased*, 1852.

M. 3511. Cast of an immature imperfect right tympanic, not improbably belonging to this variety. The original, which is from Woodbridge, is preserved in the Museum of Practical Geology; it differs from the tympanic of *B. insignis* by its narrower inferior surface, more convex anterior border, and its slighter degree of inflation, in all of which respects it resembles No. M. 3509.

Made in the Museum. 1886.

The originals of the following series of casts (purchased, 1883) *were obtained from the Antwerp Crag, and are referred to this species by Van Beneden in the 'Ann. Mus. R. Hist. Nat. Belg.'* vol. iv. pt. 2, *to which memoir the following references apply; all the specimens are preserved in the Brussels Museum.*

M. 1247. Cast of a part of the mandibular ramus; pl. xix.

M. 1250. Cast of the left periotic; pl. xix. figs. 13, 14.

M. 1251. Cast of the atlas vertebra; pl. xxi. fig. 8.

M. 1252. Cast of the left ulna; pl. xxi. fig. 7.

M. 1253. Cast of the left radius; pl. xxi. figs. 5, 6.

M. 1254. Cast of a rib; pl. xviii. fig. 6.

M. 1255. Cast of a rib; pl. xviii. fig. 7.

Balæna balænopsis (Van Beneden[1]).

Syn. *Balænula balænopsis,* Van Beneden[2].

This is a small species, approaching *B. mysticetus* in cranial and tympanic structure, its total length being estimated at about 20 feet. The atlas and seventh cervical vertebræ are separate, and the other cervicals are only anchylosed by portions of their centra. Van Beneden gives certain characters by which the tympanics can be distinguished from those of *B. insignis*; but the writer was unable to recognize these characters in the specimens in the Brussels Museum, and has accordingly referred to the present species only the smaller specimens.

Hab. Europe (Belgium and England).

[1] Bull. Ac. R. Belg. sér. 2, vol. xxxiv. p. 11 (1872).—*Balænula.*
[2] *Loc. cit.*

M. 3513. The imperfect left tympanic; from the Antwerp Crag.
By exchange with the Royal Brussels Museum of Natural History, 1886.

The originals of the following series of casts (purchased, 1883) *were obtained from the Antwerp Crag, and are preserved in the Royal Museum of Natural History at Brussels; they are figured by Van Beneden in the 'Ann. Mus. R. Hist. Nat. Belg.'* vol. iv. pt. 2, pls. i.–xvii., *to which memoir the following references apply.*

M. 1260. Cast of the left tympanic; pl. iii. figs. 1–4.

M. 1258. Cast of the right tympanic; pl. iii. figs. 5–8.

M. 1284. Cast of the left tympanic; pl. iii. figs. 9–12.

M. 1259. Cast of the right tympanic; pl. iii. figs. 13–16.

M. 1257. Cast of the left tympanic; pl. iii. figs. 17–20.

M. 1256. Cast of the left tympanic; pl. iii. figs. 21–24.

M. 1261. Cast of the right periotic; pl. iii. figs. 24–27.

M. 1327. Cast of a nasal; pl. ii. figs. 1–4.

M. 1269. Cast of the atlas vertebra; pl. viii. figs. 1–3.

M. 1328. Cast of one half of the atlas vertebra; pl. xi. figs. 4, 5.

M. 1276. Cast of one half of the atlas vertebra; pl. viii. figs. 4–6.

M. 1266. Cast of the conjoint axis and the three following cervical vertebræ; pl. ix. figs. 1–3.

M. 1278. Cast of the axis and portions of the three following cervical vertebræ of a young individual; pl. ix. fig. 6.

M. 1271. Cast of one half of the conjoint axis and five following vertebræ; pl. x. figs. 1–4.

M. 1265. Cast of one half of the conjoint axis and five following vertebræ; pl. x. fig. 5.

M. 1270. Cast of the atlas and six following vertebræ of a young individual; pl. x. fig. 6.

M. 1268. Cast of a section of the last six cervical vertebræ of a young individual; pl. x. figs. 7, 8.

M. 1264, 1267. Casts of the conjoint cervical vertebræ and section of the same; pl. xii.

M. 1329–31. Three casts of cervical vertebræ.

M. 1279. Cast of the imperfect eighth dorsal vertebra, wanting the epiphyses; pl. xiii. figs. 1, 2.

M. 1280. Cast of the imperfect eleventh dorsal vertebra, wanting one epiphysis; pl. xiii. figs. 3, 4.

M. 1281. Cast of the imperfect sixth lumbar vertebra, wanting one epiphysis; pl. xiv. figs. 5, 6.

M. 1282. Cast of the sixth caudal vertebra, wanting both epiphyses; pl. xv. fig. 3.

M. 1332. Cast of the eleventh caudal vertebra, wanting both epiphyses; pl. xv. figs. 4, 5.

M. 1333. Cast of the seventeenth caudal vertebra; pl. xv. fig. 6.

M. 1334. Cast of the twenty-first caudal vertebra; pl. xiv. figs. 8, 9.

M. 1272. Cast of the twenty-fourth caudal vertebra; pl. xiv. figs. 10, 11.

M. 1273. Cast of the left humerus; pl. xvi. fig. 7.

M. 1263. Cast of an imperfect humerus.

M. 1274. Cast of the left radius; pl. xvii. figs. 3, 4.

M. 1275. Cast of the left ulna; pl. xvii. figs. 1, 2.

M. 1283. Cast of the eighth rib of the right side; pl. ii. fig. 5.

M. 1262. Cast of the right ramus of the mandible; pl. v.

The following specimens are from the Coralline and Red Crag of Suffolk.

M. 3514. Cast of the slightly imperfect right tympanic. The original was obtained from the Red Crag, and is preserved in the Ipswich Museum; it is noticed by the present writer in the 'Quart. Journ. Geol. Soc.' vol. xliii. p. 10.
Made in the Museum, 1886.

M. 3542. The imperfect atlas vertebra; from the Coralline Crag of Sudbourn, near Orford. This specimen is noticed by Van Beneden, *op. cit.* p. 53 (where it is referred to this species), and also by the present writer in the 'Quart. Journ. Geol. Soc.' vol. xliii. p. 10. *No history.*

Some of the following specimens may belong to young individuals of B. insignis.

46078. The centrum of the seventh (?) dorsal vertebra; from the Red Crag of Boyton. Although much smaller, this specimen closely resembles the seventh dorsal of *B. insignis*, No. M. 1301. *Purchased*, 1875.

46495. The centrum of an early dorsal vertebra; from Woodbridge. This specimen is very similar to the young fifth dorsal of *B. insignis*, No. M. 1304. *Purchased*, 1875.

45843. The centrum of a very similar dorsal vertebra; from Woodbridge. *Purchased*, 1875.

46744. The centrum of a late dorsal vertebra; from Woodbridge. This specimen agrees with No. M. 1280. *Purchased*, 1875.

46743. An imperfect lumbar vertebra; from Woodbridge. *Purchased*, 1875.

48479. A middle caudal vertebra; from Woodbridge. *Presented by C. Falconer, Esq.*, 1867.

Balæna insignis (Van Beneden[1]).

Syn. *Balænotus insignis*, Van Beneden[2].

This species is larger than *B. balænopsis*. The seventh cervical is always, and the atlas generally, free, while in young individuals the axis is sometimes detached; the anchylosis of the other cervicals is less complete than in existing species, and the atlas and the sixth and seventh cervicals are much thicker; in all the cervicals distinct epiphyses are developed. The inferior surface of the tympanic is very wide, the antero-inferior angle more or less distinctly marked, and the flattening of the anterior surface extending nearly or quite to this point. The whole bone is more inflated, its involucrum thicker, aud its inferior surface wider than in *B. affinis*; but it is not always easy to be quite sure that specimens may not belong to young individuals of that species. In many of the above-mentioned features an approximation is presented to the Balænopterine section of the family.

Hab. Europe (Belgium, Italy, and England).

[1] Bull. Ac. R. Belg. sér. 2, vol. xxxiv. p. 13 (1872).—*Balænotus.*
[2] *Loc. cit.*

The originals of the following series of casts (purchased, 1883) *were obtained from the Antwerp Crag, and are preserved in the Royal Museum of Natural History, Brussels; the majority are figured by Van Beneden in the 'Ann. Mus. R. Hist. Nat. Belg.'* vol. iv. pt. 2, pls. xxiii.–xxxix., *to which memoir the following references apply.*

M. 1287–8. Casts of the associated right tympanic and periotic; pl. xxvi. figs. 1–6.

M. 1286. Cast of the left tympanic; pl. xxvi. figs. 7–11.

M. 1285. Cast of the right tympanic; pl. xxvi. figs. 12–15. This specimen is shorter and broader anteriorly than M. 1286.

M. 1291. Cast of the right frontal; pl. xxvii. fig. 1.

M. 1292. Cast of the atlas vertebra; pl. xxix.

M. 1311. Cast of the atlas vertebra; pl. xxxi. fig. 1.

M. 1322. Cast of the imperfect atlas vertebra of an older individual; pl. xxxi. fig. 2.

M. 1297. Cast of the united second, third, fourth, fifth, and sixth cervical vertebræ; pl. xxxi. fig. 3.

M. 1323. Cast of the axis vertebra; pl. xxx. fig. 1.

M. 1312. Casts of three conjoint cervical vertebræ; pl. xxxii. figs. 1, 2.

M. 1310. Cast of one half of the united cervical vertebræ; pl. xxxii. figs. 4, 5.

M. 1298. Cast of one half of the united first six cervical vertebræ; pl. xxxii. fig. 6.

M. 1295. Cast of one half of the united axis and four following cervical vertebræ; pl. xxxiii. fig. 1.

M. 1294. Cast of one half of the united axis and two following cervical vertebræ; pl. xxxiii. fig. 3.

M. 1313. Cast of the sixth cervical vertebra of the same individual; pl. xxxiii. figs. 4, 5.

M. 1293. Cast of the united first six cervical vertebræ; pl. xxxi. fig. 4, pl. xxxiii. fig. 6, and pl. xxxiv. fig. 1.

M. 1314. Cast of the centrum of the seventh cervical vertebra; pl. xxxiv. figs. 3, 4.

M. 1296–99. Casts of two specimens of the sixth cervical vertebra.

M. 1300. Cast of the first dorsal vertebra, wanting the neural arch; pl. xxxiv. fig. 5.

M. 1326. Cast of the third dorsal vertebra, wanting one epiphysis and with the neural arch incomplete; pl. xxxv. figs. 1, 2.

M. 1304. Cast of an early dorsal vertebra; pl. xxxvi. fig. 1. Described as the fifth dorsal, but very different from the next specimen.

M. 1320. Cast of the centrum and part of the arch of the fifth dorsal vertebra; pl. xxvii. figs. 4, 5.

M. 1301. Cast of the centrum of the seventh dorsal vertebra; pl. xxxvi. figs. 2, 3.

M. 1320 a. Cast of a smaller seventh dorsal vertebra, wanting one epiphysis; pl. xxxvii. figs. 1, 2.

M. 1302. Cast of the centrum of the tenth dorsal vertebra; pl. xxxvi. fig. 4.

M. 1321. Cast of the thirteenth dorsal vertebra, wanting both epiphyses and one transverse process; the centrum and transverse process are figured, pl. xxxv. figs. 3, 4.

M. 1303. Cast of the centrum of the fifth lumbar vertebra; pl. xxxvi. fig. 5.

M. 1318. Cast of the eleventh lumbar vertebra, wanting both epiphyses and transverse processes; pl. xxxviii. figs. 1, 2.

M. 1316. Cast of the fifth caudal vertebra, wanting one transverse process; pl. xxxviii. figs. 3, 4.

M. 1315. Cast of the slightly imperfect ninth caudal vertebra; pl. xxxix. fig. 3.

M. 1305. Cast of the slightly imperfect fourteenth caudal vertebra; pl. xxxix. figs. 1, 2.

M. 1307. Cast of the imperfect eighteenth caudal vertebra; pl. xxxviii. figs. 5, 6.

M. 1317. Cast of the twentieth caudal vertebra; pl. xxxix. fig. 4.

M. 1308. Cast of a smaller twentieth caudal vertebra; pl. xxxvi. figs. 6, 7.

M. 1306. Cast of the twenty-third caudal vertebra; pl. xxxix. figs. 5, 6.

M. 1309. Cast of the twenty-fourth caudal vertebra, wanting both epiphyses; pl. xxxix. figs. 7, 8.

M. 1289. Cast of the right humerus; pl. xxvii. fig. 5.

M. 1324. Cast of the right radius; pl. xxviii. figs. 4, 5.

M. 1290. Cast of the right ulna; pl. xxviii. figs. 2, 3.

M. 1325. Cast of the basihyal; pl. xxiii. fig. 2.

The following specimen is from the Antwerp Crag.

42996 d. The centrum and part of the neural arch of a comparatively late dorsal vertebra, resembling No. M. 1302, but of smaller size. *Van Breda Collection. Purchased*, 1871.

The following specimens from the Red Crag of Suffolk agree with the preceding; but it is not improbable that one or two may belong to young individuals of B. affinis.

28258. The slightly imperfect right tympanic. This specimen agrees very closely with No. M. 1285. *Purchased*, 1852.

39012. The less perfect right tympanic; from Felixstowe. This specimen resembles No. M. 1286.
Bowerbank Collection. Purchased, 1865.

27028. The slightly imperfect and partially rolled left tympanic. This bone is intermediate in form between the two preceding specimens. *Purchased*, 1851.

24865. The imperfect left tympanic; from Felixstowe.
Purchased, 1850.

36649 a. The imperfect left tympanic; from Woodbridge.
Brown Collection. Presented by Prof. Sir R. Owen, K.C.B., 1859.

39014. A more imperfect left tympanic, not improbably belonging to this species; from Felixstowe.
Bowerbank Collection. Purchased, 1865.

30261. An imperfect left tympanic, probably belonging to a small individual of this species; from Woodbridge.
Purchased, 1855.

46733. The ventral half of the atlas vertebra; from Woodbridge. This specimen agrees almost precisely with No. M. 1323. *Purchased*, 1875.

Some of the following specimens from the same deposit may perhaps belong to B. affinis *or* B. primigenia.

48481. The centrum of a first dorsal vertebra, wanting both epiphyses, which may perhaps belong to the present species; from Woodbridge. *Purchased*, 1875.

48482. The centrum of a first dorsal vertebra, closely resembling the preceding, but rather smaller; from Woodbridge. *Purchased*, 1875.

48466. The centrum and part of the arch of a middle dorsal vertebra; from Woodbridge. This specimen agrees very closely with the eighth dorsal vertebra of *B. balænopsis* (No. M. 1279), but is of considerably larger size. *Presented by C. Falconer, Esq.*, 1867.

46736. The centrum and part of the arch of the seventh (?) dorsal vertebra; from Woodbridge. This specimen closely resembles No. M. 1310. *Purchased*, 1875.

46741–2. Two centra of lumbar vertebræ; from Woodbridge. *Purchased*, 1875.

48465. An imperfect lumbar vertebra. *Presented by C. Falconer, Esq.*, 1867.

48478. The centrum of an early caudal vertebra, wanting one epiphysis, and closely resembling No. M. 1315; from Woodbridge. *Purchased*, 1878.

28270. The centrum of an early caudal vertebra; from Woodbridge. This specimen was determined by Van Beneden. *Purchased*, 1852.

37295. An imperfect caudal vertebra; from Woodbridge. *Purchased*, 1860.

Genus **PALÆOCETUS**, Seeley[1].

This genus is founded on the original of the following specimen,

[1] Geol. Mag. dec. 1, vol. ii. p. 54 (1865).

which apparently indicates a very small Whale, somewhat intermediate between the smaller fossil species of the present and following subfamilies.

Palæocetus sedgwicki, Seeley[1].

Hab. England.

41624. Cast of the second, third, and fourth cervical vertebræ. The original, which is preserved in the Woodwardian Museum, Cambridge, was obtained from the Boulder-Clay at Ely, and is regarded by its describer as having been derived from the Kimeridge Clay. The second and third vertebræ are anchylosed, but the centrum of the fourth is free. The original is described and figured by Seeley, *op. cit.* p. 54, pl. iii. *Presented by Prof. A. Sedgwick*, 1869.

B.—Balænopterine Section.

There is, at least usually, no anchylosis of the cervical vertebræ, which are thicker than in the Balænine section, and the centra of the other vertebræ are more elongated. The tympanic is long, much inflated, rounded, with the involucrum much thickened and more or less distinctly pyriform, and the notch for the eustachian canal always well marked. The tympanic varies in different individuals of the same species much less than in the Balænine section.

The large number of fossil forms described by Van Beneden from the Antwerp Crag render the determination of detached vertebræ a matter of much uncertainty; and the close affinity between some of these forms suggests that their distinction may be due to sexual or individual rather than to specific differences.

Genus **MEGAPTERA,** Gray[2].

Syn. *Megapteropsis*, Van Beneden[3].
Burtinopsis, Van Beneden[4].

The term *Burtinopsis* was applied to Balænopteroids which were regarded as intermediate between *Megaptera* and *Balænoptera*; but

[1] Geol. Mag. dec. 1, vol. ii. p. 54 (1865).
[2] Zoology of 'Erebus' and 'Terror,' p. 16 (1846).
[3] Bull. Ac. R. Belg. sér. 2, vol. xxxiv. p. 15 (1872).
[4] *Ibid.* p. 19.

a comparison of the tympanics with a large series of those of *Megaptera boops* has convinced the writer that the fossil forms should be included in the latter genus [1].

The tympanic is more inflated than in *Balænoptera*, and has the involucrum more distinctly pyriform; the eustachian part of the aperture is well-defined, and there are two well-marked ridges on the inferior surface of the adult bone, of which the outermost is the more prominent; the antero-internal extremity is usually more or less pointed, the inner surface is convex (the convexity increasing with age) and frequently has a vertical groove near the eustachian channel, and the outer surface of the involucrum is very convex. The cranium of the recent species is of moderate size, and the vertebræ are intermediate in length between those of *Balæna* and *Balænoptera*.

Megaptera affinis, Van Beneden [2].

Syn. *Megapteropsis robusta*, Van Beneden [3].

This species is of large size and closely allied to *M. boops*, but is distinguished by the bluntness of the anterior extremity of the tympanic and the greater flatness of its inner surface.

Hab. Europe (Belgium and England).

M. 3531. Cast of an imperfect right tympanic, probably belonging to an immature individual of the present species. The original was obtained from the Coralline Crag of Sudbourn, near Orford, Suffolk, and is preserved in the Museum of Practical Geology, Jermyn Street. The specimen is rather smaller than the one figured by Van Beneden in the 'Ann. Mus. R. Hist. Nat. Belg.' vol. vii. pt. 3, pl. xliii. figs. 1, 2, but agrees closely in form, and especially in the blunt anterior extremity and flattened inner wall, in both of which respects it differs from the tympanic of *M. similis* (Van Beneden). The small development of the inferior ridges indicates that the specimen is not adult. The original has been described and figured

[1] The recent tympanic figured for comparison with the fossil by Van Beneden in the 'Ann. Mus. R. Hist. Nat. Belg.' vol. vii. pt. 3, pl. lxxxix. figs. 15, 16, under the name of *Balænoptera antarctica*, really belongs to *Megaptera boops*; compare B. M. No. 2. 76. 16. 18.

[2] Bull. Ac. R. Belg. sér. 2, vol. l. p. 13 (1880).

[3] *Ibid.* vol. xxxiv. p. 15 (1872). The specific name was considered liable to lead to confusion, and was accordingly changed.

by the present writer in the 'Quart Journ. Geol. Soc.' vol. xliii. p. 11, pl. ii. figs. 4, 4 *a*.
Made in the Museum, 1886.

42996 b. A slightly imperfect early caudal vertebra of a very large Whale, apparently agreeing very closely in size and proportions with the third caudal of the present species figured by Van Beneden in the 'Ann. Mus. R. Hist. Nat. Belg.' vol. vii. pt. 3, pls. xlv., xlvi.; from the Antwerp Crag. *Van Breda Collection. Purchased*, 1871.

46738. The centrum of a middle caudal vertebra, perhaps belonging to this species; from the Red Crag of Woodbridge, Suffolk. *Purchased*, 1875.

Megaptera similis (Van Beneden [1]).

Syn. *Burtinopsis similis*, Van Beneden [2].

This species is considerably smaller than the last; the form of the tympanic appears almost indistinguishable from that of *M. boops*.

Hab. Europe (Belgium and [?] England).

39020. A left periotic, provisionally referred to this species; from the Red Crag of Woodbridge, Suffolk. The narrow elongated form of the portion containing the semicircular canals shows that this specimen (which is apparently adult) belongs to the present genus, as distinct from *Balænoptera* or *Cetotherium*; it is of much smaller size than the periotic of *M. affinis* figured by Van Beneden in the 'Ann. Mus. R. Hist. Nat. Belg.' vol. vii. pt. 3, pl. xlii. fig. 4; and since it appears too large for *M. minuta*, the probability is that it belongs to the present species. The specimen has been noticed by the writer in the 'Quart. Journ. Geol. Soc.' vol. xliii. p. 11. *Purchased*, 1865.

Megaptera minuta (Van Beneden [3]).

Syn. *Burtinopsis minutus*, Van Beneden [4].

This species is about equal in size to *Balænoptera rostrata*; the

[1] Bull. Ac. R. Belg. sér. 2, vol. xxxiv. p. 19 (1872).—*Burtinopsis.*
[2] *Loc. cit.*
[3] *Ibid.* vol. l. p. 17 (1880).—*Burtinopsis.*
[4] *Loc. cit.*

anterior extremity of the tympanic is pointed and the inner surface more convex than in the last species.

Hab. Europe (Belgium and England).

M. 3533. The imperfect right tympanic; from the Antwerp Crag. This specimen agrees precisely with the one figured by Van Beneden in the 'Ann. Mus. R. Hist. Nat. Belg.' vol. vii. pt. 3, pl. xcvii. figs. 1, 2, and also in form with the tympanic of *M. boops* (B.M. No. 1468 c).

By exchange with the Royal Brussels Museum of Natural History, 1886.

M. 3532. Cast of the slightly imperfect left tympanic of a somewhat older individual. The original was obtained from the Coralline Crag of Suffolk, and is preserved in the Ipswich Museum; it is described and figured by the present writer in the 'Quart. Journ. Geol. Soc.' vol. xliii. p. 11, pl. ii. figs. 5, 5 *a*. The specimen closely resembles the one figured by Van Beneden, *op. cit.* figs. 9-11, and also an old tympanic of *M. boops* in the British Museum (No. 2. 76. 16. 18). *Made in the Museum*, 1886.

Genus **BALÆNOPTERA,** Lacépède[1].

The skull is small, the nasals are short and rounded, and the interval between the frontal and occipital is comparatively short. The tympanic (woodcut, fig. 9) is moderately inflated, the inferior aspect forming a comparatively narrow ellipse, which is scarcely wider behind than in front; the inner surface is flattened, without a vertical groove; the notch for the eustachian canal is deep and sharply defined; the antero-inferior extremity of the inner surface is produced into a more or less well-marked ridge; and the outer surface of the involucrum is flattened; the inferior surface has two ridges corresponding to those of *Megaptera*.

Balænoptera definita (Owen[2]).

Syn. *Balæna definita*, Owen[3].
Balænodon definita, Owen[4].
Balænoptera sibbaldina, Van Beneden[5].

---

[1] Hist. Nat. d. Cétacés, Table d. Ordres &c. p. xxxvi (1804).
[2] Proc. Geol. Soc. vol. iv. p. 283 (1843).—*Balæna.*
[3] *Loc. cit.*
[4] Brit. Foss. Mamm. and Birds, table facing p. xlvi (1846).
[5] Bull. Ac. R. Belg. sér. 2, vol. l. p. 14 (1880).

Fig. 9.

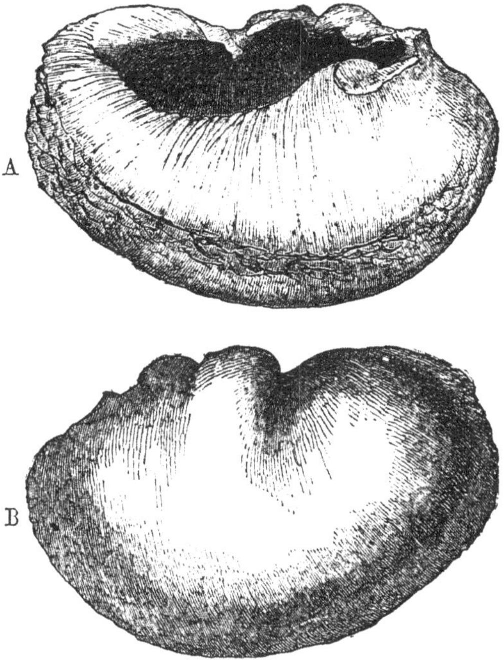

Balænoptera musculus.—The right tympanic, from the inner (A) and outer (B) aspects; recent. ½. (From the 'Proc. Zool. Soc.')

The tympanic of this species indicates a true *Balænoptera*, and is of rather larger size than the corresponding bone of *B. goropi*, from which it also differs in structure. The Belgian *B. sibbaldina* (of which the tympanic is undescribed) is almost certainly identical with this form, and is considered to be allied to the existing *B. sibbaldi*, which attains a length of 80 feet. A fine cervical vertebra from the Red Crag is preserved in the Ipswich Museum.

Hab. Europe (England and Belgium).

M. 3521. Cast of the imperfect left tympanic. The original, which is from the Red Crag and is preserved in the Ipswich Museum, is described and figured by the present writer in the 'Quart. Journ. Geol. Soc.' vol. xliii. p. 11, pl. ii. figs. 3, 3 *a*. It agrees precisely with the imperfect type tympanic figured by Owen in the 'British Fossil Mammals and Birds,' p. 531, fig. 222 (Mus. R. Coll. Surg. No. 2804), as well as with another specimen in the Museum of the Royal College of Surgeons (No. 2805). It differs from the tympanic of *B. goropi* (No. M. 3517, below) by its somewhat superior size, greater relative width, greater height of the inner wall, smaller depth of the eustachian notch, sharper posterior angle, and more gibbous involucrum.

Made in the Museum, 1886.

Balænoptera goropi (Van Beneden [1]).

Syn. *Plesiocetus goropi*, Van Beneden [2].
Balænoptera musculoides, Van Beneden [3].

This species is closely allied in size and structure to *B. musculus*, of which it may be the ancestor. The existing species attains a length of at least 70 feet; and the fossil one is very common in the Antwerp Crag.

Hab. Europe (Belgium and England).

M. 3517. The imperfect left tympanic; from the Antwerp Crag. This specimen is indistinguishable from the one figured by Van Beneden in the 'Ann. Mus. R. Hist. Nat. Belg.' vol. vii. pt. 3, pl. liv. fig. 18. *By exchange with the Royal Brussels Museum of Natural History*, 1886.

M. 3517 a. A precisely similar left tympanic; from the Antwerp Crag. *By exchange with the Royal Brussels Museum of Natural History*, 1886.

39016. A much-rolled right tympanic, agreeing very closely with the preceding specimens; from the Red Crag of Woodbridge, Suffolk. Noticed by the present writer in the 'Quart. Journ. Geol. Soc.' vol. xliii. p. 12.

Purchased, 1865.

The following specimens from the Antwerp Crag are provisionally referred to this species, although some of them may belong to Cetotherium brialmonti.

43601. The centrum and part of the neural arch of the first dorsal

[1] Bull. Ac. R. Belg. sér. 2, vol. xxxiv. p. 15 (1872).—*Cetotherium*; the specific name is misprinted *garopi*.

[2] *Loc. cit.*

[3] *Ibid.* vol. l. p. 15 (1880). When the generic name was changed the founder also proposed a new specific one; but there is no justification for the substitution.

vertebra. This specimen agrees very closely with the one figured by Van Beneden, *op. cit.* pl. lxii. *Purchased.*

42993. The centrum of a late dorsal or early lumbar vertebra. *Van Breda Collection. Purchased,* 1871.

43601 a. The centrum of a late dorsal or early lumbar vertebra, agreeing very closely with the specimen figured by Van Beneden, *op. cit.* pl. lxiii. *Purchased.*

43601 b. The centrum of a late dorsal or early lumbar vertebra. *Purchased.*

42996 a. The centrum of a middle caudal vertebra. *Van Breda Collection. Purchased,* 1871.

Some of the following specimens from the Crag of Suffolk may belong to Cetotherium brialmonti.

48475. The nearly complete axis vertebra, wanting the hinder epiphysis; from the Red Crag of Woodbridge. This specimen differs from the axis of *Cetotherium brialmonti* (compare No. 46734, page 43) by its much thinner centrum. *Purchased,* 1875.

48476. The centrum of the third cervical vertebra; from the Red Crag of Woodbridge. In its large size this specimen differs from the corresponding vertebra of *C. brialmonti* and agrees with that of the present species. *Purchased,* 1875.

48477. The centrum of the first dorsal vertebra; from Woodbridge. *Purchased,* 1875.

46739. The imperfect centrum of a late lumbar or early caudal vertebra; from Woodbridge. This specimen closely resembles the eleventh lumbar of *C. brialmonti* figured by Van Beneden, *op. cit.* vol. ix. pt. 4, pl. ix. figs. 3, 4. *Purchased,* 1875.

48464. The centrum of an early caudal vertebra; from the Coralline Crag. This specimen agrees very nearly with the second caudal of *C. brialmonti* figured by Van Beneden, *op. cit.* pl. viii. fig. 4. *Presented by C. Falconer, Esq.,* 1867.

46740. The centrum of a very similar caudal vertebra; from Wood bridge. *Purchased,* 1875.

46751. A late caudal vertebra; from Woodbridge.
Purchased, 1875.

35042. A later caudal vertebra, apparently belonging either to this or the next genus; from the Red Crag of Bawdsey.
Presented by S. V. Wood, Esq., 1860.

Balænoptera borealis, Lesson[1].

The usual length of this species is from 40 to 48 feet, but it occasionally reaches 52 feet[2].

Hab. North Atlantic.

49977. The third cervical vertebra, wanting portions of the lateral arches; dredged from the North Sea. *Purchased*, 1867.

M. 425. The centrum of a cervical vertebra; from the Pleistocene (Elephant-bed) of Brighton, Sussex.
Enniskillen Collection. Purchased, 1882.

46291. The centrum of a dorsal vertebra; dredged from the North Sea. *Owles Collection. Purchased*, 1874.

46286–7. The centra of two lumbar vertebræ; dredged from the North Sea. *Owles Collection. Purchased*, 1874.

46284. The centrum of an early caudal vertebra; dredged from the North Sea. *Owles Collection. Purchased*, 1874.

46285. The centrum of a later caudal vertebra; dredged from the North Sea. *Owles Collection. Purchased*, 1874.

46288. The centrum of a smaller caudal vertebra; dredged from the North Sea. *Owles Collection. Purchased*, 1874.

46289. The centrum of a late caudal vertebra; dredged from the North Sea. *Owles Collection. Purchased*, 1874.

33454. An early caudal vertebra; from the eastern coast.
Layton Collection. Purchased, 1858.

The following specimens are provisionally referred to this species.

7342 a. Part of a mandibular ramus; from the Pleistocene (Elephant-bed) of Brighton. This specimen is described

[1] Hist. Nat. Cétacés, p. 342 (1828).
[2] See Collett, Proc. Zool. Soc. 1886, p. 264.

by Mantell in the 'Medals of Creation,' vol. ii. p. 824, and referred to *Balæna mysticetus* [1]; but the occurrence of the cervical vertebra No. M. 425 in the same bed renders it more probable that it belongs to the present species. *Mantell Collection. Purchased*, 1836.

M. 424. Fragment of a rib: from the Pleistocene of Brighton. *Enniskillen Collection. Purchased*, 1882.

33451. The imperfect centrum of a middle caudal vertebra; from the Norfolk coast. *Layton Collection. Purchased*, 1858.

33452. The imperfect centrum of a later caudal vertebra; from the Norfolk coast. *Layton Collection. Purchased*, 1858.

33462. An imperfect caudal vertebra; dredged from the eastern coast. *Layton Collection. Purchased*, 1858.

33457. The rolled centrum of a caudal vertebra; from the eastern coast. *Layton Collection. Purchased*, 1858.

The following specimen is larger than any of the preceding, and may perhaps belong to B. musculus.

46283. The centrum of a lumbar vertebra; dredged in the North Sea. *Owles Collection. Purchased*, 1874.

Balænoptera borealina, Van Beneden [2].

This species is smaller than *B. goropi*, and closely allied to the existing *B. borealis.*

Hab. Europe (Belgium and England).

M. 3518. The imperfect right tympanic; from the Antwerp Crag. This specimen agrees with the one figured by Van Beneden in the 'Ann. Mus. R. Hist. Nat. Belg.' vol. vii. pt. 3, pl. lxviii. fig. 17. *By exchange with the Royal Brussels Museum of Natural History*, 1886.

M. 3522. Cast of the imperfect left tympanic. The original, which is much water-worn, was obtained from the Red Crag of Suffolk, and is preserved in the Ipswich Museum; it agrees precisely with the Belgian specimen, and is noticed by the present writer in the 'Quart. Journ. Geol. Soc.' vol. xliii. p. 12. *Made in the Museum*, 1886.

[1] See Owen, Brit. Foss. Mamm. and Birds, p. 542.

[2] Bull. Ac. R. Belg. sér. 2, vol. l. p. 15 (1880).

39018. The much-worn right tympanic; from the Red Crag of Woodbridge, Suffolk. This and the following specimens are noticed by the writer in the passage cited above.
Purchased, 1865.

39017. The more worn left tympanic; from Woodbridge.
Purchased, 1865.

39021. The left periotic; from the Red Crag of Woodbridge. This specimen exactly resembles the one figured by Van Beneden, *op. cit.* fig. 10. *Purchased*, 1865.

27603. The left periotic; from Woodbridge. *Purchased*, 1852.

39019. The right periotic; from Woodbridge. *Purchased*, 1865.

27434. An imperfect right periotic, provisionally referred to this species; from Woodbridge. *Purchased*, 1851.

Balænoptera emarginata (Owen[1]).

Syn. *Balæna emarginata*, Owen[2].
Balæna gibbosa, Owen[3].
Balænodon emarginata, Owen[4].
Balænodon gibbosa, Owen[5].
Balænoptera rostratella, Van Beneden[6].

The tympanic named *B. gibbosa* by Owen differs only from the type specimen of the present species by a character dependent upon age; and both specimens agree precisely with the Belgian examples forming the so-called *B. rostratella* of Van Beneden. The species is allied to *B. rostrata* (of which the usual length is from 25 to 30 feet, and specimens exceeding 33 feet are of very rare occurrence[7]), but appears of rather larger size.

Hab. Europe (England and Belgium).

39016 a. The much-worn left tympanic; from the Red Crag of Woodbridge, Suffolk. This specimen agrees exactly with the type tympanic figured by Owen in the 'British Fossil Mammals and Birds,' p. 533, fig. 224 (Mus. R. Coll.

[1] Proc. Geol. Soc. vol. iv. p. 283 (1843).—*Balæna.*
[2] *Loc. cit.* [3] *Loc. cit.*
[4] Brit. Foss. Mamm. and Birds, table facing p. xlvi (1846).
[5] *Ibid.*
[6] Bull. Ac. R. Belg. sér. 2, vol. l. p. 16 (1880).
[7] See Collett, Proc. Zool. Soc. 1886, p. 264.

Surg. No. 2822); it is noticed by the writer in the 'Quart. Journ. Geol. Soc.' vol. xliii. p. 12. *Purchased*, 1865.

M. 3519. The imperfect left tympanic, agreeing with the last specimen and also with the one figured by Van Beneden in the 'Ann. Mus. R. Hist. Nat. Belg.' vol. vii. pt. 3, pl. lxxviii. fig. 7; from the Antwerp Crag. *By exchange with the Royal Brussels Museum of Natural History*, 1886.

48977. The left periotic; from the Red Crag of Woodbridge. This specimen agrees precisely with the one figured by Van Beneden, *op. cit.* pl. lxxviii. fig. 18. *Purchased*, 1875.

Balænoptera, sp.

The following specimen apparently indicates a form allied to *B. emarginata*, but of rather smaller dimensions.

Hab. Australia.

M. 3505. The much-rolled imperfect left tympanic; from Tertiary (? Miocene) strata at Cheltenham, Victoria. The length of this specimen is 0,078; it differs from any of the tympanics figured by M'Coy in the 'Prodromus of the Palæontology of Victoria' (Geol. Surv. Victoria), dec. vi. pl. liv. (1879) under the name of *Cetotolites*, which apparently belong to *Cetotherium*. *Purchased*, 1886.

Balænoptera (?) juddi, Seeley[1].

The generic position of this comparatively small species must be regarded as uncertain.

Hab. England.

M. 133. (*Fig.*) The eighth (?) caudal vertebra; from the Upper Eocene (Lower Oligocene) Brockenhurst beds of Roydon, Hampshire. This specimen is the type, and is described and figured by Seeley in the 'Quart. Journ. Geol. Soc.' vol. xxxvii. pp. 709–712, figs. 1–3.

Presented by Prof. J. W. Judd, 1881.

Balænoptera (?), sp.

The following specimen, which is only provisionally referred to

[1] Quart. Journ. Geol. Soc. vol. xxxvii. p. 709 (1881).

the present genus, apparently indicates a species agreeing nearly in size with *B. borealina*.

Hab. Australia.

M. 3506. The hinder part of the much-rolled left tympanic; from Tertiary (? Miocene) strata at Cheltenham, Victoria.
Purchased, 1886.

Genus **CETOTHERIUM**, Brandt[1].

Including:—*Eucetotherium*, Brandt[2].
Plesiocetopsis, Brandt[3].
Cetotheriophanes, Brandt[4].
Plesiocetus, Van Beneden[5].
Heterocetus, Van Beneden[6].
Amphicetus, Van Beneden[7].

The genus was divided by Brandt into the three subgenera mentioned above; the first of which appears to be equivalent to *Heterocetus*[8], while the second is the same as *Plesiocetus* of Van Beneden. *Amphicetus* is another division which is not of more than subgeneric value.

The tympanic is much narrower anteriorly than posteriorly, its rough inferior surface forming an isosceles triangle, and the notch for the eustachian canal being smaller and descending nearer to the inferior border of the inner wall than in *Balænoptera*. The cranium is longer than in the latter, the interval between the occipital and the frontal being greater, and the nasals more elongated and flattened; the cervical vertebræ are thicker.

A. *Plesiocetine Group.*

The mandibular condyle has no talon.

Cetotherium brialmonti (Van Beneden[9]).

Syn. *Plesiocetus brialmonti*, Van Beneden[10].

This is the largest of the Belgian species, and is somewhat infe-

[1] Bull. Ac. Imp. St. Pétersbourg, vol. i. p. 145 (1843).
[2] Mém. Ac. Imp. St. Pétersbourg, vol. xx. p. 143 (1873).
[3] *Ibid.* p. 143.
[4] *Ibid.* p. 148.
[5] Bull. Ac. R. Belg. sér. 2, vol. xxxiv. p. 15 (1872).
[6] *Ibid.* vol. l. p. 21 (1880).
[7] *Ibid.* vol. l. p. 20 (1880).
[8] *Ibid.* vol. l. p. 21 (1880).
[9] *Ibid.* vol. l. p. 18 (1880).—*Plesiocetus.*
[10] *Loc. cit.*

rior in size to *Balænoptera goropi*; the parietals are longer than in *C. dubium*.

Hab. Europe (Belgium and England).

M. 3527. The imperfect left tympanic; from the Antwerp Crag. This specimen closely resembles the one figured by Van Beneden in the 'Ann. Mus. R. Hist. Nat. Belg.' vol. ix. pt. 4, pl. ii. fig. 1. *By exchange with the Royal Brussels Museum of Natural History*, 1886.

43600. The right humerus; from the Antwerp Crag. This specimen resembles the one figured by Van Beneden, *op. cit.* pl. iv. *Purchased.*

46734. The centrum of the axis vertebra; from the Red Crag of Woodbridge, Suffolk. This specimen agrees very closely with the one figured by Van Beneden, *op. cit.* pl. vii. fig. 4. *Purchased*, 1875.

Cetotherium dubium, Van Beneden[1].

Syn. *Plesiocetus dubius*, Van Beneden[2].
Plesiocetopsis dubius, Brandt[3].

This species is intermediate in size between *C. brialmonti* and *C. burtini*, many of the bones being very difficult to distinguish from those of the latter, and also resembling those of *Megaptera similis*. The atlas and axis are less massive than in *C. burtini*. Two worn tympanics in the Museum of the Royal College of Surgeons (No. 2832 A, B) indicate the occurrence of the species in the Red Crag.

Hab. Europe (Belgium and England).

M. 3526. The slightly imperfect left tympanic; from the Antwerp Crag. This specimen agrees precisely with the one figured by Van Beneden in the 'Ann. Mus. R. Hist. Nat. Belg.' vol. ix. pt. 4, pl. xiv. fig. 14.
By exchange with the Royal Brussels Museum of Natural History, 1886.

M. 3526 a. A rather more imperfect example of the left tympanic; from the Antwerp Crag.
By exchange with the Royal Brussels Museum of Natural History, 1886.

[1] Bull. Ac. R. Belg. sér. 2, vol. xxxiv. p. 18 (1872).
[2] *Ibid.* vol. l. p. 18 (1880).
[3] Mém. Ac. Imp. St. Pétersbourg, vol. xx. p. 146 (1873).

M. 3525. The imperfect right tympanic; from the Antwerp Crag.
By exchange with the Royal Brussels Museum of Natural History, 1886.

30261 a. An imperfect left periotic probably belonging to this species or to *C. burtini*; from the Red Crag of Woodbridge, Suffolk. This specimen agrees very closely with the periotic of the latter species figured by Van Beneden, *op. cit.* pl. xxviii. figs. 2–4; it is noticed by the present writer in the 'Quart. Journ. Geol. Soc.' vol. xliii. p. 12. *Purchased*, 1855.

36601. Part of a similar right periotic; from Woodbridge.
Purchased, 1855.

27434 a. An imperfect left periotic, of rather smaller size; from Woodbridge. *Purchased*, 1851.

The following specimens may probably be referred either to the present species or to C. burtini.

42996 c. The centrum of an early lumbar vertebra; from the Antwerp Crag. *Purchased.*

43599. The centrum of an early caudal vertebra; from the Antwerp Crag. This specimen closely resembles the fourth caudal of the present species figured by Van Beneden, *op. cit.* pl. xix. fig. 5. *Purchased.*

43599 a. The imperfect centrum of a caudal vertebra; from the Antwerp Crag. *Purchased.*

46735. The centrum of the third (?) cervical vertebra; from the Red Crag of Woodbridge, Suffolk. This specimen corresponds with the third cervical of the present species figured by Van Beneden, *op. cit.* pl. xviii. fig. 3.
Purchased, 1875.

Cetotherium burtini, Van Beneden[1].

Syn. *Plesiocetus burtini*, Van Beneden[2].
Plesiocetopsis burtini, Brandt[3].

This species is rather smaller than *C. dubium*, its length being

[1] Bull. Ac. R. Belg. sér. 2, vol. xxxiv. p. 18 (1872).
[2] *Ibid.* vol. l. p. 19 (1880).
[3] Mém. Ac. Imp. St. Pétersbourg, vol. xx. p. 146 (1873).

estimated at about thirty feet. The tympanics and vertebræ appear very difficult to distinguish from those of *C. hupschi*.

Hab. Europe (Belgium).

M. 3529. The imperfect left tympanic; from the Antwerp Crag. This specimen closely resembles the one figured by Van Beneden in the 'Ann. Mus. R. Hist. Nat. Belg.' vol. ix. pt. 4, pl. xxix. fig. 8. *By exchange with the Royal Brussels Museum of Natural History*, 1886.

M. 3524. The imperfect right tympanic; from the Antwerp Crag. *By exchange with the Royal Brussels Museum of Natural History*, 1886.

Cetotherium hupschi, Van Beneden [1].

Syn. *Plesiocetus hupschi*, Van Beneden [2].
Plesiocetopsis hupschi, Brandt [3].

This species is estimated to have been nearly equal in size to *Balænoptera rostrata*.

Hab. Europe (Belgium, France [4], Portugal [4], and [?] England).

43599. The imperfect centrum of an atlas vertebra belonging either to this species or to *C. burtini*; from the Antwerp Crag. This specimen is slightly larger than the one figured by Van Beneden in the 'Ann. Mus. R. Hist. Nat. Belg.' vol. ix. pt. 4, pl. xxvi. figs. 2, 3. *Purchased.*

43602. The centrum of a lumbar and another of a caudal vertebra; from the Antwerp Crag. *Purchased.*

32732. The centrum of a late lumbar vertebra; from the Pliocene (?) of Xabregas, near Lisbon, Portugal. This specimen is referred to the present species by Van Beneden in the 'Bull. Ac. R. Belg.' sér. 2, vol. l. p. 19. *Presented by J. S. Valentine, Esq.*, 1857.

The specific reference of the following specimens from the Red Crag of Suffolk is provisional.

28267. The imperfect centrum of the axis vertebra; from Woodbridge. This specimen agrees very closely with No. 43599, and doubtless belongs to the same species. *Purchased*, 1852.

[1] Bull. Ac. R. Belg. sér. 2, vol. xxxiv. p. 17 (1872)
[2] *Ibid.* vol. l. p. 19 (1880).
[3] Mém. Ac. Imp. St. Pétersbourg, vol. xx. p. 144 (1873).
[4] Van Beneden, *op. cit.* vol. l. p. 19.

28269. The imperfect centrum of an early caudal (?) vertebra; from Woodbridge. *Purchased*, 1852.

44114. Part of the right temporal; from Woodbridge. *Purchased*, 1873.

B. *Heterocetine Group.*

The mandibular condyle has a distinct talon.

Cetotherium brevifrons, Van Beneden[1].

Syn. *Heterocetus brevifrons*, Van Beneden[2].
Plesiocetopsis brevifrons, Brandt[3].

The length of this species probably did not exceed twenty feet; the frontal is short, and the cranium elevated. The type specimens are preserved in the Brussels Museum, but have not been figured.

Hab. Europe (Belgium and (?) England).

M. 3528. The slightly imperfect right tympanic; from the Antwerp Crag. This specimen agrees exactly with one of the type tympanics (No. 1371) in the Brussels Museum. *By exchange with the Royal Brussels Museum of Natural History*, 1886.

M. 3523. A worn and imperfect left tympanic, not improbably belonging to this species; from the Antwerp Crag. This specimen resembles some tympanics in the Brussels Museum labelled "*Heterocetus sprangi*, Van Beneden." *By exchange with the Royal Brussels Museum of Natural History*, 1886.

45992. The centrum of the axis vertebra; from Woodbridge, Suffolk. This specimen agrees exactly with the examples in the Brussels Museum; there is a similar specimen from the Red Crag in the Ipswich Museum. *Purchased*, 1875.

28265-6. The imperfect atlas and axis vertebræ, of rather smaller size than the preceding; from the Red Crag of Woodbridge. These specimens probably belong to a small individual of the present or following species. *Purchased*, 1852.

[1] Bull. Ac. R. Belg. sér. 2, vol. xxxiv. p. 17 (1872).
[2] *Ibid.* vol. l. p. 22 (1880).
[3] Mém. Ac. Imp. St. Pétersbourg, vol. xx. p. 145 (1873).

M. 3567. The centrum and left transverse process of a dorsal vertebra, not improbably belonging to this species; from the Antwerp Crag. *Purchased.*

Cetotherium affine (Van Beneden [1]).

Syn. *Heterocetus affinis*, Van Beneden [2].

This species is rather smaller than the preceding, and has shorter frontals and a lower cranium.

Hab. Europe (Belgium).

M. 3530. The nearly perfect right tympanic of a young individual, agreeing with the specimens in the Brussels Museum which are referred to this species; from the Antwerp Crag.
By exchange with the Royal Brussels Museum of Natural History, 1886.

Genus **HERPETOCETUS**, Van Beneden [3].

This genus is characterized by the remarkably inflated and egg-like tympanic (in which the involucrum has a very peculiar shape), and by the presence of a long talon to the mandibular condyle; the latter feature indicating affinity with the *Physeteridæ*. The tympanic is more like that of *Megaptera* than that of *Cetotherium*, but is distinguished by the abrupt outer border of the involucrum, and the filling-up of a great part of the cavity by bone.

Herpetocetus scaldiensis, Van Beneden [4].

This is the typical and only species, and is of comparatively small size.

Hab. Europe (Belgium and England).

M. 3534. The imperfect right tympanic; from the Antwerp Crag. This specimen agrees exactly with the type examples figured by Van Beneden in the 'Ann. Mus. R. Hist. Nat. Belg.' vol. vii. pt. 3, pl. cv.
By exchange with the Royal Brussels Museum of Natural History, 1886.

[1] Bull. Ac. R. Belg. sér. 2, vol. 1. p. 21 (1880).—*Heterocetus.*
[2] *Loc. cit.*
[3] *Ibid.* vol. xxxiv. p. 20 (1872).—Amended from *Erpetocetus.*
[4] *Loc. cit.*

M. 3520. Cast of the imperfect right tympanic. The original is from the Red Crag of Felixstowe, Suffolk, and is preserved in the Museum of Practical Geology, Jermyn Street; it agrees almost exactly with the last specimen, and is noticed by the present writer in the 'Quart. Journ. Geol. Soc.' vol. xliii. p. 13. *Made in the Museum*, 1886.

Generically undetermined Specimens.

A. *The following specimens probably belong to the Balænopterine Section.*

a. *From Australia.*

43997–8. Fragment of a mandibular ramus and the imperfect centrum of a lumbar (?) vertebra; from the Pleistocene (?) of Mordillioc, Hobson's Bay, Victoria. These specimens indicate a species of medium size. *Purchased*, 1872.

b. *From North America.*

M. 3568. The centrum of the sixth cervical vertebra; from Calvert Cliffs, Maryland, U.S.A. This specimen agrees in size with the corresponding vertebra of *Balænoptera rostrata*, but is relatively thicker.
Bean Collection. Purchased, 1859.

B. *The subfamily of the following specimens, which are all from the Red Crag of Suffolk, is not determined.*

26482. Fragment of the hinder part of the cranium.
Presented by Rev. Prof. Henslow, 1851.

36437, 37728–9, 44113. Four fragments of the cranium.
Purchased.

M. 1614. Fragment of the cranium; from Felixstowe.
Presented by C. Westendarp, Esq., 1884.

33395. The proximal extremity of the left ulna.
Layton Collection. Purchased, 1858.

48472. Part of the proximal extremity of a radius.
Presented by C. Falconer, Esq., 1867.

43999. Head of an anterior rib. *Purchased*, 1872.

Suborder *ARCHÆOCETI.*

Teeth differentiated into groups; nasals long.

Family ZEUGLODONTIDÆ.

Dentition:—I. $\frac{3}{3}$, C. $\frac{1}{1}$, Pm. and M. $\frac{5}{5}$. The cheek-teeth (woodcut, fig. 10) have two distinct roots, and compressed pointed crowns with denticulated cutting-edges. The cranium is elongated and depressed; the brain-cavity small; the temporal fossæ and the sagittal crest are large; the rostrum is long and largely composed laterally of the premaxillæ; and the nasals are elongated, flat and narrow. All the cervical vertebræ are free, and the lumbars elongated.

Genus **ZEUGLODON,** Owen[1].

All the known forms of the family may be provisionally referred to this genus, which has been recorded from the Eocene of North America, England, and Egypt. A tooth and caudal vertebra are figured in the accompanying woodcut.

Fig. 10.

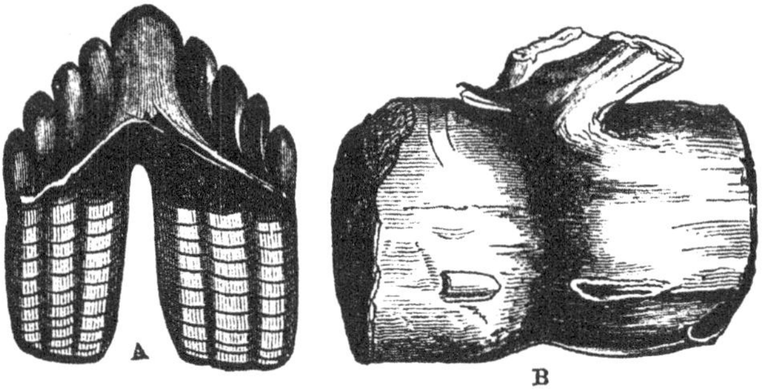

Zeuglodon cetoides.—A, molar tooth; B, caudal vertebra: reduced. From the Middle Eocene of North America. (After Lyell.)

[1] Lond. & Edin. Philos. Mag. vol. xiv. p. 302 (1839). The earlier name *Basilosaurus*, Harlan, was withdrawn by its author.

Zeuglodon cetoides, Owen[1].

Syn. (?) *Zeuglodon hydrarchus*, Carus[2].
(?) *Zeuglodon brachyspondylus*, Müller[3].
Zeuglodon macrospondylus, Müller[4].
Basilosaurus cetoides, Leidy[5].

This is the type species. The smaller form known as *Z. hydrarchus* (= *brachyspondylus*) is regarded by Leidy (*op. cit.*) and Dames[6] as being the female of the large *Z. cetoides* (*macrospondylus*), the same view being taken by Gaudry[7].

Hab. North America.

The following specimens belong to the large typical form.

35590–91. A perfect and a broken anterior tooth; from the Eocene of Clarke County, Alabama, U.S.A.
Presented by Prof. J. W. Mallet, 1859.

35592–95. Four imperfect cheek-teeth; from Clarke County.
Presented by Prof. J. W. Mallet, 1859.

40980. The greater portion of an anterior tooth, and an incomplete cheek-tooth; from the Eocene of Clarksville, Alabama.
Presented by Sir Charles Lyell, Bart., 1868.

23628. Cast of a complete unworn cheek-tooth. The original is from the Middle (?) Eocene of Claiborne, Alabama, and is preserved in the Museum at Berlin. It is figured in its original imperfect condition by Müller, *op. cit.* pl. xii. fig. 2. *Purchased.*

25444. Cast of the germ of part of a cheek-tooth. The original is from Alabama, and is figured by Müller, *op. cit.* pl. xxiii. fig. 5. *Purchased.*

25445. Cast of an imperfect worn cheek-tooth. The original is from Alabama. *Purchased.*

[1] Trans. Geol. Soc. ser. 2, vol. vi. p. 69 (1841).
[2] Nova Acta Ac. Cæs. Leop.-Car. vol. xxii. pt. 2, p. 373 (1847).
[3] 'Foss. Reste d. Zeuglodonten von Nordamerica,' p. 8 (1849).
[4] *Loc. cit.*
[5] Journ. Ac. Nat. Sci. Philad. ser. 2, vol. vii. p. 427 (1869).
[6] Sitz. k. preuss. Ak. Wiss. 1883, vol. i. pp. 134, 135.
[7] 'Les Enchaînements, etc.—Mammifères Tertiaires,' p. 39.

25446. Cast of the root of a cheek-tooth. The original is from Alabama. *Purchased.*

25442. Cast of fragment of a mandibular ramus, containing an imperfect anterior tooth. The original is from Alabama. *Purchased.*

25443. Cast of fragment of a mandibular ramus, containing the basal part of a cheek-tooth. The original is from Alabama. *Purchased.*

35641. Fragment of a jaw containing the root and the base of the crown of an anterior tooth; from Clarke County. *Presented by Prof. J. W. Mallet,* 1859.

35642. Fragment of a mandibular ramus, showing the broken bases of several cheek-teeth; from Clarke County. *Presented by Prof. J. W. Mallet,* 1859.

35597. An imperfect lumbar vertebra; from Clarke County. *Presented by Prof. J. W. Mallet,* 1859.

35596. A lumbar vertebra; from Clarke County. *Presented by Prof. J. W. Mallet,* 1859.

8142. One half of a very large lumbar vertebra; from Alabama. *Mantell Collection. Purchased,* 1836.

20707. Two imperfect lumbar vertebræ; from the Brazos River, San Felipe de Austin, Texas. *Purchased,* 1847.

40987. Fragment of a rib; from Alabama. *Presented by Sir C. Lyell, Bart.,* 1869.

The following smaller specimens belong to the so-called Z. hydrarchus.

M. 3543. Cast of the cranium. The original is from Claiborne, and is preserved in the Museum at Haarlem; it is figured by Carus, *op. cit.* figs. 39, A, B (as *Z. hydrarchus*), and by Müller, *op. cit.* pl. xxvi. (as *Z. brachyspondylus*), and also in Gaudry's ‘Enchaînements—Mamm. Tert.’ p. 39, fig. 30

(as *Z. cetoides*). In the original specimen the crowns of the teeth are broken off, but these have been restored in the cast. *Purchased.*

M. 3544. Cast of a humerus. The original is from Claiborne, and is preserved in the Museum at Haarlem. *Purchased.*

40985. Cast of an imperfect late dorsal vertebra; from Alabama. This specimen very closely resembles the one figured by Müller, *op. cit.* pl. xix. fig. 2.
Presented by Sir C. Lyell, Bart., 1868.

35645. An imperfect early caudal vertebra, wanting both epiphyses; from Alabama. *Presented by Prof. J. W. Mallet*, 1859.

40981. The imperfect centrum of a vertebra.
Presented by Sir C. Lyell, Bart., 1868.

25447. Cast of a tympanic. The original is from Alabama.
Purchased.

41846. A series of late caudal vertebræ, provisionally referred to this species; from the phosphate beds (Eocene) of the Ashley River, South Carolina.
Presented by Cowlan Gravely, Esq., 1869.

Zeuglodon, sp.

Hab. Malta. The occurrence of remains of this genus in the Miocene of Malta is recorded by Leith-Adams in his 'Nile Valley and Malta,' pp. 265, 269 (1870).

40708. The imperfect centrum of a late lumbar (?) vertebra of a large form; from the Miocene of Malta.
Purchased, 1867.

24597. The imperfect centrum of a lumbar vertebra of a small form; from the same formation. *Purchased*, 1850.

Suborder *ODONTOCETI.*

Teeth are always present after birth; the functional ones are generally numerous, but sometimes are reduced to a single pair, and are occasionally wanting; baleen is absent. The cranium is more or less asymmetrical, and the nasals are reduced to nodules which do not roof over the narial passages. The tympanic is not anchylosed to the periotic, and is of a more open structure than in the Mystacoceti.

Family PHYSETERIDÆ.

No functional teeth in the upper jaw; mandibular teeth various, and often reduced in number. The bones of the cranium are raised so as to form an elevated crest or prominence behind the nares. The anterior facette of the periotic for articulation with the tympanic is quite smooth; and the posterior tympanic surface of the former is broad, and carries a median longitudinal ridge.

Subfamily PHYSETERINÆ.

The mandibular teeth are numerous and implanted in a long groove partly divided by imperfect septa.

Genus **PHYSETER,** Linn.[1]

There are from twenty to twenty-five mandibular teeth, which have no enamel; the mandible is very long and narrow, the symphysis occupying more than half its length. The atlas is free, but all the other cervical vertebræ are fused into a solid mass.

Physeter macrocephalus, Linn.[2]

Hab. All tropical and temperate seas.

39264. The ventral half of the atlas vertebra; from the Pleistocene (?) of Uruguay, South America. As this specimen is indistinguishable from the corresponding bone of the existing Cachalot, it may be at least provisionally referred to the same species.

Presented by Captain J. Parish, R.N., 1865.

[1] Syst. Nat. ed. 12, vol. i. p. 107 (1766). [2] *Loc. cit.*

Genus **PHYSETERULA,** Van Beneden[1].

This genus appears closely allied to *Physeter*, but the length of the symphysis is only one third that of the mandible.

Physeterula dubusi, Van Beneden[2].

This is the type species, and is about one third the size of medium individuals of *Physeter macrocephalus*, but with relatively longer and more slender teeth.

Hab. Europe (Belgium and [?] England).

49966. One half of a longitudinally bisected tooth, provisionally referred to this species; from the Red Crag of Woodbridge, Suffolk. This specimen apparently agrees with the teeth of the type mandible figured by Van Beneden in the 'Bull. Ac. R. Belg.' sér. 2, vol. xliv. pl. facing p. 856; it is noticed by the writer in the 'Quart. Journ. Geol. Soc.' vol. xliii. p. 14, under the name of *Homocetus villersi*, to which form the type mandible was referred by Du Bus. *Purchased*, 1879.

Genus **EUCETUS,** Du Bus[3].

Syn. *Dinoziphius*, Van Beneden[4].

The teeth of this genus closely resemble those of *Physeter*, being only distinguished by their subcircular cross section and the small size of the inferior aperture of the pulp-cavity. Apart from the question of enamel, the teeth differ from the type of *Balænodon* by the fusiform dentine core, and the frequent presence of an axis of osteodentine of a nodular structure, and lighter in colour than the dentine. The periotic provisionally assigned to the genus is nearer to that of *Hyperoodon* than to that of *Physeter*; and this reference, if correct, would indicate the right of the present genus to stand, unless it be founded on worn teeth of *Scaldicetus*.

Eucetus amblyodon, Du Bus[5].

Syn. *Dinoziphius ramdoncki*, Van Beneden[6].

This is the type and only species; teeth from the Red Crag in the

[1] Bull. Ac. R. Belg. sér. 2, vol. xliv. p. 851 (1877). [2] *Loc. cit.*

[3] *Ibid.* vol. xxiv. p. 572 (1867).

[4] Quoted by Gervais in the 'Ostéographie des Cétacés,' p. 344 (1867–79). The specific name is misspelt.

[5] *Loc. cit.* [6] *Loc. cit.*

Ipswich Museum are nearly as large as those of full-grown speci mens of *Physeter macrocephalus*.

Hab. Europe (Belgium and England).

The majority of the following teeth apparently belong to this form.

46688. The greater part of a tooth, with a considerable portion of the coating of cement broken away; from the Red Crag of Woodbridge, Suffolk. This and the following specimen agree precisely with the type-teeth figured by Gervais and Van Beneden in the 'Ostéographie des Cétacés,' pl. xx. figs. 29–32. *Purchased*, 1875.

39009. The greater part of a tooth, wanting considerable portions of the cement-layer; from the Red Crag of Suffolk. *Purchased*, 1865.

36655. The basal portion of a tooth which has been transversely bisected. The cement is of great thickness, and the dentine core (dark brown) comparatively small; the inferior aperture of the pulp-cavity is very small. *Brown Collection. Presented by Prof. Sir R. Owen, K.C.B.*, 1859.

49457. The lower half of a tooth, wanting the cement-layer of one side; from the Red Crag of Suffolk. *Purchased*, 1867.

49979. Fragment of the upper half of a longitudinally-split tooth; from the Red Crag of Suffolk. *Wetherell Collection. Purchased*, 1871.

44606. A smaller imperfect tooth, which has been longitudinally bisected; from the Red Crag of Felixstowe. *Purchased*, 1873.

45998. Transverse section of the basal part of a tooth; from Felixstowe. *Purchased*, 1874.

47404. Two transverse sections of a tooth; from Felixstowe. In the larger specimen the osteodentine occupies a consider able space, and shows the characteristic nodular structure. *Purchased*, 1876.

27854. The slightly imperfect left periotic, provisionally referred (*Fig.*) to this species[1]; from Felixstowe. This specimen (fig. 11)

[1] This specimen can only have belonged either to the present form or to *Scaldicetus* or *Balænodon*; its large size and the circumstance that teeth of *Eucetus* appear of much more common occurrence than those of the two latter is in favour of the former reference.

has been described and figured by the writer in the 'Quart. Journ. Geol. Soc.' vol. xliii. p. 13, woodcut fig. 2. In its

Fig. 11.

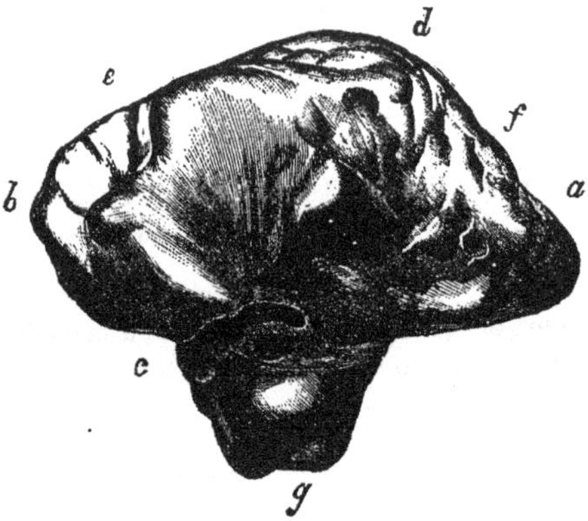

? *Eucetus amblyodon*.—The left periotic; from the Red Crag of Felixstowe. $\frac{2}{3}$. *a*, anterior articular facette for tympanic; *b*, posterior articular ridge for do.; *c*, hollow for accessory ossicle; *d*, *e*, *f*, unnamed ridges and hollows; *g*, capsule of semicircular canals. (From the 'Quart. Journ. Geol. Soc.')

somewhat produced posterior extremity it agrees much more nearly with the corresponding bone of *Hyperoodon* than with that of *Physeter*, but the median tympanic ridge is broader and lower; the anterior articular facette for the tympanic has been partially abraded, and the accessory ossicle on the posterior half of the tympanic face is wanting.

Presented by John Brown, Esq., 1852.

The following rolled and imperfect teeth from the Red Crag may probably be referred either to the present or one of the allied forms.

28255. The dentine cores of two teeth; from Felixstowe.
Purchased, 1852.

37724. The dentine core of a tooth with a fragment of cement on one side. *Purchased*, 1863.

39024. Part of the dentine core of a tooth.
Bowerbank Collection. Purchased, 1865.

36653. Part of a tooth, wanting some of the cement.
Brown Collection. Presented by Prof. Sir R. Owen, K.C.B., 1859.

44044. An imperfect tooth; from Woodbridge. *Purchased*, 1873.

27518. The greater part of a rolled tooth; from Felixstowe. *Purchased*, 1852.

27435. A rolled tooth; from Felixstowe. *Purchased*, 1851.

Genus **PHYSETODON**, M'Coy [1].

The teeth of this genus are described as being very similar to those of *Physeter*, but distinguished by the small size of the aperture of the pulp-cavity at the base of the root; in this respect they resemble the teeth of *Eucetus*, and further observations are required to prove their generic distinction.

Physetodon baileyi, M'Coy [2].

This is the type and only species.
Hab. Australia.

43996. The imperfect upper half of a tooth; from the Lower Pliocene of Mordialloc, Hobson's Bay, Victoria. This specimen agrees very closely with the type teeth figured by M'Coy, *op. cit.* pl. lv. *Purchased*, 1872.

Genus **SCALDICETUS**, Du Bus [3].

The teeth are somewhat smaller than those of *Physeter macrocephalus*, and are distinguished by a coat of grooved enamel on the summit of the crown. The section of the tooth is subcircular; the cement is abundant and much thicker in the middle than at the two extremities, thus causing the tooth to be fusiform; the dentine core is subcylindrical throughout the greater part of the tooth; and the central axis of dentine very slender, darker in colour than the overlying layer, and without a nodular structure. In young specimens the pulp-cavity is open and conical in shape.

[1] Palæontology of Victoria (Geol. Survey of Victoria), dec. vi. p. 19 (1879).
[2] *Loc. cit.*
[3] Bull. Ac. R. Belg. sér. 2, vol. xxiv. p. 568 (1867).

Scaldicetus carreti, Du Bus[1].

The type species.

Hab. Europe (Belgium).

M. 3658. Cast of a tooth. The original, which is from the Antwerp Crag, is preserved in the Museum at Brussels. The pulp-cavity is still open.

Presented by E. T. Newton, Esq., 1887.

Genus BALÆNODON, Owen[2].

Of uncertain affinities. Identified by the present writer in the 'Quart. Journ. Geol. Soc.' vol. xliii. p. 14, with the preceding genus, but later observations tend to render this very doubtful; it has been identified by Gervais with *Hoplocetus*.

Balænodon physaloides, Owen[3].

Hab. Europe (England).

27862. (*Fig.*) The lower portion of an imperfect tooth; from the Red Crag of Felixstowe, Suffolk. This specimen is the type, and is described and figured by Owen in the 'British Fossil Mammals and Birds,' p. 525, figs. 119, 226, 227, where it is regarded as a complete section of a tooth and the external layer described as cement and the axis as dentine.

Presented by John Brown, Esq., 1852.

Genus PHYSODON, Gervais[4].

Syn. *Palæodelphis*, Du Bus[5].

The teeth of this genus indicate animals much inferior in size to the Cachalot. The whole tooth is subcylindrical and frequently curved; the crown is comparatively short and coated with a finely grooved enamel, and the layer of cement is generally of considerable thickness; there is no constriction at the base of the crown, which is larger than in *Hoplocetus*. The teeth sometimes present a considerable resemblance to those of some of the larger Seals, but are readily distinguished by the thickness of the cement, which frequently splits off in the characteristic Cetacean manner, and exposes the dentine core marked by the peculiar horizontal lines.

[1] Bull. Ac. R. Belg. sér. 2, vol. xxiv. p. 568 (1867).
[2] British Fossil Mammals and Birds, p. 536 (1846).
[3] *Loc. cit.*
[4] Bull. Soc. Géol. France, sér. 2, vol. xxix. p. 101 (1872).
[5] Bull. Ac. R. Belg. sér. 2, vol. xxxiv. p. 503 (1872).

Physodon grandis (Du Bus [1]).

Syn. *Palæodelphis grandis*, Du Bus [2].

The teeth are elongated, somewhat slender, markedly curved superiorly, and slightly compressed laterally; the enamel is very finely grooved, and the pulp-cavity is scarcely ever completely closed inferiorly.

Hab. Europe (Belgium and England).

The following specimens are from the Red Crag of the Eastern Counties; they agree so closely with the specimen figured by Van Beneden and Gervais in their 'Ostéographie des Cétacés,' pl. xx. fig. 21, *that at least the majority of them may be pretty safely referred to the same species.*

27855. A tooth, with part of the cement chipped off; from Felixstowe, Suffolk. This and the following specimens are noticed by the present writer in the 'Quart. Journ. Geol. Soc.' vol. xliii. p. 14.
Presented by John Brown, Esq., 1852.

28256. The greater part of a rolled and partially-worn tooth; from Woodbridge, Suffolk. *Purchased*, 1852.

49979. The upper half of a rolled, although unworn tooth; from Suffolk. *Purchased*, 1875.

44045. The upper half of a partially-worn tooth; from Suffolk. *Purchased*, 1873.

44040. The crown of a partially-worn tooth; from Suffolk. The base has been cut and polished. *Purchased*, 1873.

44615–6, 43404, 43404 a. Four specimens of the crowns of teeth; from Suffolk. *Purchased.*

Physodon fusiformis (Du Bus [3]).

Syn. *Palæodelphis fusiformis*, Du Bus [4].

The teeth are fusiform, comparatively thick in the middle, and pointed at the two extremities, with a nearly cylindrical transverse section; in some examples the crown is curved, but others are

[1] Bull. Ac. R. Belg. sér. 2, vol. xxxiv. p. 503 (1872).—*Palæodelphis.*
[2] *Loc. cit.*
[3] *Loc. cit.—Palæodelphis.*
[4] *Loc. cit.*

straight throughout their length; the grooving of the enamel is coarser than in the preceding species.

Hab. Europe (Belgium and (?) England).

44109. A rolled tooth, with the inferior extremity broken off and a portion of the cement chipped away, provisionally referred to this species; from the Red Crag of Woodbridge, Suffolk. In the form of the dentine core this specimen agrees very closely with the one figured by Van Beneden and Gervais in the 'Ostéographie des Cétacés;' but as the enamel (if ever present) has been at least in great part worn away, it is not certain that it may not have belonged to *Eucetus*; it is noticed by the writer in the 'Quart. Journ. Geol. Soc.' vol. xliii. p. 14. *Purchased*, 1873.

Genus **HOPLOCETUS**, Gervais [1].

As already mentioned, this genus was considered by its founder [2] as identical with *Balænodon*. The teeth have very short crowns, covered with a thin coat of grooved enamel; the root is extremely long in proportion to the crown, generally tumid in the middle, and the layer of cement of great thickness; the crown is separated from the root by a constriction. The genus occurs in the Tertiaries of both Europe and America.

Hoplocetus crassidens, Gervais [3].

The teeth are fusiform, with a slight arcuation and a flattening of the root on one side; the root is of great relative thickness. The species was originally described from the Middle Miocene of Romans (Drôme), France.

Hab. Europe (France, Germany [4], Malta, and England).

24596. A tooth, wanting the upper part of the crown and the extremity of the root; from the Middle or Lower Miocene of Malta. This specimen agrees precisely with the type specimen figured by Gervais in the 'Zoologie et Paléontologie Françaises,' pl. xx. fig. 10, and the 'Ostéographie des Cétacés,' pl. xx. fig. 26. *Purchased*, 1850.

[1] Zool. et Pal. Françaises, 1st ed. vol. i. p. 161 (1848–52).
[2] Ostéographie des Cétacés, p. 345.
[3] Zool. et Pal. Françaises, *loc. cit.* (1848–52).
[4] See Probst, Jahresh. Ver. Nat. Württ. 1886, p. 106

24596 a. The unworn crown of a smaller tooth; from Malta. *Purchased*, 1850.

24596 b. The root of a tooth; from Malta. *Purchased*, 1850.

M. 1936. A small tooth, perhaps belonging to this species; locality unknown. *Presented by Prof. Sir R. Owen, K.C.B.*, 1884.

The following rolled specimens from the Red Crag of Suffolk agree very closely in general characters with the teeth of the present form, and may have been derived from Miocene deposits.

28256. A tooth, in a partially rolled condition. The characteristic constriction at the base of the crown is well marked. This and the following specimens are noticed by the writer in the 'Quart. Journ. Geol. Soc.' vol. xliii. p. 14. *Purchased*, 1852.

27518. A very similar tooth. *Purchased*, 1852.

27435. An imperfect tooth, with a more decided curvature. *Purchased*, 1851.

28981. A very similar imperfect tooth, in a much rolled condition. *Purchased*, 1854.

Hoplocetus borgerhoutensis, Du Bus[1].

In this species the root of the tooth is much longer, more slender, and much more curved inferiorly than in *H. crassidens*; the striations on the enamel of the crown are less strongly marked[2].

Hab. Europe (Belgium and England).

37944. A partially-worn tooth, with the root corroded; from the Antwerp Crag. This specimen is smaller than the one figured by Gervais in the 'Ostéographie des Cétacés,' pl. xx. fig. 28, but is otherwise similar. *Purchased*, 1863.

47987. An unworn tooth, probably belonging to this species; apparently from the Antwerp Crag. *Presented by Hon. R. Marsham*, 1877.

[1] Bull. Ac. R. Belg. sér. 2, vol. xxxiv. p. 502 (1872).

[2] In the Brussels Museum teeth resembling Gervais's figure of *H. borgerhoutensis* are labelled (?) *Palæodelphis minutus*, Du Bus.

28980. A tooth with the crown broken off; from the Red Crag of Suffolk. This specimen agrees precisely with the one figured by Gervais and already referred to.

Purchased, 1854.

44617. The imperfect root of a tooth, agreeing with the preceding specimen; from the Red Crag of Suffolk.

Purchased, 1873.

Hoplocetus curvidens, Gervais[1].

In this species the roots of the teeth are nearly as much curved as in *H. borgerhoutensis*, but are much thicker, and the groovings on the enamel are more distinct.

Hab. Europe (Belgium and [?] England).

39024. The upper half of a much-rolled tooth, provisionally referred to this species; from the Red Crag of Felixstowe, Suffolk. This specimen appears to agree very closely with the type tooth figured by Gervais in the 'Ostéographie des Cétacés,' pl. xx. fig. 25, and shows a remnant of the enamel with its basal constriction.

Bowerbank Collection. Purchased, 1865.

49979. A similar fragment of a rather smaller rolled tooth; from the Red Crag of Suffolk.

Wetherell Collection. Purchased, 1871.

The following specimens agree in size and characters with the vertebræ of Kogia breviceps, *and may therefore be probably referred either to the present or one of the allied genera of small Physeteroids, unless they indicate the occurrence of* Kogia *itself.*

48487. A caudal vertebra in a rolled condition and wanting both epiphyses; from the Red Crag of Suffolk. This specimen agrees very closely with the seventh and eighth caudals of *Kogia*. *Purchased*, 1875.

48488. A rolled sixth (?) caudal vertebra, wanting both epiphyses; from the Red Crag. This specimen is almost indistinguishable from the sixth caudal of *Kogia*.

Purchased, 1875.

[1] Zool. et Pal. Françaises, 1st ed. vol. i. p. 161 (1848–52).

Subfamily ZIPHIINÆ.

In existing genera all the mandibular teeth are rudimentary, with the exception of one or occasionally two pairs.

Genus **HYPERODON,** Lacépède[1].

There are a pair of small conical teeth concealed during life by the gum at the extremity of the mandible. The upper ends of the premaxillæ rise suddenly behind the nares to the vertex, expanding laterally, their outer edges curving backwards, and their anterior surfaces arching forwards over the nares. There are large longitudinal crests on the maxillæ at the base of the rostrum. The posterior portion of the periotic is shortened, the median ridge on the tympanic aspect of the same very strongly developed, the accessory ossicle large and rounded, and the anterior tympanic facette slightly concave. All the cervical vertebræ are fused together.

Hyperoodon rostratus (Müller[2]).

Syn. *Balæna rostrata*, Müller[3].

Hab. North Atlantic.

38140. A considerable part of the left ramus of the mandible; found in digging the main sewer in East Ham Marshes, Essex.
Presented by the Commissioners of the Metropolitan Board of Works, 1864.

Hyperoodon, sp.

Hab. Europe (England and [?] Belgium[4]).

M. 3536. Cast of the right periotic. The original, which was obtained from the Red Crag of Suffolk and is preserved in the Museum at Ipswich, is described and figured by the present writer in the 'Quart. Journ. Geol. Soc.' vol. xliii. p. 14, pl. ii. fig. 6. This specimen, in which the accessory ossicle

[1] Hist. Nat. des Cétacés—Table d. Ordres &c., p. xliv (1804).

[2] Zool. Dan. Prod. p. 7 (1776).—*Balæna.*

[3] *Loc. cit.*

[4] The cervical vertebræ of a species of *Hyperoodon* from the Antwerp Crag are noticed by Van Beneden in the Bull. Ac. R. Belg., sér. 2, vol. x. p. 407 (1860).

on the posterior half of the tympanic aspect is still attached, apparently presents no characters by which it can be distinguished from the periotic of *H. rostratus* in the Museum of the Royal College of Surgeons (No. 2899), and indicates the occurrence either of that or of a closely allied species in the Pliocene.

Made in the Museum, 1886.

Genus **CHONEZIPHIUS,** Duvernoy[1].

Including *Ziphirostrum*, Du Bus[2].
Aporotus, Du Bus[3].
Ziphiopsis, Du Bus[4].

This genus agrees with *Mesoplodon* in having the premaxillæ in contact with the intervening bones throughout the length of their inner surfaces, and in only showing a very small portion of the vomer on the inferior aspect of the rostrum; but differs in the non-ossification of the mesethmoid cartilage in the rostrum, whereby there exists a fistular excavation throughout the latter. The soldering of the inner surfaces of the premaxillæ is incomplete in some forms. The interorbital space is flat; and there are a pair of fossettes, of which the right is the larger, in the narial region. The rostrum is comparatively short and thick. The posterior portion of the periotic is rather longer than in *Hyperoodon*, the median ridge on the tympanic aspect of the same rather less strong, but the accessory ossicle of nearly equal relative size. The genus occurs in the Tertiaries of both Europe and North America, and may be regarded as connecting *Hyperoodon* with *Mesoplodon*.

Choneziphius planirostris (Cuvier[5]).

Syn. *Ziphius planirostris*, Cuvier[6].

This is the type species, and is of considerably larger size than the existing *Mesoplodon australis*. The type cranium is figured in the accompanying woodcut (fig. 12), and it seems doubtful if the second cranium described by Cuvier under the same name, but

[1] Ann. Sci. Nat., Zool. sér. 3, vol. xv. p. 61 (1851).
[2] Bull. Ac. R. Belg. sér. 2, vol. xxv. p. 622 (1867).
[3] *Ibid.* p. 626.
[4] *Ibid.* p. 628.
[5] Ossemens Fossiles, 2nd ed. vol. v. pt. i. p. 356 (1823).—*Ziphius.*
[6] *Loc. cit.*

Fig. 12.

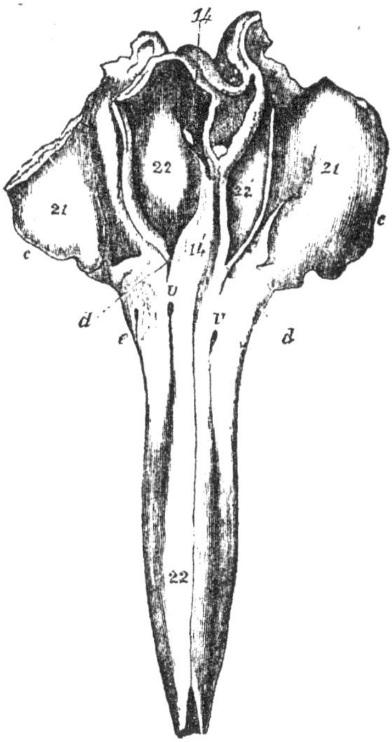

Choneziphius planirostris.—The anterior portion of the cranium; from the Antwerp Crag. Nos. as in fig. 13. $\frac{1}{4}$. (From the Mon. Pal. Soc.)

referred to a new species by Owen[1] under the name of *Ziphius cuvieri*, is really distinct[2].

Hab. Europe (Belgium, England, and Italy[3]).

M. 3560. Cast of the anterior portion of the cranium with the rostrum. The original was obtained from the Antwerp Crag, and is preserved in the Royal Museum of Natural History at Brussels; it closely resembles the type specimen.
No history.

49654. The cranial rostrum with a fragment of the interorbital region; dredged about twenty miles west of Southwold, Suffolk, having been derived from the Red Crag.
Presented by Dr. C. R. Bree, 1878.

[1] Crag Cetacea (Mon. Pal. Soc.), p. 6 (1870).
[2] See Lankester, Quart. Journ. Geol. Soc vol. xxvi. p. 507.
[3] Capellini, Atti R. Ac. Linc. ser. 4, vol. i. p. 18 (1885).

M. 3559. Cast of the cranial rostrum, with a fragment of the interorbital region. The original is in the collection of Dr. Bree of Colchester, and was dredged in the North Sea, having been derived from the Red Crag.
Made in the Museum.

44176. The cranial rostrum, with a fragment of the interorbital region; from the Red Crag of Suffolk. *Purchased*, 1873.

39459. The imperfect anterior extremity of the cranial rostrum of a large individual; from the Red Crag of Suffolk.
Purchased, 1865.

39509. The cranial rostrum; from the Antwerp Crag.
Purchased, 1865.

33459. The centrum of a first caudal vertebra which may belong to this species; probably from the Coralline Crag. This specimen is somewhat larger than the first caudal of *Mesoplodon australis*, but agrees very closely in character.
Layton Collection. Purchased, 1858.

M. 3535. Cast of the left periotic, provisionally referred to this species. The original was obtained from the Red Crag of Woodbridge, Suffolk, and is preserved in the Museum of Practical Geology, Jermyn Street; it is described and figured by the present writer in the 'Quart. Journ. Geol. Soc.' vol. xliii. p. 14, pl. ii. fig. 7. It has lost the accessory ossicle on the posterior half of the tympanic aspect; and is intermediate in size and structure between the periotics of *Hyperoodon* and *Mesoplodon*, although nearer in respect of structure to the latter. The specimen approaches *Hyperoodon* in the relative shortness of the posterior extremity, the strong development of the median articular ridge on the tympanic aspect of the same portion, and the large size of the cavity for the accessory ossicle; the anterior articular fossette for the tympanic is also less markedly concave than in *Mesoplodon*. The specimen accords well in size with the present species, to which it is referred as being the one of most common occurrence in the Crag. *Made in the Museum*, 1886.

Choneziphius planus (Owen[1]).

Syn. *Ziphius planus*, Owen [2].

This species, which is equal in size to the preceding, is referred to the present genus by Gervais in the 'Ostéographie des Cétacés,' p. 418. The type specimen is not sufficiently perfect to admit of determining whether the form described by Lankester [3] under the name of *C. packardi* may not be specifically the same.

Hab. England.

38498. The cranial rostrum; from the Red Crag of Suffolk. This (*Fig.*) is the type specimen, and is described and figured by Owen, *op. cit.* p. 16, pl. xi. fig. 1, under the name of *Ziphius planus.* It is also figured by Gervais in the 'Ostéographie des Cétacés,' pl. xxvii. fig. 16. *No history.*

Genus **MESOPLODON,** Gervais [4].

Syn. *Dioplodon*, Gervais [5].
Belemnoziphius, Huxley [6].

The mandible has a single pair of laterally-compressed pointed teeth, generally situated at some distance behind the apex. The narial region of the cranium resembles that of *Hyperoodon*, but the nasals are narrower and more deeply sunk between the premaxillæ. There are no maxillary tuberosities, and the rostrum is very long and narrow; in the latter the premaxillæ are in contact with the adjacent bones, and soldered throughout the whole length of their inner surfaces; in old specimens the mesethmoid cartilage completely ossifies and forms a well-defined band on the superior surface of the rostrum, which is solid throughout, and only shows a small portion of the vomer on its inferior aspect. The posterior portion of the periotic is produced and pointed, the articular ridge on the tympanic aspect of the same rather low, the accessory ossicle small and oval, and the anterior articular facette for the tympanic deeply concave. Two or three of the cervical vertebræ are united.

[1] Crag Cetacea (Mon. Pal. Soc.), p. 16 (1870).—*Ziphius.*

[2] *Loc. cit.*

[3] Quart. Journ. Geol. Soc. vol. xxvi. p. 502 (1870). This name is of later date than *C. planus.*

[4] Ann. Sci. Nat., Zool. sér. 3, vol. xiv. p. 16 (1850). For a full synonymy see Flower, 'Trans. Zool. Soc.' vol. viii. p. 208.

[5] Zool. et Pal. Françaises, 1st ed. vol. ii. Exp. No. 40, p. 4 (1848–52).

[6] Quart. Journ. Geol. Soc. vol. xx. p. 388 (1864). The term was applied to all the Crag forms included by Owen in *Ziphius.*

Mesoplodon longirostris (Cuvier [1]).

Syn. *Ziphius longirostris*, Cuvier [2].
Dioplodon becani, Gervais [3].
Ziphius medilineatus, Owen [4].
Dioplodon medilineatus, Gervais [5].

The specimens from the English Crag, described by Owen as *Z. medilineatus*, are undoubtedly specifically identical with the Belgian *D. becani* [6], and the latter appears to be the same as *Z. longirostris* [7]. The locality of the type rostrum of the latter (which is preserved in the Paris Museum) is uncertain.

Fig. 13.

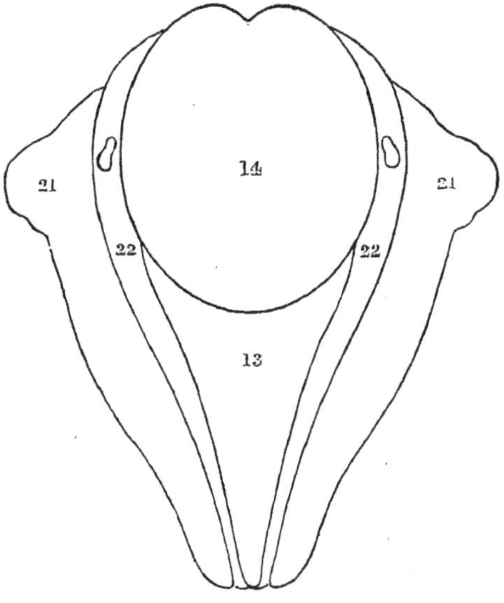

Mesoplodon longirostris.—Diagrammatic transverse section of the cranial rostrum. 13, vomer; 14, mesethmoid; 21, maxilla; 22, premaxilla. ¼. (From the Mon. Pal. Soc.)

The species is about equal in size to the existing *M. australis*. The rostrum is long and narrow, its mesethmoid element (fig. 13)

[1] Ossemens Fossiles, 2nd ed. vol. v. pt. 1, p. 357 (1823).—*Ziphius.*
[2] *Loc. cit.*
[3] Zool. et Pal. Françaises, 1st ed. vol. ii. Exp. No. 38, p. 2 (1848–52).
[4] Crag Cetacea (Mon. Pal. Soc.), p. 22 (1870).
[5] Ostéographie des Cétacés, p. 422 (1868–79).
[6] See Gervais, *op. cit.* pp. 421–422.
[7] See Du Bus, Bull. Ac. R. Belg. sér. 2, vol. xxiv. p. 570; Gervais (*loc. cit.*) had some doubt as to the identity of the two forms.

appearing largely on the superior surface and exhibiting a distinct median longitudinal groove; at its hinder extremity the mesethmoid is flat, and does not project above the plane of the adjacent bones, while the antero-posterior gibbosity near the middle is comparatively small; in advance of the gibbosity the rostrum is slender, and presents a subcircular cross section.

Hab. Europe (Belgium, England, and Italy[1]).

All the following specimens are from the Red Crag of Suffolk.

37722. The hinder part of the cranial rostrum. This specimen is (*Fig.*) the type of *Z. medilineatus*, and is figured by Owen in the 'Crag Cetacea' (Mon. Pal. Soc.), pl. iv. fig. 3.
(*From the Saul Collection.*) *Purchased*, 1863.

46924. Cast of the cranial rostrum, of which the original was obtained from Woodbridge. This specimen agrees very closely with the preceding, although the median groove has been partly obliterated by rolling. The gibbosity in the middle of the rostrum is but slightly marked, and the anterior extremity subcylindrical. *Purchased*, 1875.

M. 1612. A very similar cranial rostrum; from Felixstowe. The median groove extends over nearly two thirds the length of the specimen. This and the preceding specimen appear indistinguishable from the type rostrum of *D. becani*, figured by Gervais in the 'Zool. et Pal. Françaises,' pl. xxxviii. fig. 4, and in the 'Ostéographie des Cétacés,' pl. xxvii. fig. 7.
Presented by C. Westendarp, Esq., 1884.

M. 1613. The anterior extremity of the rostrum; from Felixstowe.
Presented by C. Westendarp, Esq., 1884.

46680. The cranial rostrum; from Woodbridge. This specimen is indistinguishable from No. M. 1612. *Purchased*, 1875.

42514. The hinder part of the cranial rostrum; from Sutton.
Purchased, 1871.

30272. The hinder part of a small cranial rostrum, probably belonging to this species. *Purchased*, 1855.

[1] See Capellini, Mem. Ac. Sci. Ist. Bologna, ser. 4, vol. vi. p. 298 (1885).

The majority of the undermentioned specimens belong either to the present or some of the following species; they were all obtained from the Red Crag of Suffolk.

M. 3537. Cast of the left periotic. The original (woodcut, fig. 14)

Fig. 14.

Mesoplodon longirostris.—The left periotic; from the Red Crag of Woodbridge. ¼.

is from Woodbridge, and is preserved in the Museum of Practical Geology, Jermyn Street; it is described and figured by the present writer in the 'Quart. Journ. Geol. Soc.' vol. xliii. p. 15, pl. ii. fig. 8. This specimen agrees almost precisely with the periotic of *Mesoplodon australis*, and shows the marked concavity of the anterior tympanic facette, which is wanting in the periotic of *Choneziphius*; the accessory ossicle on the posterior portion of the tympanic surface has disappeared.

Made in the Museum, 1886.

43526. The middle portion of the mandible, showing the hinder extremity of the symphysis and portions of the rami. The specimen agrees very closely with the corresponding portion of the mandible of *M. australis*. *Purchased*, 1872.

45993. The eighth (?) caudal vertebra. This specimen is rather smaller than the eighth caudal of *M. australis*, but otherwise agrees very closely. *Purchased*, 1874.

43183. An early caudal vertebra, much perforated by mollusks; from Falkenham.

Wetherell Collection. Purchased, 1871.

44036. A rolled middle caudal vertebra. This specimen indicates a smaller individual than an adult *M. australis*.

Purchased, 1873.

48489. A very similar vertebra. *Purchased*, 1875.

29613. A middle caudal vertebra, belonging to a different species from the two preceding specimens. *Purchased*, 1852.

48483. A caudal vertebra. *Purchased*, 1875.

48484. A caudal vertebra, of a more elongated type. *Purchased*, 1875.

48485. A caudal vertebra. *Purchased*, 1875.

28272. A caudal vertebra. *Purchased*, 1852.

Mesoplodon tenuirostris (Owen[1]).

Syn. *Ziphius tenuirostris*, Owen[2].
Dioplodon tenuirostris, Gervais[3].

This species appears closely allied to the preceding, but the median groove on the superior surface of the rostrum is apparently wanting, the posterior portion of the latter is wider, the gibbosity placed rather more anteriorly, and the vertical thickness in advance of this point somewhat greater. Whether these characters are specific or merely individual or sexual cannot be determined.

Hab. Europe (England and Italy[4]).

27601. The cranial rostrum, imperfect anteriorly; from the Red Crag of Suffolk. This specimen is the type, and is figured by Owen in the 'Crag Cetacea' (Mon. Pal. Soc.), pl. v. figs. 1, 2, and by Gervais in the 'Ostéographie des Cétacés,' pl. xxvii. fig. 8. *Purchased*, 1852.

39458. The cranial rostrum, imperfect anteriorly; from the Red Crag of Felixstowe, Suffolk. *Purchased*, 1865.

Mesoplodon gibbus (Owen[5]).

Syn. *Ziphius gibbus*, Owen[6].
Dioplodon gibbus, Gervais[7].

In this species the mesethmoid band is very prominent at the base of the rostrum, and the gibbosity in the middle is strongly

[1] Crag Cetacea (Mon. Pal. Soc.), p. 24 (1870).—*Ziphius*.
[2] *Loc. cit.*
[3] Ostéographie des Cétacés, p. 422 (1868–79).
[4] See Capellini, Mem. Ac. Sci. Ist. Bologna, ser. 4, vol. vi. p. 296 (1885)
[5] Crag Cetacea (Mon. Pal. Soc.), p. 17 (1870).—*Ziphius*.
[6] *Loc. cit.*
[7] Ostéographie des Cétacés, p. 421 (1868–79).

marked; there is no median groove, and it does not diminish in width to any great extent between its root and the median gibbosity. The rostrum is somewhat shorter than in *M. longirostris*.

Hab. Europe (England and Italy[1]).

27439. The cranial rostrum, imperfect anteriorly; from the Red (*Fig.*) Crag of Suffolk. This specimen is the type, and is figured by Owen in the 'Crag Cetacea' (Mon. Pal. Soc.), pl. ii. fig. 2, and pl. iii. fig. 3. *Purchased*, 1851.

39003. The hinder two thirds of the cranial rostrum; from the Red Crag of Felixstowe, Suffolk.
Bowerbank Collection. Purchased, 1865.

M. 3562. Cast of the complete cranial rostrum. The original is from the Red Crag of Shotley, Suffolk, and is preserved in the Norwich Museum; it agrees precisely with the type specimen. *Purchased.*

Mesoplodon angustus (Owen[2]).

Syn. *Ziphius angustus*, Owen[3].
Dioplodon angustus, Gervais[4].

This species is allied to *M. gibbus*, but distinguished by the narrower mesethmoid band, the greater constriction of the latter between its root and the gibbosity, and the smaller size and more anterior position of the gibbosity itself; a median groove is sometimes present.

Hab. Europe (England).

M. 3561. The cranial rostrum, imperfect anteriorly; from the Red (*Fig.*) Crag of Suffolk. This is the type specimen, and is figured by Owen in the 'Crag Cetacea' (Mon. Pal. Soc.), pl. iii. figs. 1, 2. *No history.*

30273. The cranial rostrum; from the Red Crag of Felixstowe, Suffolk. This specimen, which is imperfect anteriorly, agrees very closely with the type. *Purchased*, 1855.

49662. The cranial rostrum, wanting the extremity and much perforated by mollusks; from the Red Crag of Woodbridge, Suffolk. This specimen also agrees with the type in general characters, but shows a median mesethmoid groove.
Purchased, 1879.

[1] See Capellini, Mem. Ac. Sci. Ist. Bologna, ser. 4, vol. vi. p. 295 (1885).
[2] Crag Cetacea (Mon. Pal. Soc.), p. 19 (1870).—*Ziphius.* [3] *Loc. cit.*
[4] Ostéographie des Cétacés, p. 422 (1868–79).

Mesoplodon angulatus (Owen [1]).

Syn. *Ziphius angulatus*, Owen [2].
Dioplodon angulatus, Gervais [3].

In this species the mesethmoid band is broad and prominent, without a median groove or anterior gibbosity, and with its lateral surfaces bevelled away. Since, however, the type specimen is much rolled, some of these characters may be due to this cause.

Hab. Europe (England and (?) Belgium).

27433. The cranial rostrum, imperfect anteriorly; from the Red (*Fig.*) Crag of Suffolk. This is the type specimen, and is figured by Owen in the 'Crag Cetacea' (Mon. Pal. Soc.), pl. iv. figs. 1, 2. *Purchased*, 1851.

Mesoplodon compressus (Huxley [4]).

Syn. *Belemnoziphius compressus*, Huxley [5].
(?) *Ziphius compressus*, Owen [6].
Dioplodon compressus, Gervais [7].

If the *Ziphius compressus* of Owen be not identical with the *Belemnoziphius compressus* of Huxley, the specific name of the former ought to be changed. Owen's type specimen is characterized by the great vertical depth of the rostrum, the mesethmoid band being broader than in *M. angustus*, and showing a median groove in its posterior half. Additional specimens are required to indicate the full affinities of this form.

Hab. Europe (England).

27027. The cranial rostrum, imperfect at the two extremities; (*Fig.*) from the Red Crag. This is the type specimen of Owen's *Ziphius compressus*, and is figured in the 'Crag Cetacea' (Mon. Pal. Soc.), pl. v. fig. 3. *Purchased*, 1851.

[1] Crag Cetacea (Mon. Pal. Soc.), p. 20 (1870).—*Ziphius.*
[2] *Loc. cit.*
[3] Ostéographie des Cétacés, p. 422 (1868–79).
[4] Quart. Journ. Geol. Soc. vol. xx. p. 388 (1864).—*Belemnoziphius.*
[5] *Loc. cit.*
[6] Crag Cetacea (Mon. Pal. Soc.), p. 25 (1870).
[7] Ostéographie des Cétacés, p. 422 (1868–79).

Family PLATANISTIDÆ.

In the existing genera the jaws are long and narrow, with numerous simple teeth, the length of the mandibular symphysis exceeding half the length of the ramus. All the cervical vertebræ are free, and in the hinder dorsals the tubercular and capitular attachments of the ribs are blended. In *Inia* the transverse processes of the lumbar vertebræ are of great antero-posterior width.

The two following genera are usually referred to the present family[1] (see Probst, Jahresh. Ver. Nat. Württ. 1886, p. 117), although *Schizodelphis* approaches the delphinoid genus *Steno*.

Genus **CHAMPSODELPHIS,** Gervais[2].

The lumbar vertebræ of this genus have extremely wide transverse processes, indicating affinity with *Inia*; the teeth also resemble those of the latter genus.

Champsodelphis, sp.

Hab. Russia.

M. 40514. An early lumbar vertebra, wanting one epiphysis and otherwise slightly imperfect; from the Lower Pliocene of the Crimea. In general characters, and especially the great antero-posterior diameter of the transverse processes, this specimen agrees very closely with the fourth lumbar vertebra of *Inia*. *Purchased.*

Genus **SCHIZODELPHIS,** Gervais[3].

The form of the rostrum presents a great resemblance to that of the delphinoid genus *Steno*, but the mandibular symphysis is considerably longer.

Schizodelphis sulcatus, Gervais[4].

Syn. *Delphinorhynchus sulcatus*, Gervais[5].
Delphinus pseudodelphis, Gervais[6].

This is the type species, and is of nearly the same size as the existing *Steno*.

Hab. Europe.

[1] Frequently termed the *Delphinorhynchidæ*.
[2] Zool. et Pal. Françaises, 1st ed. vol. i. p. 152 (1848–52).
[3] Mém. Ac. Montpellier, vol. v. p. 126 (1861).
[4] Bull. Soc. Géol. France, sér. 2, vol. x. p. 312 (1853).—*Delphinorhynchus*.
[5] *Loc. cit.*
[6] Ext. Proc.-Verb. Ac. Montpellier, 1849, p. 11.

32730. The greater portion of a mandibular symphysis, provisionally referred to this species; from the Upper Tertiary of Xabregas, near Lisbon, Portugal. Ten teeth remain on either side, although some of their crowns are broken off.
Presented by J. S. Valentine, Esq. About 1857.

Family SQUALODONTIDÆ.

The cranium resembles in essential characters that of the *Delphinidæ*, and differs entirely from that of the *Archæoceti*, although the teeth are differentiated into groups as in the latter.

Genus **SQUALODON**, Grateloup[1].

Syn. *Pachyodon*, Meyer[2].
Phocodon, Agassiz[3].
(?) *Arionus*, Meyer[4].
Delphinoides, Pedroni[5].
Crenidelphinus, Laurillard[6].
Stereodelphis, Gervais[7].
Delphinodon, Leidy[8].
Rhizoprion, Jourdan[9].
Macrophoca, Leidy[10].

Fig. 15.

Three of the lower true molars of *Squalodon*. Miocene. (After Scilla.)

1 Act. Ac. R. Sci. Bordeaux, 1840, p. 208.
2 Neues Jahrb. 1838, p. 414. This name is the earliest, but was preoccupied.
3 In Valentin's Rep. d'Anat. et Physiol. 1841, p. 236.—*Teste* Brandt.
4 Neues Jahrb. 1841, p. 315.
5 Comptes Rendus, vol. xxi. p. 1181 (1845).
6 Dict. Univ. d'Hist. Nat. vol. iv. p. 636 (1846).
7 Zool. et Pal. Françaises, 1st ed. vol. i. p. 152 (1848–52).
8 Journ. Ac. Nat. Sci. Philad. ser. 2, vol. vii. p. 424 (1869).
9 Ann. Sci. Nat., Zool. sér. 4, vol. xvi. p. 369 (1861).
10 Proc. Ac. Nat. Sci. Philad. 1856, p. 220.

The teeth may be formulated as I. $\frac{3}{3}$, C. $\frac{1}{1}$; simple cheek-teeth (premolars) $\frac{4}{4}$, double-rooted cheek-teeth (true molars) $\frac{7}{7}$. The true molars (woodcut, fig. 15) differ from those of *Zeuglodon* by the denticulations of the crown being more developed on the posterior than on the anterior border. The genus has been recorded from the Tertiaries of Europe, Australia, and North America.

Squalodon grateloupi (Pedroni[1]).

Syn. *Delphinoides grateloupi*, Pedroni[2].
Rhizoprion bariensis, Jourdan[3].
Squalodon bariensis, Zittel[4].

This may be regarded as the type species.
Hab. Europe (France, Germany, &c.).

40343. Cast of the imperfect skull (fig. 16). The original was obtained in 1854 from the Middle Miocene of Barie, near

Fig. 16.

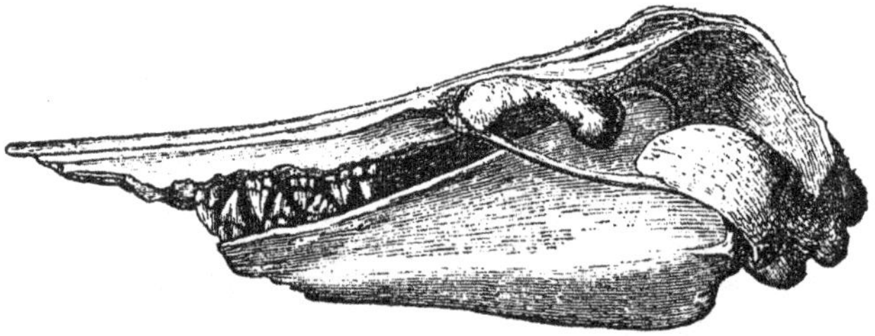

Squalodon grateloupi.—The imperfect skull; from the Middle Miocene of Barie (Drôme), France. $\frac{1}{8}$. (After Gaudry's 'Les Enchaînements,' &c.)

St. Paul, Trois-Châteaux (Drôme), France, and is preserved in the Museum of Natural History at Lyons. It is described and figured by Jourdan in the 'Ann. Sci. Nat., Zool.' sér. 4, vol. xvi. p. 369, pl. x., under the name of *Rhizoprion bariensis*, of which it is the type; and also by Gaudry in 'Les Enchaînements, &c.—Mammifères Tertiaires,' p. 30, fig. 19; and by Depéret in the 'Arch. Mus. Lyon,' vol. iv. pls. xxv.^{bis}, xxv.^{ter} (1887). *Purchased*, 1867.

1 Comptes Rendus, vol. xxi. p. 1181 (1845).—*Delphinoides.*
2 *Loc. cit.*
3 Ann. Sci. Nat., Zool. sér. 4, vol. xvi. p. 372 (1861).
4 Palæontographica, vol. xxiv. pt. 6, p. 233 (1877).

49676. Cast of the cranium (imperfect posteriorly) and the mandibular symphysis, with the crowns of the teeth broken off. The original was obtained from the Middle Miocene of Bleichenbach, Lower Bavaria; and is described and figured by Zittel in the 'Palæontographica,' vol. xxiv. pt. 6, p. 233, pl. xxxiii., under the name of *Squalodon bariensis*. *Purchased*, 1879.

M. 1335. A restored plaster-model of the preceding specimen, showing the crowns of all the teeth. *Purchased*, 1883.

Squalodon catulli (Molin [1]).

Syn. *Pachyodon catulli*, Molin [2].

This species is distinguished from the preceding by slight dental characters. The *Squalodon* teeth from Baltringen were described and figured by Van Beneden in the 'Bull. Ac. R. Belg.' sér. 2, vol. xli. p. 474, under the name of *Squalodon servatus* (Meyer); but according to Probst, Jahresh. Ver. Nat. Württ. 1885, pp. 49–67, the type skull of this species (*Arionius servatus*, Meyer) shows dental alveoli unlike those of other Squalodons, and the true *Squalodon* teeth are referred to the Italian species (see Zigno, 'Mem. R. Ist. Veneto,' vol. xx. p. 24 [1876]).

Hab. Europe (Italy and Germany).

43633. An anterior tooth; from the Middle (?) Miocene of Baltringen, near Ulm, Würtemberg. This specimen agrees very closely with the tooth figured by Van Beneden, *op. cit.* fig. 4 (as *S. servatus*), and with the two teeth figured by Probst, *op. cit.* pl. i. figs. 2, 3. *Purchased*, 1859.

43633 a. The root of an anterior tooth, probably belonging to this species; from Baltringen. *Purchased*, 1859.

Squalodon atlanticus, Leidy [3].

Syn. *Macrophoca atlantica*, Leidy [4].

This species is nearly equal in size to *S. antverpiensis*, Van Beneden, but is distinguished by the presence of denticulations on the anterior border of the teeth; the enamel is very rugose.

Hab. North America (Miocene).

[1] Sitz. k. Ak. Wiss. Wien, vol. xxxv. p. 117 (1859).—*Pachyodon.*
[2] *Loc. cit.*
[3] Proc. Ac. Nat. Sci. Philad. 1856, p. 220.—*Macrophoca.*
[4] *Loc. cit.*

35592. An anterior left lower true molar, probably belonging to this species; from the Miocene (?) of North America. The specimen shows minute denticulations on the anterior border, and has very rugose enamel; it agrees in size with the teeth figured by Leidy in the 'Journ. Ac. Nat. Sci. Philad.' ser. 2, vol. vii. pl. xxviii.

Presented by Prof. J. W. Mallet, 1859.

Family DELPHINIDÆ.

The teeth are simple, and usually numerous in both jaws; the length of the mandibular symphysis may be very small, and never exceeds one third of that of the ramus. The anterior facette on the periotic (fig. 17) for articulation with the tympanic is deeply grooved, and the posterior tympanic surface of the former is comparatively narrow, and its ridge for articulation with the free border of the tympanic ill-defined and situated close to one edge.

Genus **MONODON,** Linn.[1]

The dentition is practically restricted to a single pair of maxillary teeth, of which the left one in the male is usually developed into an immense cylindrical tusk without enamel. The cervical vertebræ are usually separate.

Monodon monoceros, Linn.[2]

This species is recorded by Owen[3] from the later deposits of the Eastern coast, and has also been obtained from the Norfolk Forest-bed[4].

Hab. Arctic Seas.

24576. Two lumbar and one caudal vertebræ; from the Pleistocene of Kotzebue Sound, Eschscholtz Bay, Alaska.

Presented by Capt. Kellet and Lieut. Wood, R.N., 1850.

Genus **DELPHINAPTERUS,** Lacépède[5].

Syn. *Beluga*, Gray[6].

This genus is mainly distingished from *Monodon* by its dentition. The teeth, which vary from $\frac{8}{8}$ to $\frac{10}{10}$, and occupy the anterior three

[1] Syst. Nat. ed. 12, vol. i. p. 105 (1766). [2] *Loc. cit.*
[3] British Fossil Mammals and Birds, p. 521.
[4] See Newton, Geol. Mag. dec. ii. vol. viii. p. 316 (1881).
[5] Hist. Nat. d. Cétacés, p. xli (1804).
[6] Spicilegia Zoologica, vol. i. p. 2 (1828).

fourths of the rostrum and the corresponding part of the mandible, are rather small, conical, and pointed, when unworn. The cranium is rather narrow and elongated, and depressed; the rostral is nearly equal to the cranial portion in length, and is triangular, broad at the base, and gradually narrowing to the apex, where it is sometimes deflected. The cervical vertebræ are free.

Delphinapterus leucas (Pallas[1]).

Syn. *Delphinus leucas*, Pallas[2].
Delphinapterus beluga, Lacépède[3].
Beluga catodon, Gray[4].

Hab. Arctic Seas.

46290. An imperfect sixth caudal vertebra; dredged from the North Sea. This specimen agrees precisely with the corresponding vertebra of the existing form.
Owles Collection. Purchased, 1874.

Delphinapterus (?) brocchii (Balsamo-Crivelli[5]).

Syn. *Delphinus brocchii*, Balsamo-Crivelli[6].

The teeth are of great relative stoutness. The species is referred to the present genus by Brandt in the Mém. Ac. Imp. Sci. St. Pétersbourg, sér. 7, vol. xx. p. 241.
Hab. Italy.

47035. Four teeth from the Lower Pliocene (Plaisancien) of Orciano, Tuscany, Italy. These specimens agree precisely with the teeth figured by Capellini in the 'Mem. Ac. Sci. Ist. Bologna,' ser. 2, vol. iii. pl. ii. *Purchased*, 1875.

The following specimens were associated with the preceding, but indicate smaller individuals which, if belonging to the same species, must have been immature. The tympanics correspond in size with those of Lagenorhynchus acutus.

47036. Two periotics and two imperfect tympanics.
Purchased, 1875.

[1] Reise etc. vol. iii. p. 85 (1776).—*Delphinus.*
[2] *Loc. cit.*
[3] Hist. Nat. d. Cétacés, p. 243 (1804).
[4] Zool. 'Erebus' and 'Terror,' p. 29 (1846).
[5] Giorn. I. R. Ist. Lombardo, vol. ii. p. 132 (1842), or vol. iii. p. 302 of Giorn. e Bibliot. Ital. of same Society (1842).—*Delphinus.*
[6] *Loc. cit.*

47037. Three vertebral centra. *Purchased*, 1875.

47038. The imperfect right humerus and four specimens of the radius and ulna. *Purchased*, 1875.

Genus **ORCA,** Gray[1].

Teeth usually $\frac{(12-13)}{(12-13)}$, occupying nearly the whole length of the rostrum, and very large and stout, with conical recurved crowns and large roots, expanded laterally and flattened. The centra of the first, second, and occasionally the third cervical vertebræ are united, while the rest are free.

Orca citoniensis, Capellini[2].

This species is of considerably smaller size than *O. gladiator*, and has thirteen teeth in each jaw; it was originally described from the Pliocene of Cetona, Italy.

Hab. Europe (Italy and (?) England).

M. 3516. The cast of an unworn tooth, provisionally referred to this species. The original, which is from the Red Crag of Suffolk, and is preserved in the Ipswich Museum, is described and figured by the writer in the 'Quart. Journ. Geol. Soc.' vol. xliii. p. 15, pl. ii. fig. 10. The specimen agrees exactly with the teeth figured by Capellini, *op. cit.* pls. i A., ii.; it also closely resembles the teeth of *Pseudorca*, but is distinguished by its more compressed root. *Made in the Museum*, 1886.

M. 3515. Cast of a right periotic, belonging to the same species as the preceding specimen. The original was obtained from the Red Crag of Woodbridge, Suffolk, and is preserved in the Museum of Practical Geology, Jermyn Street; it is described and figured by the writer, *op. cit.* pl. ii. fig. 9. It exhibits the grooved anterior tympanic articular facette characteristic of the present family (as distinguished from the *Physeteridæ*), and agrees very closely in characters with the tympanic of *O. gladiator* (Mus. R. Coll. Surg. no. 2982), although of considerably smaller dimensions, its length being 0,063. In the latter respect it agrees with the tooth already noticed. *Made in the Museum*, 1886.

1 Zool. 'Erebus' and 'Terror,' p. 33 (1846).

2 Mem. Ac. Sci. Ist. Bologna, ser. 4, vol. iv. p. 670 (1883).

Genus GLOBICEPHALUS, Lesson[1].

Syn. *Globiocephalus*, Gray[2].
Globiceps, Flower[3].

The teeth ($\frac{8-12}{8-12}$ in number) are confined to the anterior part of the jaws, and are small, conical, curved, and sharp-pointed when unworn. The cranium is broad and depressed, and the rostral not longer than the cranial portion; the upper surface of the cranium is broad and flat. The centra of the anterior five or six cervical vertebræ are united. The length of the centra of the lumbar and earlier caudal vertebræ is about equal to their width.

Globicephalus melas (Traill[4]).

Syn. *Delphinus melas*, Traill[5].
Delphinus globiceps, Cuvier[6].
Globiocephalus svineval, Gray[7].
Globiceps melas, auct.

Hab. Apparently cosmopolitan.

39483. An imperfect cranium; from the marshes near Barking, Essex. *Presented by C. P. Lane, Esq.*, 1865.

39484. The anchylosed series of cervical vertebræ; from near Barking. *Presented by C. P. Lane, Esq.*, 1865.

33458. The centrum of a lumbar vertebra, provisionally referred to this species; dredged off the eastern coast. Both epiphyses are wanting.
Layton Collection. Purchased, 1858.

Globicephalus uncidens (Lankester[8]).

Syn. *Delphinus uncidens*, Lankester[9].
Delphinus orcoides, Lankester[10].

The teeth figured by Lankester as *D. orcoides*[11] appear to belong

[1] Nouv. Tab. Règne Animal, Mammifères, p. 200 (1847). Amended from *Globicephala*. The reason for adopting this name in place of *Globiceps* is given by Flower in the 'Proc. Zool. Soc.' 1884, p. 418.
[2] Zool. 'Erebus' and 'Terror,' p. 32 (1846).
[3] Proc. Zool. Soc. 1883, p. 508.
[4] Nicholson's Journ. vol. xxii. p. 81 (1809).—*Delphinus*. [5] *Loc. cit.*
[6] Ann. du Muséum, vol. xix. p. 14 (1812). [7] *Op. cit.* p. 32 (1846).
[8] Ann. & Mag. Nat. Hist. ser. 3, vol. xiv. p. 356 (1864).—*Delphinus*.
[9] *Loc. cit.* [10] *Loc. cit.*
[11] Some error occurs in the description of the teeth made the type of *D. orcoides*, since they are stated to agree in size with those of *Orca* and *Pseudorca*, whereas they really accord in this respect with *Globicephalus*.

to the same species as the smaller type tooth. The close resemblance existing between the remains noticed below and the corresponding elements of the skeleton of *G. melas* indicates the near affinity of the two forms. Some unnamed vertebræ in the Brussels Museum probably belong to this species.

Hab. Europe (England and [?] Belgium).

46754. An anterior tooth, agreeing with one of the type-specimens figured by Lankester, *op. cit.* pl. viii. fig. 12, and corresponding in size with the first two teeth of *G. melas*; from the Red Crag. *Purchased*, 1875.

43402. Two hinder teeth; from the Red Crag. These specimens correspond with the teeth figured by Lankester, *op. cit.* pl. viii. figs. 14–18 as *D. orcoides,* and are slightly larger than the hinder teeth of *G. melas,* although otherwise indistinguishable [1]. *Purchased*, 1872.

44632. A very similar tooth, but with a straighter root; from the Red Crag of the Butley River, Suffolk. *Purchased*, 1873.

43403, 44609–10, 46753. Four teeth of the same type as the preceding; from the Red Crag of Suffolk. *Purchased.*

36657. The associated left periotic and tympanic; from the Coral-
(*Fig.*) line Crag of Orford, Suffolk. The periotic (woodcut, fig. 17) has been described and figured by the present

Fig. 17.

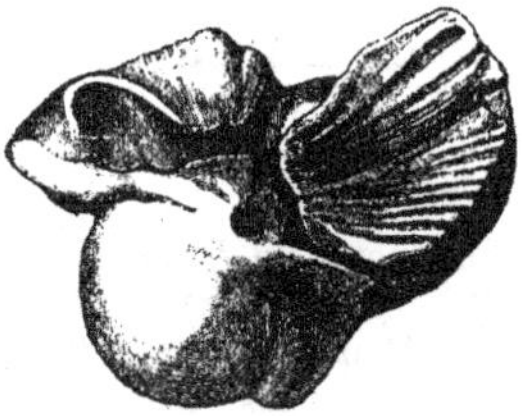

Globicephalus uncidens.—The left periotic; from the Coralline Crag of Orford. 1/1.

[1] The crowns of these teeth present considerable resemblance to the canines of certain Seals; their Cetacean nature is, however, indicated by the frequently irregular root, by the absence of a distinct curve in the whole tooth and of antero-posterior ridges on the crown. Seeing, moreover, that molars of Seals are not recorded from the Crag, it would be very strange to find such a large number of canines.

writer in the 'Quart. Journ. Geol. Soc.' vol. xliii. p. 16, pl. ii. fig. 11, and corresponds closely with the specimen figured by Lankester, *op. cit.* pl. viii. figs. 2, 3. Both specimens, which have not been rolled, agree precisely in size with the corresponding bones of *G. melas*[1].

Brown Collection. Presented by Prof. Sir R. Owen, K.C.B., 1859.

28264, 29598. Two imperfect specimens of the tympanic of opposite sides; from the Red Crag of Woodbridge, Suffolk.

Purchased, 1852.

40137–8. Two specimens of the right periotic, which have been bisected to exhibit the semicircular canals; from the Red Crag of Suffolk. *Purchased*, 1866.

49978. Five rolled specimens of the periotic; from the Red Crag of Suffolk. *Wetherell Collection. Purchased*, 1871.

27434, 28250, 28264, 28995, 29594, 49978. Several rolled specimens of the periotic; from the Red Crag of Suffolk.

Purchased.

28271. The centrum and part of the neural arch of the fifth lumbar vertebra; from Woodbridge. This specimen, which cannot be distinguished from the corresponding vertebra of *G. melas*, is noticed by the writer in the 'Quart. Journ. Geol. Soc.' vol. xliii. p. 16. *Purchased*, 1852.

46745. The imperfect centrum of a lumbar vertebra, provisionally referred to the present species; from Woodbridge.

Purchased, 1875.

37298. A similar imperfect centrum; from Woodbridge.

Purchased, 1863.

43200. The imperfect centrum of an early caudal vertebra; from Woodbridge. *Purchased*, 1872.

The two following specimens (one of which is immature) apparently indicate an allied but smaller form.

33504. The centrum of an early lumbar vertebra; from the Norwich Crag. Both epiphyses are wanting

Wigham Collection. Purchased, 1859.

33505. The centrum of a late lumbar vertebra; from the Norwich Crag. *Wigham Collection. Purchased*, 1859.

[1] British Museum, No. 76.2.16.21.

Genus **TURSIOPS,** Gervais [1].

This genus is closely allied to *Delphinus*, but has larger and less numerous teeth.

Tursiops tursio (Bonnaterre [2]).

Syn. *Delphinus tursio*, Bonnaterre [3].
Delphinus truncatus, Montagu [4].
Tursio truncatus, Gray [5].

The teeth vary in number from $\frac{21}{21}$ to $\frac{25}{25}$.

Hab. Atlantic Ocean.

20274. The centrum and portions of the neural arch and transverse processes of a lumbar vertebra, provisionally referred to this species; from the Pleistocene of Grays, Essex. This specimen is indistinguishable from the ninth lumbar of the existing form. *Purchased*, 1846.

The following specimen apparently indicates an allied form.

35042 a. The centrum of a middle caudal vertebra; from the Coralline Crag of Ramsholt, Suffolk.
Presented by S. V. Wood, Esq., 1850.

Generically undetermined Specimens.

The following specimens indicate Dolphins agreeing in size with the existing *Lagenorhynchus acutus*. They have been alluded to by the writer in the 'Quart. Journ. Geol. Soc.' vol. xliii. p. 16, where it is suggested that they may be specifically identical with one or both of two Delphinoids from the Antwerp Crag, of which remains are preserved in the Brussels Museum under the names of *Delphinus wasii* and *D. delannoyi*, Van Beneden (the generic name being employed in the Linnean sense). There are, however, no periotics or tympanics in the Brussels collection.

30265. Four periotics in a somewhat water-worn condition; from the Coralline Crag of Beccles Old Abbey, Suffolk.
Purchased, 1855.

27434. Several rolled periotics of a similar type; from the Red Crag near Woodbridge, Suffolk. *Purchased*, 1851.

[1] Hist. Nat. Mamm. vol. ii. p. 323 (1855).
[2] Cétologie, p. 21 (1789).—*Delphinus*.
[3] *Loc. cit.*
[4] Mem. Wern. Soc. vol. iii. p. 75 (1821).
[5] Cat. Seals and Whales Brit. Mus. p. 258 (1866).

28264. Similar specimens; from Woodbridge. *Purchased*, 1852.

29596. Similar specimens; from Woodbridge. *Purchased*, 1855.

29603–5. Similar specimens; same locality. *Purchased*, 1855.

36649. Similar specimens; same locality.
Brown Collection. Presented by Prof. Sir R. Owen, K.C.B., 1859.

49978 a. Similar specimens; same locality.
Wetherell Collection. Purchased, 1871.

28264 a. An imperfect tympanic; from Woodbridge.
Purchased, 1852.

29599. An imperfect tympanic; same locality. *Purchased*, 1855.

36649 a. An imperfect tympanic; same locality.
Brown Collection. Presented by Prof. Sir R. Owen, K.C.B., 1859.

Order EDENTATA.

Family MEGATHERIIDÆ.

The members of this family present characters intermediate between the *Bradypodidæ* and *Myrmecophagidæ*, combining the skull

Fig. 18.

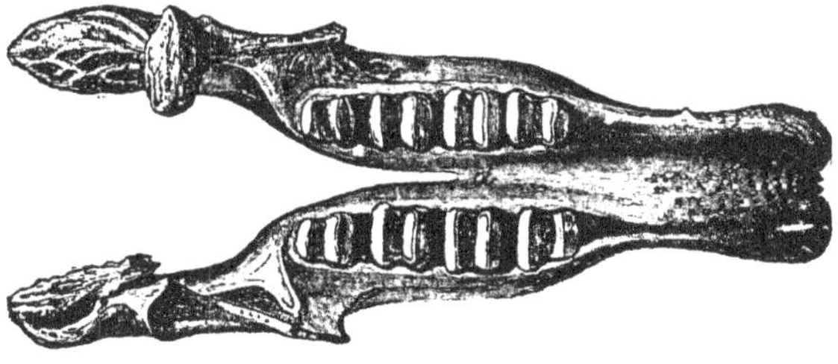

Megatherium americanum.—The mandible; from the Pleistocene of Buenos Ayres. Much reduced.

and dentition of the former with the structure of the limb-bones and vertebræ of the latter. One species of *Scelidotherium* approximates, however, in cranial characters to the *Myrmecophagidæ*. None of them appear to have been arboreal. The number of the teeth is usually $\frac{5}{4}$; and there may or may not be a diastema between the first and second teeth. There is no third trochanter to the femur; and the odontoid process of the axis vertebra has a facette on its ventral surface.

Genus **MEGATHERIUM**, Cuvier[1].

In the type species the dentition is $\frac{5}{4}$, but in the nominal *M. gervaisi*, H. Gervais and Ameghino[2], there are said to be only three lower teeth. The teeth consist of square prisms wearing into transverse ridges through the presence of two vertical plates of hard dentine; they are similar in structure, and have no diastema; the hinder part of the mandibular ramus forms a deep descending plate. The external trochlear ridge of the astragalus for the tibia is very prominent, and is raised much above the internal tibial protuberance; there is a deep navicular cup in the astragalus. The second, third, and fourth digits of the manus and the third in the pes have claws, the third being the longest. The second and third phalangeals of the third digit are anchylosed in both the manus and the pes. There is no entepicondylar foramen in the humerus.

Megatherium americanum, Cuvier[3].

Syn. *Megatherium cuvieri*, Desmarest[4].
Megatherium australe, Oken[5].
Bradypus giganteus, Pander and D'Alton[6].

This is the type species, and larger than *Rhinoceros unicornis*.

Hab. South America (Argentine Republic and Uruguay).

* Cast of the skeleton. This specimen is made up partly from casts of bones in the Museum, which were obtained in 1837 from the Pleistocene of Lujan, Buenos Ayres, Argentine Republic, and partly from those of others preserved in the Museum of the Royal College of Surgeons (No. 3441), which were obtained by Sir Woodbine Parish from the bed of the Rio Salada, one of the tributaries of the Rio Plata, south of the city of Buenos Ayres. The restored skeleton is figured in pls. i. and xxvii. of Owen's 'Memoir on the *Megatherium*.' *Purchased.*

19953. The skull, with the cerotahyals attached; from Lujan.
(*Fig.*) This specimen, from which the cast of the skull in the restored skeleton is taken, is figured by Owen, *op. cit.* pls. xii.–xvi., the figure of the mandible being copied in woodcut, fig. 18. *Purchased*, 1846.

1 Table Elém. de l'Hist. Nat. p. 146 (1798).
2 'Mammifères Fossiles de l'Amérique Méridionale,' p. 136 (1880).
3 In Shaw's 'General Zoology,' vol. i. p. 165 (1800).
4 Mammalogie, p. 365 (1822).
5 In Holl's 'Handbuch der Petrefactenkunde,' p. 27 (1829).
6 Riesenfaulthier, &c. pl. i. (Bonn, 1821).

19953 f. The slightly imperfect left ramus of the mandible, showing the aveoli of the first three teeth and the broken base of the last tooth; from Lujan. *Purchased*, 1846.

M. 3580. Casts of nine teeth. The originals of these specimens were collected by Sir W. Parish from the Rio Salado, and are preserved in the Museum of the Royal College of Surgeons (Nos. 3447–3455).
Presented by the Council of the Royal College of Surgeons, 1834.

19953 a. The superior portion of the first left upper tooth; from Lujan. *Purchased*, 1846.

19953 b. Three imperfect specimens of lower teeth; from Lujan. *Purchased*, 1846.

19953 c. The summit of a small tooth; from Lujan. *Purchased*, 1846.

43231. Three imperfect teeth; from Buenos Ayres. *Presented by Señor L. J. Fontana*, 1871.

40820. The fourth right lower tooth; from Buenos Ayres. *Presented by C. Falconer, Esq.*, 1867.

40161. Two fragments of the right and left scapula; from Buenos Ayres. *Purchased*, 1866.

19953 d. The right humerus; from Lujan. *Purchased*, 1846.

19953 e. The left humerus; from Lujan. This specimen, from (*Fig.*) which the cast in the skeleton was taken, is figured by Owen, *op. cit.* pl. xix. *Purchased*, 1846.

39393. The right humerus; from the Pleistocene of the Rio Negro, Uruguay. *Presented by D. A. Stoddart, Esq.*, 1865.

43232. The left radius; from Buenos Ayres. *Presented by Señor L. J. Fontana, Esq.*, 1871.

19953 g. The right radius; from Lujan. Figured by Owen (re- (*Fig.*) versed), *op. cit.* pl. xx. figs. 4–6. *Purchased*, 1846.

19953 h. The right radius; from Lujan. *Purchased*, 1846.

40489. The right ulna, wanting the distal extremity; from Uruguay. *Presented by W. G. Lettsom, Esq.*, 1867.

19953 i. The left ulna; from Lujan. This is apparently the speci- (*Fig.*) men figured by Owen, *op. cit.* pl. xx. figs. 1–3. *Purchased*, 1846.

19953 j. The nearly perfect right manus; from Lujan. This specimen, from which the corresponding portion of the skeleton was cast, is figured by Owen, *op. cit.* pl. xxi.
(*Fig.*)
Purchased, 1846.

19953 k. A less perfect specimen of the right manus; from Lujan.
Purchased, 1846.

19953 l. The slightly imperfect left manus; from Lujan. The corresponding portion of the skeleton is cast from this specimen. *Purchased*, 1846.

M. 3577–9. Cast of the left carpus and third and fourth metacarpals. The originals, which are from S. America, are preserved in the Museum of the Royal College of Surgeons, and are figured (with the second metacarpal) in De Blainville's 'Ostéographie'—Genus *Megatherium*, pl. iii. fig. 7 (reversed).
Presented by the Council of the Royal College of Surgeons, 1834.

19953 m. The fourth right metacarpal; from Lujan. This and the following specimen belong to a very large individual.
Purchased, 1846.

19953 n. The associated third and fourth left metacarpals of the same individual as the preceding; from Lujan.
Purchased, 1846.

19953 o. The left lunar; from Lujan. *Purchased*, 1846.

19953 p. Two specimens of the pisiform; from Lujan.
Purchased, 1846.

19946 x. The second phalangeal of a manus, provisionally referred to this species; from Buenos Ayres. *Purchased*, 1846.

41148–9. Two fragments of the innominate; from Uruguay.
Purchased, 1868.

40490. Fragment of the innominate; from Uruguay.
Presented by W. G. Lettsom, Esq., 1867.

40496. Fragment of the innominate; from Uruguay.
Presented by W. G. Lettsom, Esq., 1867.

19953 q. The left femur; from Lujan. This appears to be the specimen figured by Owen, *op. cit.* pl. xxiii. fig. 1.
(*Fig.*)
Purchased, 1846.

19953 r. The right femur; from Lujan. *Purchased*, 1846.

41618. The imperfect distal half of the right femur; from Rio Grande do Sul, Brazil. *Presented by J. Tennent, Esq.*, 1867.

19946. A patella; from Lujan. Figured (reversed) by Owen, *op.*
(*Fig.*) *cit.* pl. xxiii. fig. 2. *Purchased*, 1846.

19953 s. The right tibia and fibula, probably belonging to the same individual as No. 19953 *q*; from Lujan. *Purchased*, 1846.

19953 t. The right tibia; from Lujan. *Purchased*, 1846.

19953 u. The right tibia and fibula, wanting the external malleolus, of a smaller individual; from Lujan. *Purchased*, 1846.

39387. The distal extremity of the left tibia; from the Rio Negro, Uruguay. *Presented by D. A. Stoddart, Esq.*, 1865.

19953 v. The right pes; from Lujan. This specimen, from which
(*Fig.*) the corresponding portion of the skeleton is cast, is figured by Owen, *op. cit.* pls. xxv., xxvi. *Purchased*, 1846.

19953 w. The slightly imperfect left pes; from Lujan. The corresponding portion of the skeleton is cast from this specimen. *Purchased*, 1846.

19953 x. The right calcaneum and astragalus; from Lujan. *Purchased*, 1846.

19953 y. The right calcaneum and astragalus; from Lujan. *Purchased*, 1846.

19953 z. The right astragalus; from Lujan. *Purchased*, 1846.

19953 a′. The right astragalus of a small individual; from Lujan. *Purchased*, 1846.

19953 b′. A left astragalus, agreeing in size with the preceding; from Lujan. *Purchased*, 1846.

19953 c′. The left navicular; from Lujan. *Purchased*, 1846.

19953 d′. The left navicular; from Lujan. *Purchased*, 1846.

19953 e′. The right cuboid; from Lujan. *Purchased*, 1846.

19953 f′. The left cuboid; from Lujan. *Purchased*, 1846.

19953 g′. The atlas vertebra; from Lujan. *Purchased*, 1846.

19953 h′. The atlas vertebra of an immature individual; from Lujan. *Purchased*, 1846.

19953 i′. An imperfect cervical vertebra; from Lujan. *Purchased*, 1846.

19953 j′. Two middle dorsal vertebræ; from Lujan. *Purchased*, 1846.

19952. A middle dorsal vertebra; from Lujan. *Purchased*, 1846.

39390. A late dorsal vertebra; from the Rio Negro.
Presented by D. A. Stoddart, Esq., 1865.

19953 k′. The associated caudal vertebræ, wanting the first two of the series; from Lujan. The corresponding vertebræ of the skeleton are cast from this specimen. *Purchased*, 1846.

19953 l′. Twelve associated caudal vertebræ and detached chevron bones; from Lujan. *Purchased*, 1846.

19953 m′. The left clavicle; from Lujan. *Purchased*, 1846.

19953 n′. Thirteen sternal bones; from Lujan. *Purchased*, 1846.

39389. A sternal bone; from the Rio Negro.
Presented by D. A. Stoddart, Esq., 1865.

16404. A first rib; from Uruguay.
Presented by W. G. Lettsom, Esq., 1867.

19953 o′. Eight ribs; from Lujan. *Purchased*, 1846.

37584–625. An associated series of the bones of a fœtus, comprising the occipital, frontal, numerous vertebræ, ribs, clavicle, scapula, humerus, elements of the innominate, femur, tibia, and ungual phalangeal; from the Rio Plata.
Bravard Collection. Purchased, 1854.

Megatherium mirabile, Leidy[1].

This species is very imperfectly known, and its distinctive characters have not yet been made clear.

Hab. North America (Georgia, South Carolina, and Texas).

27503. A left astragalus, provisionally referred to this species; from the Pleistocene of Texas. This specimen is considerably larger than the corresponding bone of *M. americanus*, and shows structural differences which are certainly of specific value. *Hastings Collection. Purchased*, 1855.

Megatherium lundi, H. Gervais & Ameghino[2].

This species is founded on an astragalus which is less than one third the dimensions of the corresponding bone of *M. americanum*, from which it is said to differ in form.

Hab. South America (Argentine Republic).

43251 a. A left astragalus, probably belonging to this form; from

[1] Smiths. Contrib. Knowl. vol. vii. art. 5, p. 49 (1855).

[2] 'Mammifères Fossiles de l'Amérique Méridionale,' p. 138 (1880).

the Pleistocene of Buenos Ayres, Argentine Republic. This specimen, of which the transverse diameter is only 0,110, differs from the corresponding bone of *M. americanum* by the higher position of the navicular cup, which is also narrower; in these respects the present specimen is intermediate between *M. americanum* and *Mylodon*, in which the cup is absent.

Presented by Señor L. J. Fontana, 1871.

Genus **SCELIDOTHERIUM**, Owen[1].

Including *Platyonyx*, Lund[2].

According to Burmeister ('Monatsb. k. preuss. Ak. Wiss.' 1881, pp. 374–380) *Platyonyx* is distinguished by the structure of the foot and the absence of an entepicondylar foramen to the humerus; but since the cranium of *S. chiliense* (*infrà*, p. 100) agrees with that of typical species of *Platyonyx*, whilst its humerus has an entepicondylar foramen, it appears either that the limb-bones referred to the latter belong to some other genus[3], or that the characters on which that genus is founded can only be regarded as of specific value[4].

The genus is in some respects intermediate between *Megatherium* and *Mylodon*, approaching the former in the structure of the cranium and pes, and the latter in dentition. *S. leptocephalum* is the most aberrant form, and shows signs of affinity with the *Myrmecophagidæ*. The teeth are $\frac{5}{4}$ in number and have no diastema; the upper ones have an irregular oval section and are set obliquely in the jaw, the last being the smallest; the first three lower ones are triangular, and the fourth elongated with an inner crotchet, which varies in size in different species. The cranium is long, narrow, and low, the inferior border of the mandible being frequently nearly straight, although in some species having a descending hinder portion, recalling the jaw of *Megatherium*. The astragalus wants the articular navicular cup characteristic of *Megatherium*, and varies considerably in form in the different species, that of the type species being the most unlike that of the latter genus. There are four digits to the manus[5], the second and third being provided with nearly straight

[1] Zoology of the Voyage of the 'Beagle,' pt. i. p. 73 (1840).

[2] K. Danske Vid. Selsk. Skr. vol. ix. p. 145 (1842). Separate copies of the memoir are dated earlier.

[3] Owen ('Memoir on the *Mylodon*,' p. 170, note) suggests that the limb-bones referred by Lund to *Platyonyx* may belong to *Glyptodon*.

[4] See Lydekker, Proc. Zool. Soc. 1886, p. 496.

[5] The feet are figured by Burmeister in the 'Monatsb. k. preuss. Ak. Wiss.' 1881, plate facing p. 380.

claws. In the pes the first and second digits are aborted, the third digit being furnished with a very large curved claw, like that of *Megatherium*. All the species in the Museum have an entepicondylar foramen to the humerus.

Scelidotherium leptocephalum, Owen[1].

This is the type species. The cranium is of moderate width, the nasals and maxillæ are of remarkable length, that of the former being equal to at least half of the total length of the superior surface of the cranium. The lachrymal is not very prominent, and the interdental portion of the palate of moderate width.

The portion of the mandible in advance of the teeth is much elongated, and is nearly twice the length of the dental series, its superior border being nearly horizontal. The scapula has a straight anterior border; the femur is short and wide; and the external trochlear ridge of the astragalus for articulation with the tibia depressed. In the structure of the facial part of the cranium and of the astragalus this species approximates to *Myrmecophaga*.

Hab. South America (Patagonia and Argentine Republic).

M. 3583. Cast of the imperfect skull. The original was obtained by Darwin from the Pleistocene of Punta Alta, Patagonia, and is preserved in the Museum of the Royal College of Surgeons (No. 3506). Together with the associated specimens it forms the type of the species, and is figured by Owen in the 'Zoology of the Voyage of the Beagle,' pls. xxi.–xxiii.

Presented by the Council of the Royal College of Surgeons.

37308. (*Fig.*) The cranium, imperfect anteriorly and posteriorly; from the Pleistocene of the Rio Plata, Argentine Republic. This specimen (woodcut, fig. 19), which agrees very closely with the one figured by P. Gervais in the 'Mammifères Fossiles de l'Amérique Méridionale,' pl. xi. fig. 1, is described and figured by the present writer in the 'Proc. Zool. Soc.' 1886, p. 493, pl. xlvi. The length of the nasals, which are imperfect anteriorly, is 0,270; all the teeth are *in situ*. *Bravard Collection. Purchased*, 1854.

37309. The mandible, imperfect posteriorly; associated with the preceding specimen. This specimen, which is noticed by the present writer in the paper cited, accords with the one figured by Gervais, *op. cit.*

Bravard Collection. Purchased, 1854.

[1] Zoology of the Voyage of the 'Beagle,' pt. i. p. 73 (1840).

Fig. 19.

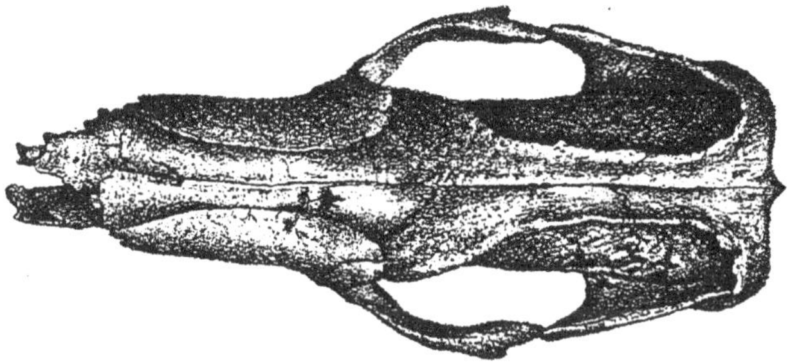

Scelidotherium leptocephalum.—Frontal aspect of the cranium; from the Pleistocene of the Argentine Republic. $\frac{1}{6}$.

32995. The cranium, imperfect anteriorly; from the Rio Plata.
(*Fig.*) This specimen is figured by Owen in the 'Phil. Trans.' for 1857, pl. viii. fig. 1; its vertical height is somewhat less than that of No. 37308, but the difference is in all probability only a sexual one.

Bravard Collection. Purchased, 1854.

32996. Fragment of the hinder portion of the right ramus of the mandible, containing the last two molars.

Bravard Collection. Purchased, 1854.

32997. Fragment of the opposite ramus of the same mandible.

Bravard Collection. Purchased, 1854.

The originals of the following casts belong to the same individual as the type skull, and are preserved in the Museum of the Royal College of Surgeons, by the Council of which body they were presented.

M. 3585. Cast of the left scapula.

M. 3584. Cast of the adjacent portions of the right scapula and humerus.

M. 3586. Cast of the left femur. Figured by Owen, *op. cit.* pl. xxv. figs. 5, 6; the length is 0,395, and the distal width 0,220.

M. 3587. Cast of the right femur, wanting the distal extremity, and with the astragalus cemented by matrix to the head.

M. 3588. Cast of the left astragalus. Described and figured by the writer in the 'Proc. Zool. Soc.' 1886, p. 494, pl. xlix. fig. 3.

M. 3590. Cast of the cervical and first seven dorsal vertebræ. Figured by Owen, *op. cit.* pl. xxi. fig. 1.

M. 3591. Cast of the sacral vertebræ.

M. 3589. Cast of four caudal vertebræ.

The following specimens were found in association with the cranium, No. 37308.

37422–7. Numerous sternal bones.

37428–9. The right and left clavicle.

37437. Part of the left scapula.

37554. The glenoidal portion of the right scapula.

37438. The left humerus.

37439. The left radius.

37430. The right radius.

37430 a. The right ulna.

37440–69. The right manus.

37310. Part of the sacrum and left innominate.

37470. Part of the right acetabulum.

37471. Part of the right femur. As far as can be determined, the bone agrees very closely with the corresponding element of the type skeleton.

37473. The right patella.

37474. The left patella.

37472. The left tibia.

37482. The left fibula.

37475. The left calcaneum.

37476. The left astragalus. This bone, although of somewhat larger size, agrees very closely in structure with the astragalus of the type skeleton (No. M. 3588), showing the comparatively slight prominence of the external trochlear ridge for the tibia; it is noticed by the writer in the 'Proc. Zool. Soc.' 1886, p. 494.

37483. The right calcaneum, astragalus, navicular, cuboid, and fifth metatarsal.

37492. The conjoint first and second phalangeals of the third digit of the pes. This bone is remarkable for the backward extension of one side of the superior surface.

37495. The imperfect terminal phalangeal of the third digit of the pes. This bone closely resembles the corresponding claw of *Megatherium*.

37489–96. Several dorsal vertebræ.

37497–506. Several caudal vertebræ.

37508–11. Numerous chevron bones.

37512–26. Several ribs of the left side.

37527–33. Several sternal ribs of the left side.

37534–50. Several ribs of the right side.

37551–3. Three sternal ribs of the right side.

The following specimens belong to the same individual as the cranium, No. 32995.

33055. The right scapula. This specimen agrees very closely with the corresponding bone of the type skeleton (No. M. 3585).

33056. The left scapula.

32998. The right humerus.

32999. The left humerus.

33000. The right ulna.

33001. The right radius.

33002. The distal extremity of the left ulna.

33003. The distal extremity of the left radius.

33004–26. The right manus. The second and third phalangeals of the second and third digits are relatively longer than in the manus figured by Burmeister in the 'Monatsb. k. preuss. Ak. Wiss.' 1881, pl. facing p. 380, fig. 1, under the name of *S. leptocephalum*, but referred below to *S. bravardi*.

33027. The left manus.

33028–34. The cervical vertebræ.

33035–50. The dorsal vertebræ.

33053. The ribs.

The following specimen probably belongs to this species.

M. 3592. The slightly imperfect axis vertebra; from the Argentine Republic. *Bravard Collection. Purchased*, 1854.

Scelidotherium bravardi, Lydekker[1].

The cranium is intermediate between that of the type species and that of *S. chiliense*, although apparently nearest to the latter. It is very narrow, the nasals and maxillæ are comparatively short, the lachrymal prominent, the interdental portion of the palate very narrow, and the premaxillæ well developed. The mandible has the portion in advance of the teeth considerably elongated and inclined distinctly upwards. The scapula is relatively broad, and the astragalus characterized by the great prominence of the external trochlear ridge for articulation with the tibia, and is intermediate in form between the corresponding bone of the type species and that of *Megatherium*.

The type skeleton indicates a smaller animal than the male of *S. leptocephalum*, but the larger limb-bones provisionally referred to the present species are fully as large as those of the type species. The skeleton, of which the feet are figured by Burmeister in the 'Monatsb. k. preuss. Ak. Wiss.' 1881, plate facing p. 380, agrees with the above-mentioned larger bones.

The structure of the facial region of the skull and of the astragalus indicates that this species departs much less widely from the *Megatherium* type than is the case with *S. leptocephalum*.

Hab. South America (Argentine Republic and Brazil).

37626. (*Fig.*) The cranium, wanting part of the nasals and left maxillary region and all the teeth; from the Pleistocene of the Banks of the Rio Plata, Argentine Republic. This specimen, which together with the associated remains is the type, is figured by Owen, under the name of *S. leptocephalum*, in the 'Phil. Trans.' for 1857, pl. viii. fig. 2, and (apparently) in a restored condition in pl. ix. fig. 1; it is also described and figured by the present writer in the 'Proc. Zool. Soc.' 1886, p. 494, pl. xlvii. *Bravard Collection. Purchased*, 1854.

[1] Proc. Zool. Soc. 1886, p. 494.

The following specimens, with the exception of No. 40674 *a*, *were found in association with the cranium.*

37649. (*Fig.*) The mandible, wanting all the teeth. This specimen is figured, under the name of *S. leptocephalum*, by Owen *op. cit.* pl. viii. figs. 4 (teeth restored) and 5, and apparently in pl. ix. figs. 2, 3 (the teeth being restored in fig. 2[1]); it is also described and figured by the present writer, *op. cit.* p. 494, pl. xlix. fig. 1. The difference between this specimen and the mandible of *S. leptocephalum* may be seen by comparing Owen's figure (pl. viii. fig. 4) with the one given by P. Gervais in the 'Mammifères Fossiles de l'Amérique Méridionale,' pl. xi. fig. 1.

37630. The right scapula. This specimen is relatively wider than the corresponding bone of *S. leptocephalum.*

40674 a. The right humerus. This bone is longer than the humerus of the type species (No. 37438), and differs in several details, especially the greater vertical depth of the inner portion of the distal expansion, and the greater width of the bar over the entepicondylar foramen.

37631. The right ulna.

37632. The right radius. This bone is smaller than the radius of *S. leptocephalum* (No. 37439).

37633. The left radius.

37634–42. Several bones of the manus.

37643. The right calcaneum. This bone differs from the calcaneum of the type species (No. 37483) by the deep concavity on the plantar surface.

37644. The imperfect right astragalus. This specimen is readily distinguished from the corresponding bone of the type species by the great prominence of the external trochlear ridge for the tibia, and agrees in this respect with the larger specimen (No. 18620 *k*) noticed below under the head of the present species.

37645. A metatarsal.

37646–7. Two sesamoids.

[1] In the description of the plate this figure is referred to the upper jaw.

The following specimens are of larger size than the preceding, but are provisionally regarded as belonging to the male of the same species; they are all from the caverns of Minas Geraes, Brazil, and belong to the Claussen Collection. Purchased, 1845.

18627. The left scapula. This bone differs from the scapula of *S. leptocephalum* by its superior size and greater relative width, as well as by the greater stoutness of the acromio-coracoidal bar and the smaller size of the foramen. In relative width it agrees with the type scapula of the present species, but differs by the more convex anterior border—a character which may be only sexual.

18627 a. The right scapula, belonging to the same individual as the preceding.

18649 c. The left ulna, wanting the distal epiphysis.

18652 a. The left ulna. This bone is relatively longer than the ulna of the type species.

18620 a^{14}. The distal half of a right ulna, perhaps belonging to this form. The distal extremity differs slightly from that of No. 18652 *a*.

18620 a^{15}. The proximal half of the left radius.

18979. The proximal half of the right radius.

18620. The right femur. This bone is longer than the type femur of *S. leptocephalum*, its extreme length being 0,448 and its distal width 0,216. The concavity on the anterior border is relatively longer than in the type species.

18932. The right femur.

18932 a. A crushed specimen of the left femur.

18652 a^{25}. The right patella. This specimen differs from the corresponding bone of *S. leptocephalum* (No. 37473) by its greater length and flatness, as well as by the smaller proportionate size of the articular facette. It agrees precisely with the corresponding bone of the hind limb of the present form, figured by Burmeister in the 'Monatsb. k. preuss. Ak. Wiss.' 1881, plate facing p. 380, fig. 3, under the name of *S. leptocephalum*.

18620 a^{16}. The left fibula.

18621. The left tibia. This bone is longer than the tibia of *S. leptocephalum* (No. 37472), and differs by the form of the distal articular surface, the facette for the external trochlea of the astragalus being in a higher plane than that for the internal tuberosity, and the facette for the latter relatively smaller; the facette adjacent to the fibula is smaller and placed at an obtuse angle to the other two facettes, instead of being nearly in the same plane. There is a foramen on the internal border below the head, which is wanting in *S. leptocephalum*.

18744. The right tibia. The facettes for the astragalus are slightly different from those of the preceding specimen, but the other characters are similar.

18620 k. (*Fig.*) The left astragalus, belonging to the same individual as the tibia, No. 18621. This bone, which has lost the extremity of the internal tibial tuberosity, is described and figured by the present writer, *op. cit.* p. 495, pl. xlix. fig. 4; it differs from the type astragalus of *S. leptocephalum* by the great projection of the external trochlear ridge above the level of the internal tuberosity, and is thereby intermediate in character between that bone and the astragalus of *Megatherium*, although wanting the navicular cup of the latter. It agrees precisely with the corresponding bone of the foot figured by Burmeister in the 'Monatsb. k. preuss. Ak. Wiss.' 1881, plate facing p. 380, fig. 2, under the name of *S. leptocephalum*.

18935. The conjoint first and second phalangeals of the third digit of the pes. This bone differs from the corresponding phalangeal of *S. leptocephalum* (No. 37492) by the absence of the backward projection of one side of the superior surface.

18620 x. The conjoint first and second phalangeals of the third digit of the pes, imperfect on the palmar aspect.

Scelidotherium chiliense, Lydekker[1].

This species is characterized by the extreme breadth and shortness of the nasals and maxillæ, and the want of production of the anterior part of the mandible. The length of the nasals does not

[1] Proc. Zool. Soc. 1886, p. 496.

exceed one third the total length of the cranium, the lachrymal is prominent, the interdental portion of the palate very narrow, and the premaxilla probably aborted. The length of the portion of the mandible in advance of the teeth scarcely exceeds that of the whole dental series, and the superior border of the symphysis is but slightly curved upwards. The astragalus is intermediate between that of *S. leptocephalum* and *S. bravardi*, but nearest to the latter. There is an entepicondylar foramen to the humerus.

The cranium of this species comes very close to that of the smaller *S. brongniarti*, from Brazil, figured by Lund in the 'K. Danske Vid. Selsk. Skr.' vol. ix. pl. xxviii., under the name of *Platyonyx*, but is distinguished by the much greater breadth of the facial portion. In its extremely short nasals this species comes nearest of all to *Mylodon*.

Hab. South America (Chili).

M. 2819. (*Fig.*) The cranium, incomplete anteriorly and posteriorly; from the Pleistocene of Tamarugal, in the province of Tarapaca, a district formerly belonging to Peru, but now annexed to Chili[1]. This specimen is described and figured by the present writer in the 'Proc. Zool. Soc.' 1886, p. 496, pl. xlviii., and, together with the associated remains, forms the type of the species. The length of the superior surface is 0,400, and that of the nasals 0,150.

Presented by Don Modesta Basadre, 1886.

M. 2820. A less perfect cranium, wanting the nasals; from Tamarugal. *Same history.*

M. 2818. Fragment of the middle portion of the cranium; from Tamarugal. *Same history.*

M. 2821. (*Fig.*) The symphysis and portions of the rami of the mandible; from Tamarugal. This specimen is described and figured by the present writer, *op. cit.* p. 496, pl. xlix. fig. 2. The length of the symphysis is 0,124, and the interval between the latter and the first tooth 0,012. *Same history.*

The following bones were found in association with the preceding specimens, and were presented at the same time.

M. 2827. The imperfect right scapula.

[1] See Proc. Zool. Soc. 1886, pp. 395–6, where a map of the vicinity is given.

M. 2828. The right scapula of a young individual, wanting the epiphyses.

M. 2829. The left humerus. The entepicondylar foramen is well shown.

M. 2830. The slightly imperfect right humerus of a rather smaller individual.

M. 2831. The left humerus of a considerably smaller individual.

M. 2832. The imperfect left ulna.

M. 2833. The right and left radius, probably belonging to a single individual.

M. 2834. Three specimens of the acetabulum.

M. 2835. The left femur. This bone is of a long and narrow type; its extreme length being 0,405, and its distal width 0,196; the concavity on the anterior border is very short.

M. 2836. The right and left femur, probably belonging to a single individual.

M. 2837. The left tibia.

M. 2838. The right tibia, belonging to a rather smaller individual.

M. 2839. The imperfect right tibia.

M. 2840. The proximal epiphysis of the right tibia.

M. 2841. The right and left calcaneum.

M. 2842. The left astragalus. This bone is intermediate in structure between the astragalus of *S. leptocephalum* and that of *S. bravardi.*

M. 2822. Ten dorsal vertebræ.

M. 2823. Four more or less imperfect ribs.

M. 2824. The anterior two thirds of the sacrum, with the last lumbar vertebra anchylosed to it.

M. 2825. The posterior half of the sacrum.

M. 2826. The last two vertebræ of the sacrum.

Scelidotherium tarijense, H. Gervais and Ameghino[1].

The symphysis and that portion of the mandible in advance of the teeth is very short and the superior border much bent upwards, so that nearly the whole of it is above the level of the dental alveoli; the descending process of the zygoma is not digitated, and the nasals appear to have been of considerable length.

Hab. South America (Bolivia and Brazil).

18649 a. The nearly complete left ramus of a mandible, probably belonging to a female of this species; from a cavern in Minas Geraes, Brazil. This specimen, in which all the teeth are wanting, agrees very closely in contour with the mandible of the type male skull from Bolivia, figured by P. Gervais (without specific name) in the 'Mammifères Fossiles de l'Amérique Méridionale' (Castelnau's Voyage), pl. xi. fig. 2, but is of more slender type. In addition to the difference in the structure of the symphysis, it differs from the mandible of *S. chiliense* (No. M. 2821) by the form of the dental alveoli. Noticed by the present writer in the 'Proc. Zool. Soc.' 1886, p. 496, note 2.
Claussen Collection. Purchased, 1845.

18933. A right femur, which may not improbably belong to the present species; from a cavern in Minas Geraes. This bone is longer and narrower than the femur of *S. leptocephalum*; its extreme length being 0,430, and its distal width 0,185. *Claussen Collection. Purchased,* 1845.

SPECIFICALLY UNDETERMINED SPECIMENS.

Unless otherwise stated, these specimens are from the caverns of Minas Geraes, Brazil, and belong to the Claussen Collection. Purchased, 1845.

M. 3600. Part of the cranium and left ramus of the mandible of a young individual; from the Pleistocene of Buenos Ayres, Argentine Republic. The lower teeth are much wider at the base than at the summit.
Bravard Collection. Purchased, 1854.

M. 3603. Fragment of the right maxilla with the last three teeth, apparently belonging to a young individual; from the Pleistocene of the La Plata, Argentine Republic. The teeth

[1] 'Mammifères Fossiles de l'Amérique Méridionale,' p. 148 (1880).

are slightly larger than those of the type maxilla of the so-called *S. minutum*, figured by Lund in the 'K. Danske Vid. Selsk. Skr.' vol. viii. pl. iii. fig. 9.

Bravard Collection. Purchased, 1854.

43252. The mandible of a young individual; from the Pleistocene of Buenos Ayres. The teeth agree very closely with those of the preceding specimen, and are wider at the base than at the summit. *Presented by Señor L. J. Fontana*, 1871.

18620 h. The second left metacarpal.

18620 a^{1,2,3}. The three phalangeals associated with the preceding specimen.

18620 a^{4}. The third right metacarpal. This specimen differs very markedly from the corresponding bone of *S. leptocephalum*.

18620 a$^{5-7}$. The three phalangeals, associated with the preceding specimen.

18652. The conjoint first and second phalangeals of the third digit of the pes.

18652 x. The fifth metatarsal.

18620 a. The left scaphoid.

18620 b. The left lunar.

18620 c. The left cuneiform.

18620 e. The left trapezoid.

18620 f. The left magnum.

18620 g. The left unciform.

18620 i. The right cuneiform.

18620 j. The right pisiform.

18620 l. The left navicular.

18620 m. The left cuboid.

18620 n. The left ectocuneiform.

18620 z. The fifth left metatarsal.

18620 p. The third left metatarsal.

18620 q. The left mesocuneiform.

18620 r. The left entocuneiform.

18620 s. The right navicular.

18620 t. The right cuboid.

18620 u. The right ectocuneiform.

18620 v. The right mesocuneiform.

18620 w. The third right metatarsal.

18620 a–8. The first phalangeal of the fourth digit of the manus.

18620 a–20. A small cervical vertebra.

18638 a. A caudal vertebra.

18638 b. A third metatarsal.

18638 c. Various phalangeals.

18638 d, e. Ungual phalangeals.

18638 f. Fragment of a rib.

18629. Fifth metatarsal.

18630. Three metacarpals.

18631. The left cuboid.

18649 r. Three lumbar vertebræ.

18649 s. A caudal vertebra.

18633. Numerous vertebræ, mostly imperfect.

There are numerous other detached vertebræ and bones of the feet, &c., which have not been catalogued.

Genus **MYLODON,** Owen[1].

Syn. *Glossotherium,* Owen[2].
Orycterotherium, Harlan[3].
Eubradys, Leidy[4].
Including:—*Grypotherium,* Reinhardt[5].
Lestodon, P. Gervais[6].
Pseudolestodon, H. Gervais and Ameghino[7].

---

[1] Zoology of the Voyage of the 'Beagle', pt. 1, p. 63 (1840).
[2] *Op. cit.* p. 57.
[3] Amer. Journ. ser. 1, vol. xliv. p. 69 (1843).
[4] Proc. Ac. Nat. Sci. Philad. vol. vi. p. 241 (1853, vol. dated 1854).
[5] K. Danske Vid. Selsk. Skr. ser. 5, vol. xii. p. 353 (1879).
[6] Mammifères Fossiles de l'Amérique Méridionale (Castelnau's Voyage), p. 47 (1855).
[7] Mammifères Fossiles de l'Amérique Méridionale, p. 158 (1880).

The forms ranged under *Pseudolestodon* connect the typical species so closely with those of the so-called *Lestodon* that it seems advisable to include the whole of them under one generic title; an analogous remark applies to the so-called *Grypotherium*[1].

Teeth $\frac{5}{4}$; those of the upper jaw are usually subtriangular or oval in section, the hinder ones being grooved internally; the first three lower teeth are very similar in structure, but the fourth is elongated and has an inner crochet. In the typical forms there is only a small interval between the first and second teeth, and the plane of wear of the first tooth in both jaws is horizontal; in other species, however, the diastema is longer and the plane of wear of the first tooth oblique, as in *Cholœpus*. The mandible has no inferior descending plate. The humerus has no entepicondylar foramen. The tibia is relatively short, and has a notch on the anterior aspect of the distal extremity for the reception of the astragalus. The astragalus, which varies considerably in the different species, approaches that of *Megatherium*, but has no navicular cup. The terminal phalangeals of both limbs are narrow and straight, and there is no anchylosis of any of the phalangeals; the first, second, and third digits of the manus, and the second and third of the pes are furnished with claws; the third digit of the manus being the longest. The scapula closely resembles that of *Scelidotherium*. There are numerous small ossifications in the dermis.

The genus approaches the nearest of any member of the family to the *Bradypodidæ*, species like *M. gracilis* and *M. armatus* being those in which this approximation is the most marked.

Group A.

The premaxillæ elongated and united anteriorly by a bony arch with the nasals; the anterior portion of the mandible equally elongated, and longer than the dental series.

Mylodon darwini, Owen[2].

Syn. *Grypotherium darwini,* Reinhardt[3].
Scelidotherium ankilosopum, Bravard[4].

This is the only species of the present group, and is nearly equal in size to *M. armatus*. The length of the nasals is 0,126, that of

[1] See Burmeister, Descript. Phys. Rep. Argentine, vol. iii. pt. 1, p. 358 (1879).

[2] Zoology of the Voyage of the 'Beagle', pt. 1, p. 68 (1840).

[3] K. Danske Vid. Selsks. Skr. ser. 5, vol. xii. p. 353 (1879).

[4] In P. Gervais, Zool. et Pal. Générales, sér. 1, p. 132 (1867–69).

the lower series of cheek-teeth 0,125, and that of the portion of the mandible in advance of the first tooth 0,152. The first tooth in each jaw is extremely minute and wears horizontally; it may be sometimes absent; the palate is narrower than in any other species. This species connects the members of the following group with *Scelidotherium*.

Hab. South America (Patagonia and Argentine Republic).

M. 3591 a. Cast of the right ramus of the mandible, showing a section of the four teeth. The original, together with the opposite ramus, was obtained by Darwin from the Pleistocene of Punta Alta, Bahia Blanca, Patagonia, and is preserved in the Museum of the Royal College of Surgeons (No. 3490). It is the type of the species, and is figured by Owen in the 'Zoology of the Voyage of the Beagle,' pt. 1, p. 69, pls. xvii. fig. 5, xviii., and xix.

Presented by the Council of the Royal College of Surgeons.

GROUP B.

The premaxillæ short and separated from the nasals; the anterior portion of the mandible equally short, its length being much less than that of the dental series.

Mylodon harlani, Owen[1].

Syn. *Orycterotherium missouriense*, Harlan[2].
Eubradys antiquus, Leidy[3].

This is the type species. The first and second teeth in the lower jaw are approximated, and the plane of wear of the former was probably horizontal.

Hab. North America (Kentucky, Missouri, Mississippi, South Carolina, and Georgia).

7375. Cast of a portion of the right ramus of the mandible, showing the alveolus of the first tooth and the three posterior teeth. The original, which is the type, was obtained from 'Big-Bone-Lick,' Kentucky, and is preserved in the Museum of the Lyceum of Natural History, New York; it is described and figured by Harlan in his 'Medical and Physical Researches,' p. 334, pl. xv. figs. 1–4 (1835), under the name of *Megalonyx laqueatus*; by Owen in the 'Zoology of the

[1] Zoology of the Voyage of the 'Beagle,' pt. 1, p. 68 (1840).
[2] Amer. Journ. ser. 1, vol. xliv. p. 69 (1843).
[3] Proc. Ac. Nat. Sci. Philad. vol. vi. p. 241 (1853; vol. dated 1854).

Voyage of the Beagle,' pt. 1, p. 68, pl. xvii. figs. 3, 4; and Leidy in the 'Smiths. Contrib. Knowl.' vii. art. 5, p. 47, pl. xiv. figs. 1, 2, the teeth being figured in pl. xvi. fig. 19 of the latter memoir.

Mantell Collection. Purchased, 1836.

40853. The terminal phalangeal of a lateral digit; from the Pleistocene of Skiddaway, near Savannah, Georgia.

Presented by C. Falconer, Esq., 1867.

Mylodon robustus, Owen [1].

This species is about one third smaller than *Megatherium americanum*. The interval between the first and second teeth is comparatively small, and the plane of wear of these teeth horizontal; the length of the last lower tooth is moderate, and the palate of great relative width.

Hab. South America (Argentine Republic and Uruguay).

40483. The cranium, wanting both zygomatic arches, and with some of the teeth imperfect; from the Pleistocene of Gutierrez, near Paysandu, Uruguay. This specimen agrees precisely with the type cranium figured by Owen, *op. cit.* pls. ii.–iv. The length inferiorly is 0,436.

Presented by W. G. Lettsom, Esq., 1867.

M. 3591 b. Cast of the right astragalus. The original (which is figured in De Blainville's 'Ostéographie,' Genus *Megatherium*, pl. iii. fig. 18) was obtained from the Argentine Republic, and is in the Museum of the Royal College of Surgeons.

Presented by the Council of the Royal College of Surgeons.

The following specimens are provisionally referred to this species, but some of them may belong to M. lettsomi.

40489. The left humerus, imperfect distally; from Uruguay. This specimen agrees very closely with the one figured by Owen, *op. cit.* pls. xi. & xii.

Presented by W. G. Lettsom, Esq., 1867.

40154. The left femur; from the Pleistocene of Buenos Ayres, Argentine Republic. This specimen agrees very closely with the one figured by Owen, *op. cit.* pls. xvii. & xviii.

Purchased, 1866.

[1] Description of the Skeleton of an Extinct Gigantic Sloth (1842).

43250. The greater number of the bones of the two hind limbs of one individual, comprising the femur (imperfect), patella, tibia, fibula, calcaneum, astragalus, and the fifth right metatarsal; from Buenos Ayres.
Presented by Señor L. J. Fontana, 1871.

40069. The right and left tibia, calcaneum, and astragalus; from Buenos Ayres. The length of the tibia is 0,240, or very nearly the same as in the specimen figured by Owen, *op. cit.* pl. xx. *Purchased*, 1866.

43251. The terminal phalangeal of a third digit; from Buenos Ayres. *Presented by Señor L. J. Fontana*, 1871.

40485. An imperfect middle cervical vertebra; from Uruguay.
Presented by W. G. Lettsom, Esq., 1867.

43248. The centrum and part of the arch of a dorsal vertebra; from Buenos Ayres. *Presented by Señor L. J. Fontana*, 1871.

M. 3571. Numerous dermal ossifications, probably belonging either to this or one of the other species of *Mylodon*; from Buenos Ayres. *No history.*

⁂ The Museum also possesses a considerable series of detached bones of the feet from Buenos Ayres, entered in the Register under No. 43251, which have not been catalogued.

Mylodon lettsomi, Owen [1].

Syn. *Pseudolestodon lettsomi*, Gervais and Ameghino [2].

This hitherto unfigured species is intermediate in size between *M. robustus* and *M. armatus*. It agrees in the position of the upper dental alveoli with the former, but is distinguished by the narrower and more convex palate, the excessive elevation of the inner wall of the hinder dental alveoli, and the narrower interval between the pterygoids.

Hab. South America (Uruguay and [?] Argentine Republic).

40486. The cranium (woodcut, fig. 20), wanting both zygomatic (*Fig.*) arches and all the teeth; from the Pleistocene of Santa Lucia, Uruguay. This is the type specimen; its length inferiorly is 0,465.
Presented by W. G. Lettsom, Esq., 1867.

[1] Quoted by H. Gervais and Ameghino in the 'Mammifères Fossiles de l'Amérique Méridionale,' p. 164 (1880). The name is misspelt.

[2] *Op. cit.*

39382. Fragment of the left maxilla, showing the alveoli of the three middle teeth; from the Pleistocene of the Rio Negro, Uruguay. The great elevation of the inner wall of the alveoli is well shown.

Presented by D. A. Stoddart, Esq., 1865.

The following specimens may perhaps belong to this species.

19946. The associated right and left femur and the right tibia and fibula; from Buenos Ayres, Argentine Republic. The femur is larger than the corresponding bone of *M. robustus*, and the tibia shorter and thicker, its extreme length being 0,196. *Purchased* 1846.

Fig. 20.

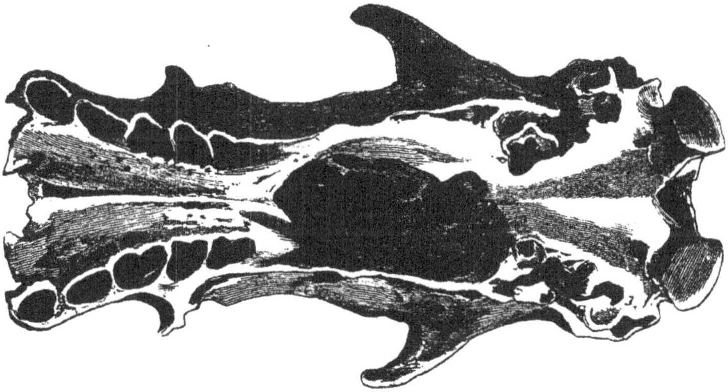

Mylodon lettsomi.—Palatal aspect of the cranium; from the Pleistocene of Uruguay. $\frac{1}{6}$.

19946 a. The right astragalus; associated with the preceding.
Purchased, 1846.

19946 b. The right calcaneum; associated with the preceding.
Purchased, 1846.

19946 c. Three terminal phalangeals; associated with the preceding.
Purchased, 1846

Mylodon armatus (P. Gervais[1]).

Syn. *Lestodon armatus*, P. Gervais[2].
Mylodon giganteus, Burmeister[3].
Lestodon giganteus, H. Gervais and Ameghino[4].

This species is nearly equal in size to *Megatherium americanum*; there is a long diastema between the first and second upper teeth; the plane of wear of the first tooth in either jaw is very oblique; the first upper tooth is triangular and large, and the corresponding lower one nearly cylindrical and small.

Hab. South America (Argentine Republic, Uruguay, and Brazil).

39377. Fragment of the occipito-parietal region of the cranium; from the Pleistocene of Uruguay.
Presented by D. A. Stoddart, Esq., 1865.

The following specimens agree very closely with the limb-bones figured by P. Gervais in the 'Mém. Soc. Géol. France,' sér. 2, vol. ix. pl. xxvi., *and are therefore provisionally referred to this species.*

39386. The distal extremity of the right femur; from Uruguay.
Presented by D. A. Stoddart, Esq., 1865.

39385. The left tibia; from Uruguay. This bone is much larger than the tibia of *M. robustus*, its extreme length being 0,326. *Presented by D. A. Stoddart, Esq.*, 1865.

41619. The imperfect left tibia; from Rio Grande do Sul, Brazil.
Presented by T. J. Tennent, Esq., 1869.

41620. The terminal phalangeal of a median (3rd) digit; associated with the preceding.
Presented by T. J. Tennent, Esq., 1869.

M. 1903. The associated left tibia, calcaneum, and astragalus; together with the metacarpal and phalangeals of the third digit of the right manus: and the neural arch of a dorsal and the centrum of a lumbar vertebra; from the Pleistocene of the Argentine Republic.
Presented by Prof. Sir R. Owen, K.C.B., 1883.

[1] Mammifères Fossiles de l'Amérique Méridionale (Castelnau's Voyage), p. 47 (1855).—*Lestodon*.
[2] *Loc. cit.*
[3] An. Mus. Buenos Aires, vol. i. p. 162 (1864–69).
[4] Mammifères Fossiles de l'Amérique Méridionale, p. 172 (1880).

Mylodon gracilis, Burmeister [1].

Syn. *Lestodon myloides*, P. Gervais [2]
Pseudolestodon myloides, H. Gervais and Ameghino [3].
Pseudolestodon gracilis, H. Gervais and Ameghino [4].

The identity of *M. gracilis* with *L. myloides* is on the authority of Burmeister (*loc. cit.*); and although this is not accepted by Gervais and Ameghino, there seems no doubt of its correctness. Burmeister's specific name, although of later date than that given by Gervais, is adopted on account of the inappropriateness of the latter in conjunction with the term *Mylodon.*

The species is of rather more slender build than *M. robustus*, the plane of the occiput is less oblique, the plane of wear of the anterior teeth very oblique, and the fore and aft edges of the first lower tooth bear a distinct ridge; the last lower tooth is much elongated, and has been compared to a figure of 8.

Hab. South America (Argentine Republic).

M. 2500. The nearly entire skeleton; from the Pleistocene of Buenos Ayres, Argentine Republic. The skull agrees precisely with the type specimen figured in De Blainville's 'Ostéographie,' Genus *Megatherium*, pl. i. figs. 8, 9, 18.
Purchased, 1885.

Genus **MEGALONYX,** Jefferson [5].

Including *Megalochnus*, Leidy [6] (= *Myomorphus*, Pomel).

In this genus, which is confined to North America, the teeth are $\frac{5}{4}$; the hinder teeth form suboval or triangular prisms, with an angle directed outwards; there is a long diastema between the first and second teeth, the first upper tooth being curved and having an oval section, with a horizontal plane of wear. There is an entepicondylar foramen to the humerus. The astragalus is very like that of *Mylodon*; the terminal phalangeals are strongly curved, and there is no anchylosis of the phalangeals.

Megalonyx jeffersoni, Desmarest [7].

Syn. *Megalonyx laqueatus*, Harlan [8].

This is the type species; the identity of the so-called *M. laqueatus*

[1] An. Mus. Buenos Aires, vol. i. p. 166 (1864–69).
[2] Mammifères Fossiles de l'Amérique Méridionale (Castelnau's Voyage), p. 47 (1855).
[3] Mammifères Fossiles de l'Amérique Méridionale, p. 160 (1880).
[4] *Ibid.* p. 164.
[5] Trans. Amer. Phil. Soc. vol. iv. p. 248 (1799).
[6] Proc. Ac. Nat. Sci. Philad. 1868, p. 180.
[7] Mammalogie, p. 366 (1822).
[8] Med. Phys. Research. p. 319 (1835).

is given on the authority of Leidy, 'Smiths. Contrib. Knowl.' vol. vii. art. 5, p. 4 (1855).

Hab. North America (Tennessee, Mississippi, Virginia, and Alabama).

41143. Cast of the cranium, wanting the zygomatic arches, and showing the caniniform teeth of both sides, and the first three molars on the right and the last three on the left side. The original was obtained from the Pleistocene of Henderson County, Kentucky; it appears to be the specimen figured by Leidy in the 'Smiths. Contrib. Knowl.' vol. vii. art. 5, pls. i.–iv. *Purchased*, 1868.

The originals of the following specimens were obtained from Big-Bone Cave, White County[1], *Tennessee, and are preserved in the Museum of the Academy of Natural Sciences of Philadelphia; they are described and figured by Harlan in his* '*Medical and Physical Researches*,' p. 319, pl. xiii. *All the casts belong to the Mantell Collection, purchased*, 1836.

7376. Cast of a portion of a tooth, including the termination of the pulp-cavity. This specimen appears to be the one figured by Owen in the 'Zoology of the Voyage of the Beagle,' pt. 1, pl. xvii. fig. 2, under the name of *M. laqueatus*, and agrees in contour with the one figured by Leidy in the 'Smiths. Contrib. Knowl.' vol. vii. art. 5, pl. vi. fig. 7, and pl. xvi. fig. 5.

7368. Cast of the left humerus, imperfect distally. A less imperfect specimen is figured by Leidy, *op. cit.* pl. ix. figs. 1–4.

7366. Cast of the immature right humerus, wanting the distal epiphysis.

7369. Cast of the immature right radius, wanting the distal epiphysis.

7374. Cast of the right ulna.

7371. Cast of the distal epiphysis of the right femur.

7377. Cast of the left tibia, wanting the distal epiphysis.

7364. Cast of the distal epiphysis of the right tibia.

7363. Cast of the right calcaneum. Figured by Harlan, *op. cit.* pl. xvi., as the ilium.

[1] This name has apparently given rise to the 'White Cave,' from which the specimens have been said to have been obtained.

7373. Cast of the right calcaneum of an immature individual.

7367. Cast of the terminal phalangeal of a third digit.

7372. Cast of the terminal phalangeal probably of an adjoining digit.

7365. Cast of a lumbar vertebra.

7363. Cast of a rib.

Genus **CŒLODON**, Lund[1].

According to Reinhardt[2] this genus agrees in the structure of the limbs with *Megalonyx*, although having teeth of the type of *Megatherium*, which are, however, only $\frac{4}{3}$ in number[3]. There is an entepicondylar foramen to the humerus; and the terminal phalangeals are curved and slender.

Cœlodon maquinensis, Lund[4].

This is the type species, and is distinguished from *C. escrivanensis*, Reinhardt[5], merely by the presence of a groove on the posterior aspect of the last upper molar. The following specimens, as being from Brazil, are provisionally referred to the type species.

Hab. South America (Brazil).

18639. Two teeth; from a cavern in Minas Geraes, Brazil.
Claussen Collection. Purchased, 1845.

The following were associated with the preceding specimens.

18638. The terminal phalangeal of the second digit of the manus. This bone agrees with the specimen figured by Reinhardt, *op. cit.* pl. iv. figs. 8, 9.

18638 a. Several caudal vertebræ.

18638 b. A third metatarsal.

[1] Ann Sci. Nat. sér. 2, vol. xi. p. 220 (1839).

[2] K. Danske Vid. Selsk. Skr. ser. 5, vol. xii. p. 259 (1878).

[3] A mandibular ramus with four teeth has been described and figured by Burmeister in the 'Sitz. k. preuss. Ak. Wiss.' 1885, pp. 567–573, pl. v., under the name of *Cœlodon*. Ameghino ('Bol. Ac. Nac. Cordova,' vol. viii. p. 394 [1886]) identifies this jaw with his previously described *Oracanthus burmeisteri*, which is more nearly allied to *Megatherium* than *Cœlodon*. See also Lütken, 'Overs. K. Danske Vid. Selsk. Forhandl.' 1886, pp. 78–84. The name *Oracanthus* is preoccupied by a genus of fossil fishes.

[4] *Loc. cit.*

[5] *Op. cit.* pls. i.–iv.

18638 c. Several first and second phalangeals.

18638 d, e. Two specimens of the terminal phalangeal of the third digit of the manus. These specimens agree with the corresponding bone figured by Reinhardt, *op. cit.* pl. iii. fig. 4.

18638 f. Part of a rib.

18640. The distal extremity of the right radius.

18641. The proximal extremity of a radius.

18642. Two fragments not improbably belonging to the humerus.

Family GLYPTODONTIDÆ.

The teeth are $\frac{8}{8}$ in number, and have two deep grooves on either side, dividing them into three nearly distinct lobes. The facial part of the cranium is extremely short, and there is a long descending maxillary process in the zygoma. The vertebral column is almost entirely anchylosed into a long tube, but has a complex joint at the base of the neck. The carapace has no movable bands, and is composed of polygonal or quadrangular bony scutes, which are usually united by suture into a solid buckler, but are occasionally (*Thoracophorus*) separate; there is usually a ventral buckler, or plastron; and the caudal region is enclosed in a complete bony sheath. The scutes of the carapace may be either tuberculated, sculptured, or plain; when sculptured the pattern is either simply radiate, or in the form of a rosette consisting of a central elevated disk, surrounded by several peripheral disks, the disks being separated by grooves which vary in depth in the different genera; the disks themselves may be either plain or sculptured. There is no third trochanter to the femur; and the feet are short and stout. The cerebrum is relatively much smaller than in the *Dasypodidæ*.

Genus **GLYPTODON**, Owen[1].

Including *Schistopleurum*, Nodot[2].

The scutes of the carapace are articulated, usually of great thickness, polygonal in form, and in the larger species with from five to seven peripheral disks[3]; the anterior lateral scutes are not

[1] In Parish's 'Buenos Ayres and La Plata,' p. 178 *b* (1838). Many copies of this work belong to an issue dated 1839.

[2] Comptes Rendus, vol. xli. p. 336 (1855).

[3] Some of the scutes mentioned under the head of species *a* have from ten to twelve peripheral disks.

elongated. In the peripheral regions the central disks are larger than in the dorsal region; the grooves between the disks are deep, and in the groove surrounding the central disk there are frequently deep pits, which apparently contained hairs; the surface of the disks is always more or less rugose. The antero-inferior angles of the carapace are not produced in advance of the nuchal border; and the peripheral row of scutes form conical protuberances. At least in those forms which have been generically separated under the name of *Schistopleurum*, the caudal sheath consists of a series of movable rings ornamented with large conical tubercles, and has no terminal tube. A ventral buckler is present. The fronto-nasal region of the cranium has a straight profile. The humerus has no entepicondylar foramen; and there are five perfect digits in the manus and four in the pes.

If Owen's reference of the caudal tube to *G. clavipes* be correct, the genus *Schistopleurum* might be retained.

Glyptodon clavipes, Owen [1].

The middle of the central disk of each scute of the carapace is concave, with its rim raised above the level of the peripheral disks, which are much smaller than the central one. The surfaces of the disks are comparatively smooth, although pitted, and the intervening grooves are wide and shallow. In the periphery of the carapace the central disks of each scute become larger and elevated above the peripheral disks. The scutes themselves are relatively larger than in *G. reticulatus*, and the type specimen is of smaller size.

The terminal tube of a caudal sheath appended to the type carapace (woodcut, fig. 21) in the Museum of the Royal College of Surgeons (No. 3584), and thus figured by Owen in pls. iv. and v. of the Catalogue of the Fossil Mammalia in that Collection as the complete caudal sheath, resembles that of *Hoplophorus*, to which genus it is considered by Ameghino [2] that it belongs. The specimen figured by Burmeister in the 'An. Mus. Buenos Aires,' vol. ii. pl. xxxvi., under the present name has been referred by Gervais and Ameghino to *Hoplophorus* [3].

Hab. South America (Argentine Republic).

[1] In Parish's 'Buenos Ayres and La Plata,' p. 178 *b* (1838).

[2] Bol. Ac. Nac. Cordova, vol. v. pp. 4, 20, 24, 25 (1883); see also H. Gervais and Ameghino, 'Mammifères Fossiles de l'Amérique Méridionale,' p. 203 (1880). Ameghino (*op. cit.* pp. 24–5) states that a carapace has been found associated with a caudal sheath like that of *G. reticulatus.*

[3] *Vide infrà*, p. 133.

M. 3611. Fragment of the dorsal region of the carapace; probably from the Pleistocene of Buenos Ayres, Argentine Republic. This specimen agrees precisely in characters with the carapace of the type specimen. The depression of the middle of each central disk and the elevation of its rim is well shown. *No history.*

Fig. 21.

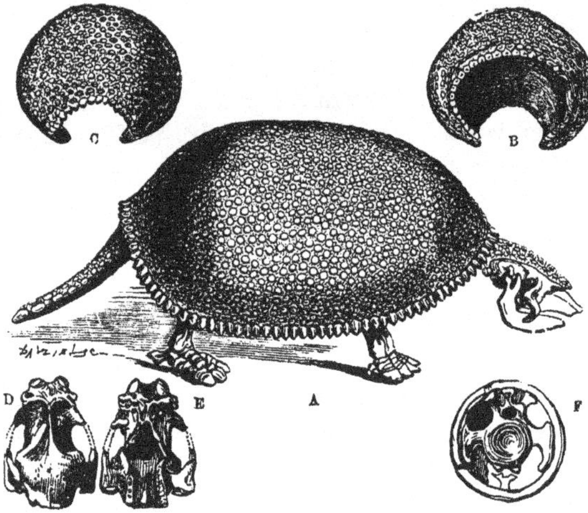

Glyptodon clavipes; from the Pleistocene of Buenos Ayres. The caudal sheath probably belongs to *Hoplophorus.* Much reduced. A. View of entire animal. B. Front end of carapace. C. Back view of same. D and E. Upper and under side of cranium. F. Section of tail, showing caudal vertebræ inside the bony sheath.

M. 3612. Two scutes from the anterior portion of the carapace; from Buenos Ayres. The central disks are of very large size. *Presented by R. Lydekker, Esq.*, 1886.

M. 3613. A scute from the peripheral region of the carapace; from Buenos Ayres. The elevation of the central disk is well shown. *Presented by R. Lydekker, Esq.*, 1886.

39266. Fragment of the carapace; from Buenos Ayres. *Presented by Captain J. Parish*, 1865.

M. 3609. Cast of a portion of the peripheral region of the carapace. The original was obtained from Buenos Ayres, and is believed to be in the Museum of the Royal College of Surgeons.
? *Presented by the Council of the Royal College of Surgeons*

M. 3614. A left femur, imperfect proximally, agreeing with the specimen figured by Burmeister in the 'An. Mus. Buenos Aires,' vol. ii. pl. xxxiv. fig. 1, and referred to the present species; from Buenos Ayres. This specimen is equal in size to the corresponding bone of *G. reticulatus* (No. 39481), but differs in the sharp angulation of the border connecting the head with the great trochanter. It apparently indicates a larger individual than the one to which the type carapace belonged, and may be specifically distinct. *No history.*

M. 3610. Casts of the bones of the left pes and of the distal extremity of the left tibia. The originals are said to have been obtained in association with the type carapace, and are preserved in the Museum of the Royal College of Surgeons. *Presented by the Council of the Royal College of Surgeons.*

Glyptodon reticulatus, Owen [1].

Syn. *Schistopleurum typum*, Nodot [2].
Glyptodon asper, Burmeister [3].
Glyptodon spinicaudus, Burmeister [4].
Schistopleurum asperum, Burmeister [5].
Glyptodon typus, H. Gervais and Ameghino [6].

From a comparison of the type specimen of *G. reticulatus* with undoubted specimens of *G. typus* (= *G. asper*), the writer has no reasonable doubt of the identity of the two [7].

The species is of large size, and has the carapace of considerable relative width; the disks of the scutes are very rough, and the intervening grooves wide and deep; not unfrequently in the dorsal region of the carapace the central disk of each scute is not larger than the peripheral ones, and the rosettes are then very indistinctly defined and the grooves form an irregular network. The central

1 Cat. Foss. Mamm. Aves Mus. R. Coll. Surg. p. 119 (1845).
2 Comptes Rendus, vol. xli. p. 336 (1855).
3 An. Mus. Buenos Aires, vol. i. p. 200 (1864–69).
4 *Ibid.* p. 75 (1864–69).
5 Descrip. Phys. Argentine Republic, vol. iii. pt. 1, p. 424 (1879).
6 Mammifères Fossiles de l'Amérique Méridionale, p. 198 (1880).
7 The proposed identification by Ameghino in the Bol. Ac. Nac. Cordova, vol. v. p. 28 (1883), of *G. reticulatus*, Owen, with *Panochthus tuberculatus*, Burmeister, and of *G. tuberculatus*, Owen, with *G. typus* (Nodot), is quite untenable, see p. 125.

disk of each scute is not unfrequently surrounded by several deep hair-pits. The specimen No. 3585 in the Museum of the Royal College of Surgeons apparently belongs to this species [1].

Hab. South America (Argentine Republic and Uruguay).

M. 3608. Cast of a portion of the dorsal region of the carapace. The original, which is the type, is preserved in the Museum of the Royal College of Surgeons (No. 3606), and was obtained from the Pleistocene near Rio Metanza, 20 miles south of Buenos Ayres, Argentine Republic; it is now somewhat smaller than the cast. A small portion of the original is figured by Owen in the 'Catalogue of the Fossil Mammalia and Aves in the Museum of the Royal College of Surgeons,' pl. v. figs. 1, 2; in the figured portion the rosettes are very indistinctly defined, but they are well marked in a portion which is not figured.

Presented by the Council of the Royal College of Surgeons.

M. 3604. The greater portion of the carapace; from the Pleistocene of Buenos Ayres. The dorsal scutes agree precisely with those of the preceding specimen; the rosettes being frequently indistinct, and the central disks often not larger than the peripheral ones. The periphery of the carapace is wanting. This specimen is noticed by Ameghino in the 'Bol. Ac. Nac. Cordova,' vol. v. p. 8, where it is identified with Burmeister's *G. asper.*

Bravard Collection. Purchased, 1854.

39479. Cast of the nearly complete carapace and caudal sheath. The original was obtained from the Pleistocene of Monte Video, Uruguay, and is preserved in the Museum at Dijon; it is figured by Nodot in the 'Mém. Ac. Sci. Dijon,' sér. 2, vol. v. pl. i. (1857), under the name of *Schistopleurum typum* (of which it is the type). The dorsal scutes agree very closely with those of the preceding specimens; in the periphery the rosettes are always distinct. Both Burmeister and Ameghino agree in identifying this specimen with the so-called *G. asper* of the former writer, of which a specimen is figured in the 'An. Mus. Buenos Aires,' vol. ii. pl. xxxvii. With this carapace and caudal sheath have been mounted casts of the skull and limb-bones of specimens in the Paris Museum, and the cast of a cephalic buckler (No. 37560) in the Museum Collection.

By exchange, 1865.

[1] Provisionally referred by Gervais and Ameghino to *G. gemmatus.*

39478. Cast of the mandible belonging to the same individual as No. 39479. The original is preserved in the Museum at Dijon, and is figured by Nodot, *loc. cit.*; it accords well with the mandible of the skull figured by Burmeister in the 'An. Mus. Buenos Aires,' vol. ii. pl. xxiv., under the name of *G. asper*. *By exchange*, 1865.

39481. Cast of the right femur belonging to the same individual as the preceding specimen. The original is preserved at Dijon. This specimen agrees very closely with the one figured by Burmeister, *op. cit.* pl. xxxiv. fig. 2, under the name of *G. asper*. *By exchange*, 1865.

39481 a. Cast of the left tibia, fibula, and pes belonging to the same individual as the preceding specimen. The originals are preserved at Dijon, and are figured by Nodot, *loc. cit.* *By exchange*, 1865.

Some of the following may belong to G. gemmatus (*Nodot*).

39403. Fragment of the dorsal region of the carapace; from the Pleistocene of the Rio Negro, Uruguay. The pits surrounding the central disks are very numerous and of large size. *Presented by D. A. Stoddart, Esq.*, 1865.

39404. Fragment of the dorsal region of the carapace; from the Rio Negro. *Presented by D. A. Stoddart, Esq.*, 1865.

M. 3605. A considerable portion of the carapace in fragments; from Buenos Ayres. *Bravard Collection. Purchased*, 1854.

37559. Two fragments of the dorsal region of the carapace; from Buenos Ayres. The pits surrounding the central disks are irregular. *Bravard Collection. Purchased*, 1854.

40493. Fragment of the carapace; from Uruguay. The rosettes are well defined, and their surfaces of extreme rugosity. *Presented by W. G. Lettsom, Esq.*, 1867.

39403 a. Fragment of a water-worn carapace; from the Rio Negro. *Presented by D. A. Stoddart, Esq.*, 1865.

Some of the following remains of large Glyptodonts may belong to the present, and others to some of the allied forms.

37560. The nearly entire cephalic buckler; from Buenos Ayres. This specimen differs from the corresponding buckler of *Panochthus* figured by Burmeister, *op. cit.* pl. xiii. fig. 3. *Bravard Collection. Purchased*, 1854.

33241. An entire tooth; from Buenos Ayres.
Presented by the Council of the Royal College of Surgeons. About 1855.

42289. Two imperfect teeth; from Buenos Ayres.
Presented by G. J. Hinde, Esq., 1870.

40067. The pelvis, sacrum, and anterior caudal vertebræ; from Buenos Ayres. This specimen is smaller than the pelvis and sacrum of *Panochthus* figured by Burmeister, *op. cit.* pl. vi. figs. 1, 2, and may therefore very probably belong to *G. reticulatus*. *Purchased*, 1866.

40076. An imperfect sacrum; from Buenos Ayres.
Purchased, 1866.

19947 a. The slightly imperfect left astragalus; from Buenos Ayres.
Purchased, 1846.

19947. The imperfect right astragalus; from Buenos Ayres.
Purchased, 1846.

40081. The slightly imperfect atlas vertebra; from Buenos Ayres.
Purchased, 1866.

Glyptodon, sp. *a*.

The humerus of this form indicates an animal fully as large as *G. reticulatus*, but the femur is somewhat smaller. The fragments of carapace found in the Minas-Geraes caverns with the following limb-bones, and therefore provisionally referred to the same species, have much smaller scutes than those of any of the preceding species.

Hab. South America (Brazil). All the following specimens were obtained from the caverns of Minas Geraes, and belong to the Claussen Collection, purchased 1845.

18963. The distal portion of the left humerus. There is no entepicondylar foramen.

18962. The distal extremity of the right humerus.

18962 a. Fragment of the sacrum.

18943. The acetabular region of the right innominate.

18750. The proximal half of the right femur.

18743. The proximal half of a somewhat smaller right femur.

18944. The distal two thirds of the right femur.

18652 m. A patella.

18620. An imperfect patella.

Some or all of the following may be specifically identical with the preceding specimens, or they may belong to the next species.

18944 c. Portion of the dorsal region of the carapace. The scutes are considerably smaller and thinner than those of *G. reticulatus*, but thicker than those of *Hoplophorus ornatus*. The central disk of each scute is much larger than the disks of the peripheral row, and the latter is frequently double. The surface of the scutes is moderately rugose. This and the following specimens are intermediate in structure between the carapace of the preceding species and that of *Hoplophorus*.

18944 d. Fragment of the anterior (?) part of a carapace. The scutes have a double row of peripheral disks, but are smaller than in the last specimen.

18944 e. Small fragments and detached scutes of carapaces, agreeing in characters with the preceding specimens. The thickness of these specimens averages 0,020.

Glyptodon euphractus (Lund[1]).

Syn. *Hoplophorus euphractus* Lund[2].

This species was referred to the present genus by Reinhardt in the 'Vid. Medd. Nat. Foren. Kjöbenhavn,' 1875, p. 165 *et seq.*; it is of considerably smaller size than the preceding, but somewhat larger than *Hoplophorus ornatus*. Its characters are still very imperfectly known.

Hab. South America (Brazil).

18645. A right astragalus, agreeing in size with the corresponding bone of the foot figured by Lund in the 'K. Danske Vid. Selsk. Skr.' vol. xii. pl. lii., under the name of *Hoplophorus euphractus*; from a cave in Minas Geraes, Brazil.

Claussen Collection. Purchased, 1845.

18941. The left half of the conjoint centra and part of the neural arches of the axis and following cervical vertebræ, pro-

[1] Ann. Sci. Nat. sér. 2, vol. xi. p. 218 (1839).—*Hoplophorus*. The name also occurs in the Overs. K. Danske Vid. Selsk. Forhandl. 1838, p. 11.

[2] *Loc. cit.*

visionally referred to this species; from Minas Geraes. This specimen is larger than the corresponding elements of *Hoplophorus meyeri* (No. 18941 *a*), and differs by the direction of the longer diameter of the articular facette for the atlas being longitudinal instead of vertical.

Claussen Collection. Purchased, 1845.

The following specimens from the caverns of Minas Geraes agree very closely with the fragment figured by Lund in the 'K. Danske Vid. Selsk. Skr.' vol. viii. pl. xi., *under the name of* Hoplophorus euphractus; *they may perhaps belong either to the present species or to* G. dubius, *Reinhardt, if the latter be really a distinct species. All the specimens belong to the Claussen Collection.*

18944 a. Several fragments from the anterior half of the carapace.

18944 b. Several fragments from the posterior half of the carapace.

Genus **DÆDICURUS,** Burmeister[1].

According to Ameghino, the carapace is composed of very thick articulated scutes without sculpture, but marked by vascular perforations, and is supposed to have been invested by a continuous horny epidermis. The caudal sheath (on which the genus was established) consists of several movable rings and a long tube with its terminal extremity enlarged into a flattened club-like expansion. This sheath is covered with coarse tubercles interspersed with large rough disks, having a radiate sculpture, and to which horny spines were probably attached. The fronto-nasal region of the cranium is flat; the humerus has an entepicondylar foramen; and there are three perfect digits in the manus and four in the pes. The genus attains a larger size than any other.

Dædicurus clavicaudatus (Owen[2]).

Syn. *Glyptodon clavicaudatus*, Owen[3].
Hoplophorus clavicaudatus, Nodot[4].
Glyptodon giganteus, Serres[5].
Panochthus giganteus, Burmeister[6].

---

[1] An. Mus. Buenos Aires, vol. ii. p. 393 (1870–74).
[2] Rep. Brit. Assoc. for 1846, Trans. of Sections, p. 67 (1847).—*Glyptodon*.
[3] *Loc. cit.*
[4] Mém. Ac. Sci. Dijon, sér. 2, vol. v. p. 101 (1857).
[5] Comptes Rendus, vol. lxii. p. 207 (1866).
[6] An. Mus. Buenos Aires, vol. ii. p. 91 (1870–74).

Dædicurus giganteus, Burmeister[1].
Dædicurus uruguayensis, H. Gervais and Ameghino[2].
Dædicurus gigas, Ameghino[3] (*ex* Bravard, MS.).

This species was originally founded upon the tube of a caudal sheath (No. 19955), and a precisely similar specimen was made the type of *D. uruguayensis*. The so-called *G. giganteus* was founded upon a pelvis, which according to Burmeister belongs to the same species as the caudal sheath; this identification being made upon the evidence of a skeleton in the museum at Buenos Ayres.

The total length of the animal is given by Burmeister as twelve feet.

Hab. South America (Argentine Republic and Uruguay).

19955. (*Fig.*) The nearly complete terminal tube of the caudal sheath, together with the enclosed vertebral column; from the Pleistocene of Buenos Ayres, Argentine Republic. This specimen is the type, and is figured by Mantell in his 'Petrifactions, &c.,' p. 359, fig. 75.
Purchased, 1846.

M. 3615. The extremity of the terminal tube of the caudal sheath; from Buenos Ayres. This specimen is of rather larger size than the preceding, but presents no structural difference. *No history.*

39482. Cast of the greater portion of the terminal tube of the caudal sheath. The original was brought by Villardebó from the Pleistocene of Uruguay, and is preserved in the Ecole Normale Supérieure of Paris; it is figured by De Blainville in his 'Ostéographie,' Genus *Glyptodon*, pl. i. figs. 4, 5, and by Nodot in the 'Mém. Ac. Nat. Sci. Dijon,' sér. 2, vol. v. pl. viii. figs. 6–8. This specimen is the type of *D. uruguayensis*, but differs in no respects from the two preceding specimens. *By exchange*, 1865.

Genus **EURYURUS**, H. Gervais and Ameghino[4].

The carapace is thick and composed of articulated subquadrangular scutes, which have no sculpture, but are simply rugose. The caudal sheath has several movable rings, and is terminated

[1] An. Mus. Buenos Aires, vol. ii. p. 394 (1870–74).
[2] Mammifères Fossiles de l'Amérique Méridionale, p. 182 (1880).
[3] Bol. Ac. Nac. Cordova, vol. v. p. 30 (1883).
[4] Mammifères Fossiles de l'Amérique Méridionale, p. 184 (1880).

by a long tube excessively compressed and ending in a point, the smaller scutes being separated from one another, and the enlarged lateral ones of great prominence. The skull and limbs are unknown.

Euryurus rudis (P. Gervais[1]).

Syn. *Glyptodon rudis*, P. Gervais[2].

This is the type and only species, and is intermediate in size between *Panochthus tuberculatus* and *Dœdicurus*[3].

Hab. South America (Argentine Republic and [?] Brazil).

M. 3616. Two scutes of the carapace; from the Pleistocene of Buenos Ayres, Argentine Republic.
Presented by Señor Florentino Ameghino, 1879.

18651. Fragment of the terminal portion of a caudal sheath, which in the extreme prominence of the large lateral tubercles agrees with the description of the caudal sheath of the present species; from a cavern in Minas Geraes, Brazil.
Claussen Collection. Purchased, 1845.

Genus **PANOCHTHUS**, Burmeister[4].

The carapace is of great thickness, and composed of quadrangular or pentagonal scutes, each of which in the dorsal region carries from thirty to fifty small tubercles, varying in size from a pea to a nut, the tubercles being separated by grooves. The caudal sheath consists proximally of several movable rings, and is terminated by a long, slightly compressed tube ornamented in the same manner as the carapace, but also bearing on its lateral surfaces a few very large tubercles ornamented with a radiate sculpture. The scutes of the carapace are mostly articulated, but a certain amount of movement exists at the peripheral extremities of some of the rows. There is a ventral buckler. The profile of the fronto-nasal region is very convex. There are four digits in each foot, and, according to Ameghino, there is an entepicondylar foramen to the humerus. All the species are of very large size.

[1] Comptes Rendus, vol. lxxxvi. p. 1361 (1878).—*Glyptodon.*

[2] *Loc. cit.*

[3] The scutes were regarded by Burmeister as belonging to this genus.

[4] An. Mus. Buenos Aires, vol. ii. p. 2 (1870–74).

Panochthus tuberculatus (Owen[1]).

Syn. *Glyptodon tuberculatus*, Owen[2].
Schistopleurum tuberculatum, Nodot[3].

This is the type species, and is described as being nearly equal in bulk to a Rhinoceros.

The specimen which may be taken as the type is the one figured by Owen in the 'Catalogue of Fossil Mammalia and Aves in the Museum of the Royal College of Surgeons,' pl. v. figs. 3–5 (No. 3609[4]), since it is probable that the unfigured specimen (No. 3607), which, intended as the type, belongs to *G. reticulatus*.

Hab. South America (Argentine Republic and ? Uruguay[5]).

M. 3606. Fragment of the peripheral portion of the carapace; from the Pleistocene of Buenos Ayres, Argentine Republic. This specimen accords closely with the above-mentioned fragment figured by Owen, and also with the corresponding portion of the complete carapace figured by Burmeister in the 'An. Mus. Buenos Aires,' vol. ii. pl. xiii.
Bravard Collection. Purchased, 1854.

39402. Fragment of a carapace, somewhat water-worn, provisionally referred to this species; from the Pleistocene of the Rio Negro, Uruguay.
Presented by D. A. Stoddart, Esq., 1865.

19954. The greater portion of the terminal tube of the caudal shield from the Pleistocene of Buenos Ayres. This specimen agrees precisely with the corresponding element figured by Burmeister, *op. cit.* pl. xvi. figs. 1, 2.
Purchased, 1846.

39401. A small fragment of a caudal sheath, doubtless belonging to the same form as No. 39402; from the Rio Negro. This specimen apparently agrees exactly with the preceding.
Presented by D. A. Stoddart, Esq., 1865.

[1] Cat. Foss. Mamm. Aves Mus. R. Coll. Surg. p. 120 (1845).—*Glyptodon.* The proposed identification of *G. tuberculatus* with *G. typus* (Nodot) is noticed under the head of *G. reticulatus.*

[2] *Loc. cit.*

[3] Mém. Ac. Sci. Dijon, sér. 2, vol. v. p. 81 (1857).

[4] See Flower, Cat. Vert. Mus. R. Coll. Surg. pt. 2, p. 691 (1884).

[5] The remains of a *Panochthus* from Uruguay have been named *P. morenoi* by Ameghino, but the two fragments from that region in the Museum do not apparently differ from the type species.

43241. The left femur; from Buenos Ayres. This specimen agrees precisely with the one figured by Burmeister in the 'An. Mus. Buenos Aires,' vol. ii. pl. viii.; the characteristic deep pit on the anterior surface above the trochlea is well shown. *Presented by Señor L. J. Fontana*, 1871.

19947 a. The right fourth metatarsal; from Buenos Ayres. This specimen agrees precisely with the corresponding bone of the foot figured by Burmeister, *op. cit.* pl. x. fig. 1; the metapodials of the present genus are readily distinguished from those of *Glyptodon* by their more elongated form. *Purchased*, 1846.

Genus **HOPLOPHORUS**, Lund[1].

The carapace is usually thin and composed of articulated quadrangular or pentagonal scutes, in which the sculpture generally consists either of a rosette composed of a large central disk, with eight or more smaller peripheral disks separated by shallow grooves, or simply of a central disk surrounded by radiations; the surface of the disks being comparatively smooth. The anterior lateral scutes, at least in some species, are elongated antero-posteriorly and of oblong shape. The peripheral scutes of the carapace, at least in many species, are not elevated into conical protuberances, but its antero-inferior angles are produced in advance of the nuchal border. The caudal sheath has several movable rings, and terminates in a long conical tube, ornamented with a number of large disks, surrounded by a series of much smaller ones; in the lateral region of the extremity some of the disks become of very large size. The fronto-nasal region of the cranium is somewhat less convex than in *Panochthus*; the humerus has an entepicondylar foramen; and there are four perfect digits in each foot[2]. The majority of the species are of moderate size, but certain specimens provisionally referred to the genus indicate a very large form. The genus comes nearer to the *Dasypodidæ* than any other member of the family, as is shown by the elongated form of the carapace, its produced antero-inferior angles, and the presence of an entepicondylar foramen to the humerus.

[1] Ann. Sci. Nat. sér. 2, vol. xi. p. 217 (1839). The name also occurs in the Overs. K. Danske Vid. Selsk. Forhandl. 1838, p. 11. The real authority for the name must, however, be taken later, as the type species (*H. euphractus*) has been shown to belong to *Glyptodon*.

[2] These characters have been verified only in some of the species.

Hoplophorus meyeri, Lund[1].

Syn. *Glyptodon gracilis,* Nodot[2].
Hoplophorus gracilis, H. Gervais and Ameghino[3].

This may be regarded as the type species; and it will be advisable to take as the type specimen the imperfect cranium figured by Lund in the 'K. Danske Vid. Selsk. Skr.' vol. xii. pl. li., under the name of *H. euphractus,* which is different from that of *H. ornatus,* although of the same generic type. The teeth figured by the same writer in vol. ix. pl. xxxv. figs. 2, 4, are also referred by Reinhardt to this species, but the specimens figured in Lund's earlier memoirs under the name of *H. euphractus* are referred by the former writer to *Glyptodon (Schistopleurum).* The characters of the species are too imperfectly known to admit of diagnosis.

Hab. South America (Brazil).

M. 3617. A small fragment of the terminal tube of a caudal sheath provisionally referred to this species; from a cavern in Minas Geraes, Brazil. The central disk in each scute is smaller and more rugose than in the caudal sheath of *H. ornatus.* *Claussen Collection. Purchased,* 1845.

M. 3618. Two fragments from the anterior lateral portion of the carapace, not improbably belonging to the present species; from Minas Geraes. The central disk of each scute is relatively smaller than in the corresponding portion of the carapace of *H. ornatus.*
Claussen Collection. Purchased, 1845.

The following specimens indicate the presence of two species of Hoplophorus *in the caves of Minas Geraes, and some of them may probably be referred to the present form. All of them were obtained from Minas Geraes, and belong to the Claussen Collection.*

18936 a. The nearly complete right humerus, showing the entepicondylar foramen. The proximal portion of this specimen differs from the imperfect humerus from Brazil (No. 40675) provisionally referred to *H. ornatus.*

[1] Overs. K. Danske Vid. Selsk. Forhandl. for 1843, p. 79 (1844)—*teste* Lütken (MS.).

[2] Mém. Ac. Sci. Dijon, sér. 2, vol. v. p. 97 (1857).

[3] Mammifères Fossiles de l'Amérique Méridionale, p. 196 (1880).

18936. The distal extremity of the right humerus of an equal-sized but different species of *Hoplophorus.* This bone differs considerably in form from the corresponding portion of the last specimen, the entepicondylar foramen being oval instead of nearly circular.

18648. The right astragalus. This specimen agrees in size with the corresponding bone of *H. ornatus.*

18648 a. The left astragalus.

18941 a. The left half of the conjoint centra and part of the neural arches of the axis and three following vertebræ. This specimen agrees very closely with the corresponding element figured by Lund in the 'K. Danske Vid. Selsk. Skr.' vol. ix. pl. xxxv. fig. 1, under the name of *H. euphractus*; and from its resemblance to the vertebræ of *H. ornatus* figured by Burmeister in the 'An. Mus. Buenos Aires,' vol. ii. pl. xix. fig. 6, may perhaps belong to the present species.

Hoplophorus ornatus (Owen[1]).

Syn. *Glyptodon ornatus*, Owen[2].
Hoplophorus burmeisteri, Ameghino[3].

Since there may possibly be a doubt as to the identity of the complete carapace figured by Burmeister in the 'An. Mus. Buenos Aires' under the present name with the fragment to which the name *G. ornatus* was applied, it will be advisable to regard the former as the type.

The features distinguishing this species are the depression of the central disks of the scutes of the carapace, and the smooth surface of the disks. There are usually from eight to ten or twelve peripheral disks in each scute, which are always separated from the disks of the adjacent scutes by shallow grooves. In the middle of the dorsal region the central disk of each scute is but slightly larger than the peripheral ones, but in the lateral region the central disk becomes very much larger. The carapace is elongated in form, its total length being about four feet. The disks on the dorsal aspect of the terminal tube of the caudal sheath are often oval, convex, subequal in size, and regularly arranged; the hinder lateral disks being saddle-shaped.

Hab. South America (Argentine Republic).

1 Cat. Foss. Mamm. Aves Mus. R. Coll. Surg. p. 119 (1845).—*Glyptodon.*
2 *Loc. cit.*
3 Quoted in the Bol. Ac. Nac. Cordova, vol. v. p. 32 (1883).

M. 481. The carapace, imperfect dorsally; from the Pleistocene of Buenos Ayres, Argentine Republic. The missing portion of the carapace, together with the caudal sheath, has been restored in plaster from the figures given by Burmeister in the 'An. Mus. Buenos Aires,' vol. ii. pl. xvii.
Purchased, 1883.

37558. Part of the terminal tube of the caudal sheath (woodcut, (*Fig.*) fig. 22); from Buenos Ayres. The four larger lateral

Fig. 22.

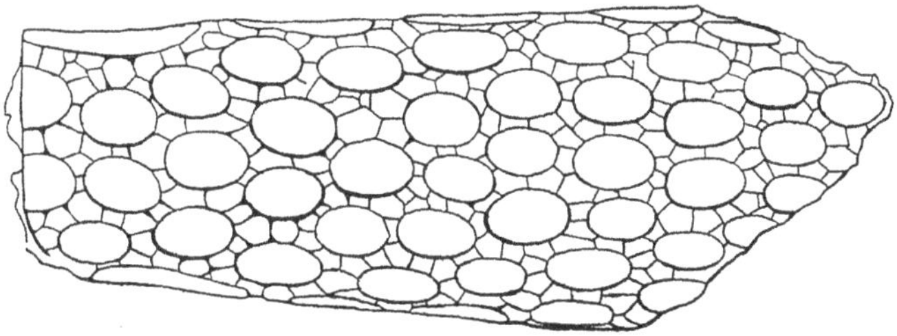

Hoplophorus ornatus.—Ventral aspect of the imperfect extremity of the terminal tube of the caudal sheath; from the Pleistocene of Buenos Ayres. $\frac{1}{3}$.

disks diminish gradually in size anteriorly, and the second and third are saddle-shaped, with the axis of the concavity directed transversely; the disks on the dorsal surface are oval and subequal.
Bravard Collection. Purchased, 1854.

37565. The imperfect occipital region of the cranium; from Buenos Ayres. This and the two following specimens accord very closely with the cranium figured by Burmeister, *op. cit.* pls. xviii. & xix. The characteristic pitting of the parietal region is well known.
Bravard Collection. Purchased, 1854.

37556. The greater portion of the right half of the cranium, wanting the palate; from Buenos Ayres.
Bravard Collection. Purchased, 1854.

37557. Fragment of the right maxillary region of the cranium; from Buenos Ayres. The eversion of the anterior border of the orbit and the deflection of the fronto-nasal region is well shown. *Bravard Collection. Purchased,* 1854.

40066. The imperfect left ramus of the mandible; from Buenos Ayres. This specimen accords with the one figured by Burmeister, *op. cit.* pl. xviii. *Purchased*, 1866.

Some of the following specimens may perhaps be specifically distinct. All are from Buenos Ayres, and, unless otherwise indicated, belong to the Bravard Collection.

40675. The left humerus, wanting the distal extremity. This bone apparently agrees very closely with the example figured by Burmeister, *op. cit.* pl. xxi. fig. 1.

43237. The left ulna, agreeing very closely with the specimen figured by Burmeister, *op. cit.* pl. xxi. fig. 3.
Presented by Señor L. J. Fontana, 1871.

32501. The imperfect sacrum and pelvis. This specimen agrees very closely in size with the sacrum figured by Burmeister, *op. cit.* pl. xx. fig. 1.

37307. The right femur. The distal half agrees very closely with the imperfect specimen of the corresponding bone figured by Burmeister, *op. cit.* pl. xxi. fig. 2. In Burmeister's restoration of the proximal half, the great trochanter is represented as not extending above the level of the head, but in the present specimen it does so to a considerable extent.

43250. The right calcaneum of a smaller form.
Presented by Señor L. J. Fontana, 1871.

43412. The three phalangeals of the second digit of the left pes. These specimens agree precisely with the corresponding digit of the foot figured by Burmeister, *op. cit.* pl. xxii. figs. 3, 4, 5.

Hoplophorus, sp. *a*.

The carapace is thin, larger than that of *H. ornatus*, and with the scutes smoother. In the dorsal region there are usually eight peripheral disks to each scute, which are polygonal, and frequently coalesce with the corresponding disks of the adjacent scutes. The intervening grooves are well-marked, the rosettes very distinct, and the peripheral disks of relatively large size; the central disks of the scutes are not depressed, and are polygonal in form. It is possible that the caudal sheath referred by Owen to *Glyptodon clavipes* may belong to this form.

Hab. South America (Argentine Republic).

M. 3607. The imperfect hinder portion of the carapace; from the Pleistocene of Buenos Ayres, Argentine Republic. The form of the posterior peripheral scutes indicates that the specimen belongs to *Hoplophorus*.
Bravard Collection. Purchased, 1854.

M. 3619. A small fragment of a carapace, agreeing in the structure of the scutes with the preceding specimen; from Buenos Ayres. *Bravard Collection. Purchased*, 1854.

43243. Fragment of a similar carapace; from Buenos Ayres.
Presented by Señor L. J. Fontana, 1871.

M. 3620. Fragments of the peripheral region of the carapace belonging to the same individual as No. M. 3607. The resemblance of these scutes to those of *H. ornatus* confirms the reference of the present form to the same genus.
Bravard Collection. Purchased, 1854.

Hoplophorus (?), sp. *b.*

The following specimens, which are provisionally regarded as belonging to a single species, indicate a form agreeing approximately in size with the preceding, but distinguished by the rugosity of the scutes and the depression of their central disk, which is nearly circular in form. Not unfrequently there is an additional row of peripheral disks between the normal peripheral rows of adjacent scutes. The grooves are much more deeply marked in the posterior than in the anterior scutes.

Hab. South America (Brazil).

The following specimens were obtained from the caverns of Minas Geraes, and belong to the Claussen Collection, purchased, 1845.

M. 3621. Fragment of the dorsal region of the posterior part of the carapace, showing twelve scutes, which are polygonal in form. In one or two places an additional row of peripheral disks may be observed. The thickness of the scutes is 0,011.

M. 3622. A smaller fragment from the posterior part of the carapace.

M. 3622 a. A still smaller fragment from the same half of the carapace.

M. 3623. Small fragment from the anterior half of the carapace, showing two complete scutes.

The two following specimens (Claussen Collection, purchased, 1845) from the Caves of Minas Geraes, appear specifically identical with the preceding.

M. 3624. Fragment from the anterior half of the carapace.

M. 3625. Fragment from the anterior half of the carapace.

Hoplophorus, sp.

This form, which may be identical with one of the preceding, was apparently somewhat larger than *H. ornatus*.

Hab. South America (Uruguay).

M. 76. (*Fig.*) The hinder portion of the terminal tube of a caudal sheath, with the enclosed vertebral column; from the Pleistocene of Monte Video, Uruguay. This specimen (woodcut,

Fig. 23.

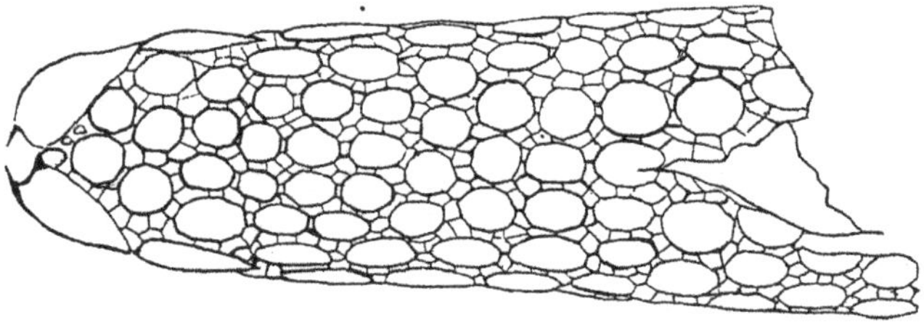

Hoplophorus, sp.—Dorsal aspect of the extremity of the terminal tube of the caudal sheath; from the Pleistocene of Uruguay. $\frac{1}{4}$.

fig. 23) is larger than the corresponding element of *H. ornatus*; the lateral enlarged disks are four in number, and mesially convex instead of saddle-shaped, while the superior disks are more nearly circular. It differs from the caudal sheath of the large form noticed on the next page, not only by its inferior dimensions, but by the equal size of the disks on the dorsal surface, and the absence of a row of small disks between the two enlarged terminal ones. *Presented by W. G. Lettsom, Esq.*, 1881.

Hoplophorus elegans, Burmeister[1].

Syn. *Hoplophorus radiatus*, Gervais and Ameghino[2] (*ex* Bravard).

Each scute is of small size and subquadrangular; the central disk is slightly rugose and concave, and the peripheral disks are

[1] An. Mus. Buenos Aires, vol. ii. p. 219 (1871).

[2] Mammifères Fossiles de l'Amérique Méridionale, p. 194 (1880).

replaced by a rugose zone, ornamented by a number of fine lines radiating from the border of the central disk.

Hab. South America (Buenos Ayres).

37559. Several associated fragments of the carapace; from the Pleistocene of Buenos Ayres, Argentine Republic. These specimens agree very closely with the fragment figured by Burmeister in the 'An. Mus. Buenos Aires,' vol. ii. pl. xx. fig. 7. *Barvard Collection. Purchased*, 1854.

43242. Several fragments of the carapace; from Buenos Ayres. *Presented by Señor L. J. Fontana*, 1871.

37559 a. Fragment of the anterior portion of a carapace, probably belonging to the present species; from Buenos Ayres. *Bravard Collection. Purchased*, 1854.

Hoplophorus (?), sp. *c*.

Syn. (?) *Glyptodon clavipes*, Burmeister[1] (*non* Owen).

The following specimens indicate a species apparently nearly or quite as large as *Glyptodon reticulatus*. The large terminal disks on the lateral surfaces of the tube of the caudal sheath are separated from one another by a row of small disks, and are followed laterally by a series of enlarged disks, which decrease gradually in size as they recede from the tip. The disks on the dorsal aspect of the tube are subcircular in shape, frequently concave, and present great variation in size; but those on the ventral surface are more regular in this respect. The scutes of the fragment of carapace are oblong, with the grooves very indistinctly marked, and with a series of hair-pits in the groove surrounding the central disk.

The Glyptodont figured by Burmeister in the 'An. Mus. Buenos Aires,' vol. ii. pl. xxxvi., under the name of *Glyptodon clavipes*, differs from the type specimen of that species figured by Owen[2] in the more elongated and less vaulted form of the carapace, and the straight inferior border of the same, and also in the more elongated scutes, in which the central disks are much less distinctly defined; in all of which respects it has a much closer resemblance to the carapace of *Hoplophorus ornatus*, to which species it is referred by Ameghino. The carapace and caudal sheath of that specimen may very probably belong to the same species as the specimens noticed below; but the cranium, which is like that of *Glyptodon*, may possibly belong to another form, and the margins of the carapace are perhaps incorrectly restored[3].

[1] An. Mus. Buenos Aires, vol. ii. pl. xxxvi. (1874). [2] *Vide suprà*, p. 115.

[3] If this cranium really belongs to the same individual as the carapace, it would apparently indicate that the species presents characters intermediate between those of *Hoplophorus ornatus* and *Glyptodon reticulatus*.

Although the present form is apparently distinct from all described species, with the exception of the so-called *Glyptodon clavipes* of Burmeister, the writer thinks it inexpedient at present to assign to it a distinct specific name.

Hab. South America (Uruguay, and (?) Argentine Republic).

40661. (*Fig.*) The extremity of the terminal tube of the caudal sheath; from the Pleistocene of Uruguay. This specimen, although of considerable larger size, apparently agrees very closely with the terminal tube figured by Burmeister, under the name of *Glyptodon clavipes*, in the 'An. Mus. Buenos Aires,' vol. ii. pl. xl. figs. 7, 8, showing the irregular arrangement of the disks on the ventral surface (fig. 7), and the large subcircular lateral ones (fig. 8), which decrease regularly in size from the hindmost one. In the terminal tube of a caudal sheath referred by Owen [1] to *G. clavipes*, but which, as already mentioned, probably belongs to *Hoplophorus*, the lateral disks are oval, and the two last are much larger than the others, while on the ventral and dorsal aspects the disks are less numerous and of a decidedly oval form. The upper surface of the present specimen is represented in the accompanying woodcut (fig. 24), where the form of the disks is well shown. Its transverse diameter is 0,144, against 0,075 in the corresponding element of *H. ornatus* (*suprà*, p. 129).

Presented by W. G. Lettsom, Esq., 1867.

Fig. 24.

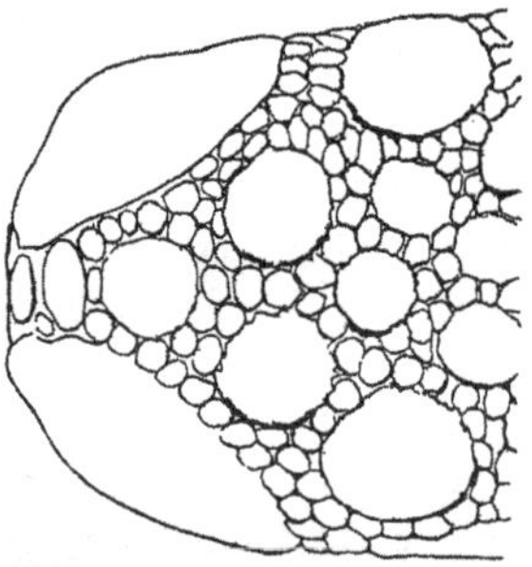

Hoplophorus (?) sp.—Dorsal aspect of the extremity of the terminal tube of the caudal sheath; from the Pleistocene of Uruguay. ½.

39405. Fragment from the anterior region of a carapace, showing nine scutes, which agrees in relative size with the pre-

[1] Cat. Foss. Mamm. Aves Mus. R. Coll. Surg. pl. i. (side view), and pl. ii. figs. 5 and 6 (dorsal and ventral views).

ceding specimen; from the Pleistocene of Uruguay. The scutes are nearly smooth, their central disks being mainly defined by a circle of hair-pits, and the peripheral ones scarcely differentiated. The scutes form oblong hexagons, and in this respect, as well as in the absence of rugosity, agree with those of *H. ornatus* and differ from those of *Glyptodon*. Their longer diameter is 0,052, and their thickness 0,031.

Presented by D. A. Stoddart, Esq., 1865.

Family DASYPODIDÆ.

The teeth in the existing genera are simply conical; the facial portion of the skull is elongated, and there is no descending maxillary process in the zygoma. The carapace always contains a certain number of movable bands, and may be entirely made up of such; the scapular and pelvic bucklers, when present, consist of polygonal articulated scutes, while the scutes of the movable bands overlap and are quadrangular. There is a cephalic, but no ventral shield; the tail is either completely cased in bone, or bears a certain number of bony tubercles. The cervical vertebræ have extremely short, broad, and depressed centra; the atlas is free, but the second, third, and frequently some of the other vertebræ are anchylosed by their centra. The humerus has an entepicondylar foramen, the femur a third trochanter, and the tibia is always anchylosed distally with the fibula; while the fore feet have long, curved claws. The cerebrum is of relatively large size.

Genus **CHLAMYDOTHERIUM,** Lund[1].

Syn. *Pampatherium*, Ameghino[2].

This genus connects the *Dasypodidæ* with the *Glyptodontidæ*, having the carapace of the former, while the dentition is stated to approach that of the latter[3].

There are nine lower teeth. The carapace has several movable bands, composed of large quadrangular scutes; the majority of the fixed scutes being either pentagonal or hexagonal.

[1] Ann. Sci. Nat. sér. 2, vol. xi. p. 217 (1839).—The name also occurs in the 'Overs. K. Danske Vid. Selsk. Forhandl.' 1838, p. 11, but with insufficient description, and misspelt *Chlamytherium*.

[2] Quoted by Gervais and Ameghino in 'Mammifères Fossiles de l'Amérique Méridionale,' p. 210 (1880).

[3] Gervais and Ameghino, *op. cit.* p. 210. Ameghino, Bol. Ac. Nac. Cordova, vol. ix. p. 205 (1886), says that the lower teeth have an elliptic cross-section.

Chlamydotherium giganteum, Lund [1].

Syn. *Chlamydotherium gigas*, Lund [2].
Chlamydotherium majus, Lund [3].

This species is described as being nearly equal in size to a Rhinoceros.

Hab. South America (Brazil).

18945 a. A scute belonging to the posterior row of the scapular buckler; from a cavern in Minas Geraes, Brazil. The length of this specimen is 0,092, and its width 0,047.
Claussen Collection. Purchased, 1845.

18646. The right astragalus; from Minas Geraes. This bone is too large to have belonged to *C. humboldti*; it agrees very closely in general structure with the astragalus of *Dasypus sexcinctus*, its transverse diameter being 0,065 against 0,010 in the latter.
Claussen Collection. Purchased, 1845

18645. A navicular, agreeing in relative size with the preceding specimen; from Minas Geraes.
Claussen Collection. Purchased, 1845.

18951. The left patella; from Minas Geraes. This specimen cannot be structurally distinguished from the patella of *Dasypus sexcinctus*; its length being 0,082, against 0,014.
Claussen Collection. Purchased, 1845.

Chlamydotherium humboldti, Lund [4].

This is the type species, and is described as being about equal in size to *Tapirus americanus*.

Hab. South America (Brazil and Argentine Republic).

18945 b. Numerous scutes from the movable bands of the carapace; from the caverns of Minas Geraes, Brazil. These scutes agree very closely in structure with those of *Dasypus sexcinctus*, but are nearly four times the size.
Claussen Collection. Purchased, 1845.

18945 c. Numerous scutes from the solid portions of the carapace and the cranium; from Minas Geraes.
Claussen Collection. Purchased, 1845.

[1] Ann. Sci. Nat. sér. 2, vol. xi. p. 217 (1839).
[2] K. Danske Vid. Selsk. Skr. vol. ix. p. 142 (1842). [3] *Op. cit.*
[4] Ann. Sci. Nat. sér. 2, vol. xi. p. 217 (1839). The name occurs in the passage cited under the head of the generic title.

18945. Several teeth, agreeing with those figured by Lund in the 'K. Danske Vid. Selsk. Skr.' vol. ix. pl. xxxiv. figs. 1–8.
Claussen Collection. Purchased, 1845.

The following specimens are also from Minas Geraes, and belong to the Claussen Collection.

18937. The distal portion of the left humerus. In characters this specimen agrees very closely with the corresponding bone of *Dasypus sexcinctus*; the transverse diameter of the distal articular surface is 0,057, against 0,016 in the latter species.

18943. An imperfect scapula. This specimen is apparently very similar in structure to the corresponding bone of *Dasypus*.

18643. Several phalangeals.

18948. The nearly complete pelvis. In general structure this specimen agrees very closely with the pelvis of *Dasypus sexcinctus*; but the sacral vacuity is less elongated, and the ischial tuberosity produced laterally instead of vertically. The diameter across the acetabula is 0,238 against 0,054 in the latter.

18941. The centrum of the conjoint axis and third cervical vertebræ. This specimen closely agrees with the corresponding bone of *D. sexcinctus*; the transverse diameter across the anterior articular surface being 0,063 against 0,020.

18941 a–42. Several imperfect vertebræ and ribs.

Genus **TOLYPEUTES,** Illiger[1].

Teeth $\frac{(8-9)}{9}$. Carapace with large scapular and pelvic bucklers, and only three movable bands, the sculpture consisting of subconcentrically arranged granulations; tail covered with large bony tubercles. Five complete digits in the manus, the third being much the largest; in the pes there are also the same number of digits, the three middle ones having broad flat nails. In the structure of the carapace this genus makes the nearest approach of any of the living forms to the *Glyptodontidæ*.

[1] Prodromus Syst. Mamm. et Avium, p. 111 (1811).

Tolypeutes conurus (I. Geoffroy[1]).

Syn. *Dasypus conurus*, I. Geoffroy[2].
Dasypus apareoides, Bravard[3].
Tolypeutes affinis-conurus, H. Gervais and Ameghino[4].

No difference of specific value has been indicated between the fossil and recent forms[5].

Hab. South America (Argentine Republic).

M. 3626. Portion of the first row of scutes of the pelvic buckler; from the Pleistocene of La Plata, Argentine Republic. This specimen, which belongs to Bravard's *D. apareoides*, cannot be distinguished from the corresponding portion of the covering of the existing race.

Bravard Collection. Purchased, 1854.

Genus DASYPUS, Linn.[6]

Syn. *Euphractus*, Wagler[7].

Teeth $\frac{(8-9)}{(9-10)}$, the first upper one being usually planted in the premaxilla. The carapace has six or seven movable bands, each scute being marked by a regular ellipse, formed of widely separated punctures; the tail has several distinct rings near the base. There are five digits in each foot; in the manus the first is very slender and has a long terminal phalangeal; the second is also slender but rather longer; the third, fourth, and fifth gradually diminish in length, and are armed with curved, compressed claws. In the pes all the digits are short.

Dasypus villosus, Desmarest[8].

Syn. *Euphractus affinis-villosus*, H. Gervais and Ameghino[9].

Since the fossil apparently presents no character by which it can be distinguished from the existing form, it is regarded as specifically the same[10].

Hab. South America (Argentine Republic and North Patagonia).

[1] Comptes Rendus, vol. xxiv. p. 575 (1847).—*Dasypus*. [2] *Loc. cit.*
[3] In P. Gervais's Zool. et Pal. Générales, sér. 1, p. 132 (1867–69).
[4] Mammifères Fossiles de l'Amérique Méridionale, p. 216 (1880).
[5] See Burmeister, Descript. Phys. Republic Argentine, vol. iii. pt. 1, p. 443 (1879).
[6] Syst. Nat. ed. 12, vol. i. p. 53 (1766). [7] Syst. d. Amphibien, p. 36 (1830).
[8] Nouv. Dict. d'Hist. Nat. ed. 2, vol. xxxii. p. 489 (1819).
[9] Mammifères Fossiles de l'Amérique Méridionale, p. 215 (1880).
[10] See Burmeister, Descript. Phys. Republic Argentine, vol. iii. pt. 1, p. 440 (1879).

M. 3627. Three scutes from one of the movable bands of the carapace; from the Pleistocene of La Plata, Argentine Republic.
Bravard Collection. Purchased, 1854.

Genus **XENURUS**, Wagler[1].

Teeth $\frac{(8-9)}{(8-9)}$. Twelve or thirteen movable bands in the carapace, the sculpture obscurely granular; tail nearly naked, with a few bony tubercles. In the manus the first and second digits are slender and have small claws, the other three have but two phalangeals; the third has an immense claw, and the fourth and fifth smaller claws of the same form. There are five short digits in the pes, of which the third is the longest.

Xenurus unicinctus (Linn.[2]).

Syn. *Dasypus unicinctus*, Linn.[3]
Dasypus tatouay, Desmarest[4].
Dasypus gymnurus, Illiger[5].
Xenurus affinis-nudicaudo, Lund[6].
Xenurus antiquus, Lund[7].

The fossil form cannot apparently be distinguished from the living Broad-banded Armadillo.

Hab. South America (Brazil).

18881. The left tibia and fibula; from a cavern in Minas Geraes, Brazil. These bones agree precisely with those of the living form; the suboval notch at the anterior extremity of the space between the two bones, so characteristic of the genus, is well shown.
Claussen Collection. Purchased, 1845.

Genus **EUTATUS**, P. Gervais[8].

Teeth $\frac{9-8}{10}$. The whole of the carapace is divided into movable bands, which are 33 in number; the free extremity of each scute

[1] Syst. d. Amphibien, p. 36 (1830).
[2] Syst. Nat. ed. 12, vol. i. p. 53 (1766).—*Dasypus.*
[3] *Loc. cit.*
[4] Mammalogie, p. 369 (1822).
[5] Maximilian, Beitr. Nat. Bras. vol. ii. p. 529 (1832).
[6] K. Danske Vid. Selsk. Skr. vol. ix. p. 197 (1842).
[7] *Ibid.* vol. viii. p. 227 (1841).
[8] Comptes Rendus, vol. lxv. p. 280 (1867).

has a straight row of from three to five pits; the tail is cylindrical and covered by small flat scutes; there are five digits to each foot, of which the first is the most slender, while the third and fourth are the stoutest.

Eutatus seguini, P. Gervais[1].

This is the type and largest species, and is somewhat superior in size to *Priodon giganteus*. The skeleton is described and figured by Gervais in the 'Mém. Soc. Géol. France,' sér. 2, vol. ix. pt. 5, pls. xxviii., xxix., and the carapace and feet by Burmeister in the 'Sitz. k. preuss. Ak. Wiss.' 1883, vol. ii. p. 1045, pl. xiii.

Hab. South America (Argentine Republic).

M. 3628. Numerous scutes of the carapace, belonging to a single individual; from the Pleistocene of La Plata, Argentine Republic. These specimens agree precisely with the scutes figured by Burmeister, *op. cit.* figs. 8, 9.
Bravard Collection. Purchased, 1854.

M. 3629. Several scutes of another individual, from La Plata.
Bravard Collection. Purchased, 1854.

M. 3630. A terminal lateral phalangeal, probably belonging to this species; from La Plata.
Bravard Collection. Purchased, 1854.

Genus **TATUSIA,** F. Cuvier[2].

Syn. *Praopus*, Burmeister[3].

Teeth $\frac{(7-8)}{(7-8)}$; very small, cylindrical, and, with the exception of the last, preceded by milk-teeth. Carapace with seven movable bands, the sculpture consisting of pits arranged in a V-shape; caudal sheath composed of distinct rings resembling those of *Glyptodon*. The manus has four perfect digits, of which the third and fourth are the longest; while there are five in the pes, of which the third is the longest.

[1] Comptes Rendus, vol. lxv. p. 280 (1867).
[2] Hist. Nat. d. Mammifères (1822).—*Tatusie.*
[3] Syst. Uebersicht Thier Brasiliens, p. 295 (1854).

Tatusia novemcincta (Linn.[1]).

Syn. *Dasypus novemcinctus*, Linn.[2]
Dasypus peba, Desmarest[3].
Dasypus septemcinctus, Linn.[4]
Dasypus octocinctus, Schreber[5].
Praopus longicaudatus, Burmeister[6].
Dasypus affinis-octocincto, Lund[7].
Praopus affinis-longicaudatus, H. Gervais and Ameghino[8].

There appears no reason for separating the fossil from the recent race.

Hab. Central and South America.

18882. The right tibia and fibula, provisionally referred to this species; from a cave in Minas Geraes, Brazil. These bones cannot be distinguished from those of the existing form. *Claussen Collection. Purchased*, 1845.

M. 3631. Several scutes of the movable bands of the carapace, provisionally referred to this species; probably from the Pleistocene of La Plata, Argentine Republic. (?) *Bravard Collection. Purchased*, 1854.

Tatusia punctata (Lund[9]).

Syn. *Dasypus punctatus*, Lund[10].
Praopus punctatus, H. Gervais and Ameghino[11].

This species is slightly larger than *T. kappleri*, Gray, of Dutch Guiana, but was probably closely allied.

Hab. South America (Brazil).

18881 a. Numerous scutes belonging to the movable bands of the carapace; from the caverns of Minas Geraes, Brazil. Some of these specimens are rather larger than the type scutes figured by Lund, *op. cit.* pl. xiv. figs. 10, 11. *Claussen Collection. Purchased*, 1845.

1 Syst. Nat. ed. 12, vol. i. p. 54 (1766).—*Dasypus*. 2 *Loc. cit.*
3 Mammalogie, p. 368 (1822).
4 Amœnitat. Acad. 3rd ed. vol. i. p. 281 (1785).
5 Säugethiere, vol. ii. pl. lxxiii. (1775). 6 *Op. cit.* p. 298.
7 K. Danske Vid. Selsk. Skr. vol. ix. p. 197 (1842).
8 Mammifères Fossiles de l'Amérique Méridionale, p. 217 (1880).
9 K. Danske Vid. Selsk. Skr. vol. viii. p. 227 (1841).—*Dasypus*.
10 *Loc. cit.*
11 Mammifères Fossiles de l'Amérique Méridionale, p. 218 (1880).

Family MANIDÆ.

Teeth are absent in all known forms. The limbs are short; there are five digits in each foot, the terminal phalangeals being long (especially in the manus), curved, and bifid at the extremity, and the third digit the longest; the proximal articular surface of the proximal phalangeal is placed distally, so that the bone is not bent back upon the metapodial, and there is no anchylosis of the phalangeals. The humerus has an entepicondylar foramen, but there is no third trochanter to the femur, and clavicles are absent. The cranium is elongated and subconical, with an incomplete zygomatic arch, and no distinct lachrymal; the ramus of the mandible is straight and very slender, and has neither coronoid nor angular process. The superior surface of the animal is covered with imbricated horny scales, in the intervals between which scattered hairs occur.

Genus **MANIS**, Linn.[1]

Including *Smutsia*, etc., Gray.

Manis gigantea, Illiger[2].

This is the largest species, and attains a length of five feet.

Hab. West Africa (recent), and Southern India (Pleistocene).

M. 2962. Cast of the terminal phalangeal of the third digit of the right manus. The original, which is preserved in the Indian Museum, Calcutta, was obtained from the Cathedral cave, Billa Surgam, Karnul district, Madras; and is described and figured by the writer in the 'Palæontologia Indica' (Mem. Geol. Surv. Ind.), ser. 10, vol. iv. p. 50, pl. viii. figs. 8, 8 *a*. *Presented by the Director of the Geological Survey of India*, 1886.

Family MACROTHERIIDÆ.

It is probable that teeth were present in some forms. The terminal phalangeals, especially in *Macrotherium sindiense*, present a considerable resemblance to those of *Manis*, but are shorter; the proximal phalangeals have their proximal articular surface directed

[1] Syst. Nat. ed. 12, vol. i. p. 52 (1766).

[2] Abh. k. Ak. Wiss. Berlin, 1811, pp. 78, 84 (1815).

dorsally, and are bent back upon the metapodials. The humerus has no entepicondylar foramen, and there is no third trochanter to the femur. The cranium is unknown. In one genus (*Ancylotherium*) the first and second phalangeals of the manus were anchylosed.

Genus **MACROTHERIUM**, Lartet[1].

This is the type genus. In the type species, teeth were apparently present. The terminal phalangeals form short, thick claws, which were of enormous size in the manus.

Macrotherium sindiense, Lydekker[2].

Syn. *Manis sindiensis*, Lydekker[3].

This species is smaller than *M. giganteum*, and, judging from the one phalangeal, apparently presents characters intermediate between *Manis* and the latter. It has been suggested that it may have been the ancestor of *Manis*.

Hab. India.

M. 2962 a. Cast of the second phalangeal of the third (?) digit of the manus. The original (fig. 25), which is the type

Fig. 25.

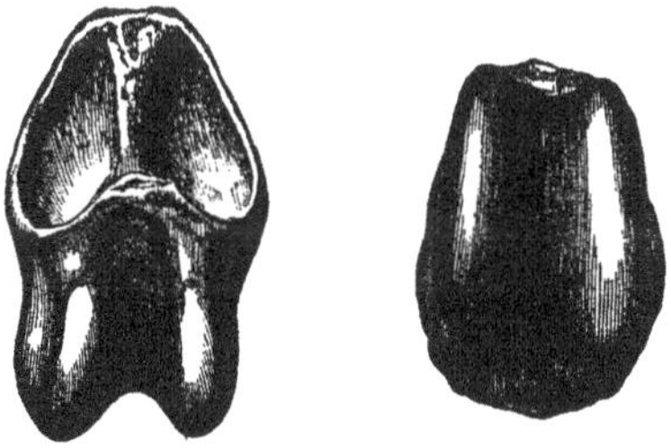

Macrotherium sindiense.—The second phalangeal of the third (?) digit of the manus, viewed from the anterior and distal aspects; from the Lower Siwaliks of Sind. 1/1.

and only known specimen, and is preserved in the Indian Museum, Calcutta, was obtained from the Lower

[1] Annuaire du Département du Gers, 1839; *teste* Lartet 'Notice sur la Colline de Sansan,' p. 22 (1851). The name does not occur in the 'Comptes Rendus,' vol. iv. p. 90 (1837), which is the passage commonly quoted as the authority.

[2] Palæontologia Indica (Mem. Geol. Surv. Ind.), ser. 10, vol. i. p. 82 (1876).—*Manis.*

[3] *Loc. cit.*

Siwaliks (Lower Pliocene or Upper Miocene) of Sind, and is figured by the writer in the 'Palæontologia Indica,' ser. 10, vol. i. pl. viii. figs. 11–14, under the name of *Manis*, and in vol. iv. p. 50, woodcut, fig. 4, of the same work, under the present generic name. *Presented by the Director of the Geological Survey of India*, 1886.

Macrotherium giganteum (Meyer[1] [ex Cuv.]).

Syn. *Pangolin gigantesque*, Cuvier[2].
Manis gigantea, Meyer[3].
Macrotherium sansaniense, Lartet[4].

Although the name *Pangolin gigantesque* or *Manis gigantea* (which was applied to a phalangeal from Eppelsheim) was really preoccupied by Illiger's term for the large existing *Manis* of West Africa, yet it seems advisable to retain the specific name in conjunction with *Macrotherium*, rather than to substitute Lartet's name of *M. sansaniense*, which was applied to the apparently identical Sansan form[5]. The species was apparently nearly as large as *Mylodon robustus*.

Hab Europe.

M. 418. Cast of the imperfect terminal phalangeal of the third (?) digit of the (probably) pes. The original, which is rather more incomplete than the cast, was obtained from the bone-sand of Eppelsheim, Hessen-Darmstadt, and is the type of the species; it is figured by Cuvier in the 'Ossemens Fossiles,' 2nd ed. vol. v. pt. 1, pl. xvi. figs. 26, 27, under the name of *Pangolin gigantesque*, and by Kaup in the 'Oss. Foss. de Darmstadt,' pt. 4, pl. ii. *bis*, figs. 6, 7, 8, as *Dinotherium giganteum*.
Egerton Collection. Purchased, 1882.

21492. The terminal phalangeal of the third (?) digit of the (probably) manus; from Eppelsheim. This specimen is considerably larger than the preceding; and is therefore, in accordance with Gaudry's provisional determination of the homologous bones of *Ancylotherium pentelici*, referred to the fore limb. *Purchased*, 1847

1 Palæologica, p. 63 (1832).—*Manis*.
2 Ossemens Fossiles, 2nd ed. vol. v. pt. i. p. 193 1823).
3 *Loc. cit.*
4 Annuaire du Département du Gers, 1839, *teste* Lartet; see Note, p. 143.
5 In Gervais's Zool. et Pal. Françaises, 1st ed. pt. 1, p. 136 (1848–52), the name *M. giganteum* is applied to the Sansan form as a misquotation from Lartet.

1751 a. Cast of the proximal phalangeal of the third (?) digit of the (probably) pes. The original was obtained from Eppelsheim, and is figured by Kaup, *op. cit.* pl. ii. *bis*, figs. 4, 5, under the name of *Dinotherium giganteum*. It is also represented in fig. 9 of the same plate in conjunction with the terminal phalangeal, as if the two were adjacent bones. *Purchased*, 1836.

Ma. 3. Cast of a metapodial; the original of which was obtained from Eppelsheim. *Purchased*, 1836.

21880. The imperfect proximal phalangeal of a digit, restored in plaster; from the Middle Miocene of Sansan (Gers), France. *Croizet Collection. Purchased*, 1848.

21879. The second phalangeal of an equal-sized digit; from Sansan. *Croizet Collection. Purchased*, 1848.

21881. A laterally crushed specimen of a similar second phalangeal; from Sansan. *Croizet Collection. Purchased*, 1848.

SERIAL POSITION UNCERTAIN.

The affinity of the following specimens, which are entered in the Museum Register under the name of *Myopotherium bravardi* [1], has not been determined.

43411–46. Numerous associated bones of the pes and two cervical vertebræ; from the Pleistocene of Buenos Ayres, Argentine Republic. *Presented by Señor L. J. Fontana*, 1872.

[1] The writer has been unable to find any published authority for this name.

Subclass METATHERIA.

Order MARSUPIALIA.

The homology of the cheek-teeth of the Marsupialia having been recently worked out by O. Thomas [1], it will suffice to mention that the tooth which is either preceded by a milk-molar, or which is homologous with the one so preceded, corresponds with the fourth premolar of the Eutheria; while the four teeth which usually occur posteriorly to the latter are reckoned as true molars, and numbered consecutively. The tooth immediately in advance of the fourth premolar corresponds to the third premolar of the Eutherian series; but in the numerous existing members of the Polyprotodont suborder in which there are three premolars, the first of these corresponds with the first (and not with the second) of the Eutherian series; judging, however, from *Plagiaulax*, it is probable that the same rule may not hold good for all those few members of the Diprotodont suborder in which three premolars are present. A peculiarity in regard to the serial homology of the two premolars, which occur in one genus of the Polyprotodont family *Dasyuridæ*, will be noticed under that head.

Suborder *MARSUPIALIA DIPROTODONTIA.*

Incisors $\frac{(1-3)}{1}$. The central pair of upper incisors and the one pair of lower incisors large and cutting. Canines absent or small. Molars with bluntly tuberculate, or transversely or longitudinally ridged crowns. There are usually but two premolars in existing forms, but four of these teeth are present in one species of the Mesozoic *Plagiaulax*. The fourth premolar is frequently of a simply secant structure, and may be considerably longer than the first true molar.

Family PHASCOLOMYIDÆ.

Dentition[2] :—I. $\frac{1}{1}$, C. $\frac{0}{0}$, Pm. $\frac{1}{1}$, M. $\frac{4}{4}$. All the teeth have persistent pulps. The incisors are large and scalpriform, with enamel only on the anterior surface. The cheek-teeth (woodcut, fig. 26) are curved, the convexity being inwards in the upper and outwards in the lower jaw; the premolar, which has no milk predecessor, is usually single-

[1] Abstract in Proc. R. Soc. vol. xlii. p. 310 (1887).

[2] It may possibly prove that there were more than a single pair of upper incisors in *Phascolonus*.

lobed, while the true molars consist of two lobes with a triangular section. The masseteric fossa of the mandible has a perforation and a deep pit, while the condyle is convex, prominent, and separated by a deep notch from the coronoid process. The limbs are of

Fig. 26.

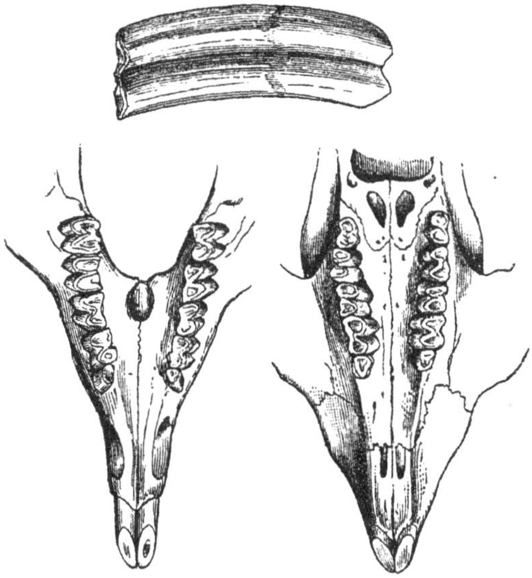

Phascolomys wombat.—Mandibular and palatal dentition, reduced, and single molar of the natural size; from Tasmania.

equal length; the humerus is very stout, and has an entepicondylar foramen; there is a well-developed olecranon to the ulna; the manus has five complete digits, but the hallux of the pes is imperfect, and the second, third, and fourth digits are of nearly equal size and partly syndactylous.

Genus **PHASCOLOMYS**, Geoffroy[1].

The upper incisors are not greatly wider than the lower, are either subquadrangular or oval in section, and their biting-surfaces are not worn very obliquely; in the lower incisors the biting-surfaces are comparatively short.

[1] Ann. d. Muséum, vol. ii. p. 365 (1803).

A. *Latifront Group.*

In the existing species the fronto-nasal suture is straight; the nasals are narrow; the enamelled surface of the upper incisors is directed anteriorly; the inferior border of the mandible is highly convex; the masseteric fossa is shallow; and $\overline{\text{pm. 4}}$ has a quadrangular cross-section.

Phascolomys magnus, Owen[1].

The species is considerably smaller than *Phascolonus gigas*, and is mainly known by portions of the cranium, the complete mandible being unrepresented in the collection. The disposition of the upper cheek-teeth and palate is after the type of *P. latifrons*; but the transverse concavity of the palate, which is most marked at the diastemal region, is much greater. The length of the upper series of cheek-teeth in the type specimen is 0,086.

Hab. Queensland and New South Wales.

39989. The palatal region of the cranium, showing all the cheek-
(*Fig.*) teeth; from the Pleistocene of Gowrie, Drayton, Queensland. This specimen is the type, and is figured by Owen in the 'Phil. Trans.' 1872, pl. xxxv. figs. 1–5, and also in pl. xl. figs. 1–5 of the 'Extinct Mammals of Australia.'
Presented by Sir D. Cooper, Bart., 1866.

38607. Portion of the right maxilla, showing the last three true molars; from the Pleistocene of Eton Vale, Darling Downs, Queensland. Described by Owen in the 'Phil. Trans.' 1872, p. 248.
Presented by Sir D. Cooper, Bart., 1864.

42613. Two upper true molars, one imperfect; from a cavern in
(*Fig.*) the Wellington Valley, New South Wales. Figured by Owen in the 'Phil. Trans.' for 1872, pl. xxxv. fig. 6, and in the 'Extinct Mammals of Australia,' pl. lx. fig. 6.
Presented by the Trustees of the Australian Museum, 1870.

47850. A small fragment of the left ramus of the mandible, containing one of the true molars; from Darling Downs. The tooth of this specimen, which is provisionally referred to the present species, is smaller than $\overline{\text{m. 1}}$ of *Phascolonus gigas*. *Presented by Dr. George Bennett*, 1875.

[1] Phil. Trans. 1872, p. 246.

The following specimens may be provisionally referred to this species.

43959. The proximal extremity of the left humerus; from Darling Downs. The longest transverse diameter is 0,063.
Presented by Dr. George Bennett, 1874.

43911. The proximal extremity of the right radius; from a cave in the Wellington Valley.
Presented by the Trustees of the Australian Museum, 1870.

46073. The first phalangeal of the second digit of the right manus; from Queensland.
Presented by Dr. George Bennett, 1874.

40062. Cast of the distal two thirds of the right femur. The original is from Queensland, and is preserved in the Museum at Sydney. It has the prominent inner ridge on the tibial trochlea characteristic of *P. platyrhinus* and *P. latifrons*.
Presented by the Trustees of the Australian Museum, 1870.

38781. The distal portion of a similar left femur; from Darling Downs. *Presented by F. N. Isaac, Esq.*, 1861.

43924. Two first phalangeals; from the caves of the Wellington Valley.
Presented by the Trustees of the Australian Museum, 1870.

Phascolomys medius, Owen [1].

This species is intermediate in size between the preceding and following ones, with which it agrees in the structure of the molars and palate; the concavity of the latter being rather less than in *P. magnus*, but greater than in *P. latifrons*. The length of the space occupied by the first three upper cheek-teeth is 0,038, against 0,0255 in *P. latifrons* and 0,055 in *P. magnus*; and that of the upper diastema 0,063 against 0,044 in the existing species. The inferior border of the mandible is strongly convex, the section of $\overline{\text{pm. 4}}$ quadrangular, and the masseteric fossa of the mandible shallow.

Hab. Queensland and New South Wales. The following specimens include the types.

[1] Phil. Trans. 1872, p. 241.

32904. The anterior portion of the cranium, showing most of the
(*Fig.*) cheek-teeth in a broken condition; from the Pleistocene of the Condamine River, Queensland. Figured by Owen in the 'Phil. Trans.' 1872, pl. xxxiii. figs. 2–6, and in the 'Extinct Mammals of Australia,' pl. lviii. figs. 2–6.

Purchased. About 1854.

39991. The anterior portion of the cranium, showing the broken
(*Fig.*) bases of the incisors, and the first four cheek-teeth in a broken condition; from the Pleistocene of Gowrie, Drayton, Queensland. Figured by Owen in the 'Phil. Trans.' 1872, pl. xxxii. figs. 2–7.

Presented by Sir D. Cooper, Bart., 1866.

47846. Part of the right maxilla, showing the four true molars and the alveolus of $\underline{\text{pm. } 4}$; from Darling Downs, Queensland.

Presented by Dr. George Bennett, 1875.

39990. The imperfect symphysis and right ramus of the mandible,
(*Fig.*) showing the broken bases of the incisors and the whole of the cheek-teeth; from Gowrie. Figured by Owen in the 'Phil. Trans.' 1872, pl. xxxiv. figs. 1, 2, and in the 'Extinct Mammals of Australia,' pl. lix. figs. 1, 2.

Presented by Sir D. Cooper, Bart., 1865.

47847. The imperfect mandible, showing the whole of the cheek-teeth; from Darling Downs. The left condyle is preserved, and the characteristic shallowness of the masseteric fossa well shown. *Presented by Dr. George Bennett*, 1875.

50069. The imperfect mandible, showing the four true molars on the left side and the alveoli of the other teeth; from Darling Downs.

Presented by G. F. Bennett, Esq., 1879.

The following specimens are provisionally referred to this species.

M. 3636. The proximal half of the left ulna; from Queensland.

No history.

46839–40. The associated calcanea of both sides, the left fourth and fifth metatarsals, the fifth right metatarsal, and a terminal phalangeal; from Queensland.

Presented by Dr. George Bennett, 1874.

38763. The seventh cervical vertebra; from Darling Downs.

Presented by F. N. Isaac, Esq., 1861.

Phascolomys latifrons, Owen[1].

The characters of this species are given under the head of the group. The length of the space occupied by the upper cheek-teeth in the specimen figured by Owen in the 'Phil. Trans.' is 0,049.

Hab. South Australia (Recent) and New South Wales (Pleistocene).

42605. Fragment of the right ramus of the mandible, showing the (*Fig.*) four premolars and the broken base of $\overline{\text{pm. 4}}$; from a cavern in the Wellington Valley, New South Wales. Figured by Owen in the 'Phil. Trans.' 1872, pl. xxii. fig. 5, and in the 'Extinct Mammals of Australia,' pl. lii. fig. 5.

Presented by the Trustees of the Australian Museum, 1870.

42609. Fragment of the right ramus of the mandible, showing the (*Fig.*) first four cheek-teeth and the alveolus of $\overline{\text{m. 4}}$; from the Wellington Valley. Figured by Owen in the 'Phil. Trans.' for 1872, pl. xix. fig. 4, pl. xxi. fig. 4, and pl. xxii. fig. 4; and also in the 'Extinct Mammals of Australia,' pl. lii. fig. 7, and pl. lv. fig. 2.

Presented by the Trustees of the Australian Museum, 1870.

42614. Fragment of the right ramus of the mandible, showing the four true molars; from the Wellington Valley.

Presented by the Trustees of the Australian Museum, 1870.

Phascolomys krefftii, Owen[2].

This species is very closely allied to *P. latifrons*, from which it is mainly distinguished by a difference in the direction of the sutures between the nasals, maxillæ, and parietals, and also by the more backward extension of the mandibular symphysis, and a difference in the contour of the inner surface of the ramus.

Hab. New South Wales.

42601. The anterior portion of the cranium, showing the broken (*Fig.*) incisors; from a cavern in the Wellington Valley, New South Wales. This is the type specimen, and is figured by Owen in the 'Phil. Trans.' 1872, pl. xvii. figs. 2, 6, and in the 'Extinct Mammals of Australia,' pl. l. figs. 2, 6.

Presented by the Trustees of the Australian Museum, 1870.

[1] Proc. Zool. Soc. 1845, p. 82.
[2] Phil. Trans. 1872, p. 178.

42602. (*Fig.*) The anterior portion of the mandible, showing the bases of the incisors and the first three cheek-teeth; from a cavern in the Wellington Valley. Figured by Owen in the 'Phil. Trans.' 1872, pl. xx. fig. 7, and pl. xxiii. fig. 4; and in the 'Extinct Mammals of Australia,' pl. lii. fig. 7, and pl. lvi. fig. 4.

Presented by the Trustees of the Australian Museum, 1870.

42610. (*Fig.*) Fragment of the left ramus of the mandible, showing two true molars; from the Wellington Valley. Figured by Owen in the 'Phil. Trans.' 1872, pl. xix. fig. 3, pl. xxii. fig. 6, and pl. xxiii. fig. 5, and in the 'Extinct Mammals of Australia,' pl. liii. fig. 3.

Presented by the Trustees of the Australian Museum, 1870.

B. *Serial position uncertain.*

Phascolomys curvirostris, Owen[1].

This species, which was apparently somewhat larger than *P. medius*, is known only by the type specimen. It is characterized by the great downward deflection of the palatal border of the premaxillæ. The enamelled surface of the upper incisor is mainly directed anteriorly, in which respect it agrees with the Latifront group.

Hab. New South Wales.

M. 2576. Cast of the premaxillæ and incisors. The original, which is the type, was obtained from a cavern in the Wellington Valley, New South Wales, and is preserved in the Australian Museum, Sydney; it is described and figured by Owen in the 'Quart. Journ. Geol. Soc.' vol. xlii. p. 1, pl. i.

Presented by Sir R. Owen, K.C.B., 1886.

C. *Platyrhine Group.*

The two existing members (*P. platyrhinus* and *P. wombat*) of this group have the fronto-nasal suture convex posteriorly; the nasals wide; $\underline{\text{pm. 4}}$ small; a tubercle on the lachrymal; the enamelled surface of the upper incisors directed mainly outwards, the inferior border of the mandible slightly convex; the masseteric fossa deep; and $\overline{\text{pm. 4}}$ ellipsoidal in section.

Phascolomys mitchelli, Owen[2].

This species is very close to *P. platyrhinus*, with which it is

[1] Quart. Journ. Geol. Soc. vol. xlii. p. 1 (1886).

[2] In Mitchell's 'Three Expeditions into Eastern Australia,' 2nd ed. vol. ii. p. 362 (1838).

identified by Murie in the 'Proc. Zool. Soc.' 1865, p. 851—an identification which, if correct, should lead to the abolition of the latter name. The distinctive characters of the present form, according to Owen, are the still smaller connexion of the nasal with the maxilla, the lesser protuberance of the lachrymal in advance of the tubercle, the smaller size of the posterior palatal foramina, the greater depth of the outer surface of the maxilla below the zygoma, and the less extension backwards of the mandibular symphysis.

Hab. New South Wales and Queensland.

42598. The middle region of the cranium, showing the true molars, (*Fig.*) and the alveoli of $\underline{\text{pm. 4}}$ and of the incisors; from a cave in the Wellington Valley, New South Wales. Figured by Owen in the 'Phil. Trans.' 1872, pl. xvii. figs. 1, 3–5, and in the 'Extinct Mammals of Australia,' pl. l. figs. 1, 3–5.
Presented by the Trustees of the Australian Museum, 1870.

32912. The palatal region of the cranium; from the Pleistocene of Australia. The whole of the cheek-dentition is shown.
Presented by Sir W. McArthur, Bart. About 1854.

43048. The palatal region of the cranium, showing the whole of the (*Fig.*) cheek-teeth; from the Pleistocene of Queensland. Figured by Owen in the 'Phil. Trans.' 1872, pl. xviii. figs. 1–4, and in the 'Extinct Mammals of Australia,' pl. li. figs. 1–4.
Presented by Dr. George Bennett, 1861.

43049. The imperfect palatal region of the cranium; from Australia. (*Fig.*) Figured by Owen in the 'Phil. Trans.' 1872, pl. xviii. figs. 5–7, and in the 'Extinct Mammals of Australia,' pl. li. figs. 5–7. *Presented by Dr. George Bennett*, 1861.

42612. The imperfect left maxilla, without the teeth, from a cave (*Fig.*) in the Wellington Valley. Figured by Owen in the 'Phil. Trans.' 1872, pl. xvii. figs. 7–8, and in the 'Extinct Mammals of Australia,' pl. l. figs. 7, 8.
Presented by the Trustees of the Australian Museum, 1870.

46915. The imperfect palatal region of the cranium, showing all the cheek-teeth; from the Pleistocene of Kirban, near Mendoran, New South Wales.
Presented by W. L. R. Gipps, Esq., 1875.

47849. Fragment of the right maxilla, with two true molars; from the Pleistocene of Darling Downs, Queensland.
Presented by Dr. George Bennett, 1875.

50070–1. The imperfect palate and mandible; from the Pleistocene of Gowrie, Queensland. The left ramus of the mandible is nearly entire. *Presented by G. F. Bennett, Esq.*, 1879.

42604. (*Fig.*) Part of the left ramus of the mandible, showing the five cheek-teeth; from a cave in the Wellington Valley. Figured by Owen in the 'Phil. Trans.' 1872, pl. xix. fig. 5, and pl. xxi. figs. 5, 6, and also in the 'Extinct Mammals of Australia,' pl. liii. fig. 5, and pl. lv. figs. 5, 6.
Presented by the Trustees of the Australian Museum, 1870.

45999. The imperfect right ramus of the mandible; from a cave in the Wellington Valley.
Presented by the Trustees of the Australian Museum, 1870.

42603. The imperfect left ramus of the mandible; from a cave in the Wellington Valley. Noticed by Owen in the 'Phil. Trans.' 1872, p. 192.
Presented by the Trustees of the Australian Museum, 1870.

42608. Fragment of the right ramus of the mandible, showing the first four cheek-teeth; from a cave in the Wellington Valley.
Presented by the Trustees of the Australian Museum, 1870.

42611. Fragment of the left ramus of the mandible, showing the last three cheek-teeth; from a cave in the Wellington Valley.
Presented by the Trustees of the Australian Museum, 1870.

42616. A lower true molar; from a cave in the Wellington Valley.
Presented by the Trustees of the Australian Museum, 1870.

38749. Part of the left ramus of the mandible, with the four true molars; from Darling Downs.
Presented by F. N. Isaac, Esq., 1861.

38750. Fragment of the right ramus of the mandible, with two true molars; from Darling Downs.
Presented by F. N. Isaac, Esq., 1861.

32892. Fragment of the left ramus of the mandible, showing the broken bases of the true molars; from the Pleistocene of the Condamine River, Queensland. *Purchased*, 1857.

50072–3. (*Fig.*) The symphysis and part of the right ramus of the mandible, showing the last three true molars and the left incisor; from Gowrie. The symphysis appears somewhat narrower than in the recent mandible of *P. platyrhinus*. Figured by Owen in the 'Extinct Mammals of Australia,' pl. liv. fig. 1. *Presented by G. F. Bennett, Esq.*, 1879.

42606. Several imperfect upper incisors, probably belonging either to this species or to *P. platyrhinus*; from the caves of the Wellington Valley.
Presented by the Trustees of the Australian Museum, 1870.

42606 a. Two imperfect mandibular incisors agreeing in size with the preceding; from the same locality. *Same history.*

Some of the following limb-bones probably belong to the present species, and others to some of the other smaller Wombats.

43951. The proximal half of the left humerus; from the same locality. *Same history.*

43951 a. The imperfect left humerus; from the same locality. *Same history.*

43951 b. The distal half of the right humerus; from the same locality. *Same history.*

40020. The associated left radius and ulna; from Queensland.
Presented by Sir D. Cooper, Bart., 1866.

43908. The proximal portion of the left ulna; from a cave in the Wellington Valley.
Presented by the Trustees of the Australian Museum, 1870.

42617. The distal half of the left femur; from the same locality. *Same history.*

32293. The distal extremity of the left femur; from the same locality. *Purchased*, 1857.

43939. Several specimens of the calcaneum; from the same locality.
Presented by the Trustees of the Australian Museum, 1870.

43939 a. Two specimens of the astragalus; from the same locality. *Same history.*

43939 b. Several metapodials and phalangeals; from the same locality. *Same history.*

Phascolomys platyrhinus, Owen[1].

The characters of this, the largest existing Wombat, are given under the head of the group. The length of the space occupied by the upper cheek-teeth is 0,053.

Hab. South Australia (Recent) and Queensland (Pleistocene).

[1] Cat. Osteol. Series Mus. Roy. Coll. Surg. vol. i. p. 334 (1853).

39272. The anterior portion of the mandible, showing all the cheek-
(*Fig.*) teeth and the broken bases of the incisors; from the Pleistocene of Queensland. Figured by Owen in the 'Phil. Trans.' 1872, pl. xx. figs. 3–5, and in the 'Extinct Mammals of Australia,' pl. liv. figs. 3–5.

Presented by Edward Hill, Esq., 1863.

M. 3450. The symphysis and greater portion of the left ramus of the mandible; from Queensland.

Presented by Dr. George Bennett.

Phascolomys thomsoni, Owen[1].

This species agrees in size with *P. wombat*; the general mandibular characters are those of *P. platyrhinus*, but the symphysis is less produced backwards; the length of the mandibular series of cheek-teeth is 0,050. The cranium is unknown.

Hab. Queensland.

38608. The imperfect right ramus of the mandible; from the Pleis-
(*Fig.*) tocene of Darling Downs, Queensland. This is the type specimen, and is figured by Owen in the 'Phil. Trans.' 1872, pl. xvii. figs. 8, 9, and pl. xxi. fig. 7, as well as in the 'Extinct Mammals of Australia,' pl. li. figs. 8, 9, and pl. lv. fig. 7. *Presented by Sir D. Cooper, Bart.*, 1864.

Phascolomys parvus, Owen[2].

This is the smallest species, the length of the series of mandibular cheek-teeth being 0,036; the mandibular characters are very similar to those of the preceding species. The cranium is unknown. This form may be merely a smaller race of *P. thomsoni*.

Hab. Queensland and New South Wales.

32893. The slightly imperfect left ramus of the mandible, showing
(*Fig.*) all the teeth; from the Pleistocene of the Condamine River, Queensland. This is the type specimen, and is figured by Owen in the 'Phil. Trans.' 1872, pl. xx. figs. 6–8, and in the 'Extinct Mammals of Australia,' pl. liv. figs. 6–8. *Purchased*, 1857.

32899. The greater part of the left ramus of the mandible, showing
(*Fig.*) all the teeth except $\overline{\text{m. 4}}$; from the Condamine River. Figured by Owen in the 'Phil. Trans.' 1872, pl. xix. figs.

[1] Phil. Trans. 1872, p. 192.
[2] *Ibid.* p. 193.

6, 7; and in the 'Extinct Mammals of Australia,' pl. liii. figs. 6, 7. *Purchased*, 1857.

32911 x. Fragment of the left ramus of the mandible, showing the last three true molars; from the Condamine River. Noticed by Owen in the 'Phil. Trans.' 1872, p. 194. *Purchased*, 1857.

43303 x. (*Fig.*) Fragment of the right ramus of the mandible, showing the last three true molars; from a cavern in the Wellington Valley, New South Wales. Figured by Owen in the 'Phil. Trans.' 1872, pl. xxxviii. figs. 5, 6, and in the 'Extinct Mammals of Australia,' pl. lxiii. figs. 5, 6. *Purchased. About* 1852.

Genus **PHASCOLONUS**, Owen [1].

Syn. (?) *Sceparnodon*, Owen [2] (*ex* Ramsay, MS.).

There is every probability that the upper incisors on which *Sceparnodon* is founded belong to *P. gigas*, and it is on this provisional reference that the latter is generically separated from *Phascolomys*.

The upper incisors are wider than the lower, are much compressed antero-posteriorly, have a bevelled cutting-edge, a concave posterior surface, and their enamelled surface directed anteriorly; in the lower incisors the biting-surface is much elongated.

The action of the upper upon the lower incisors must have been analogous to that obtaining in *Diprotodon*.

Phascolonus gigas, Owen [3].

Syn. *Phascolomys gigas*, Owen [4].
(?) *Sceparnodon ramsayi*, Owen [5].

This is the type and only known species; it was apparently about equal in size to a Tapir, but of much stouter build. The inferior border of the mandible has the great convexity characteristic of the Latifront group of *Phascolomys*, but the masseteric fossa is deeper, and has a prominent inferior ridge; $\overline{\text{pm. 4}}$ has the ellipsoidal section of the Platyrhine group, and $\underline{\text{pm. 4}}$ the complexity characteristic of the latter; in the partially compressed upper incisors *P. platyrhinus*

[1] Phil. Trans. 1872, p. 257 (as a subgenus).
[2] *Ibid.* 1884, p. 245.
[3] 'Encyclopædia Britannica,' 8th ed. vol. xvii. p. 175 (1859).—*Phascolomys.*
[4] *Loc. cit.*
[5] Phil. Trans. 1884, p. 247.

makes an approach to the present form, but differs by the twisting of the axis of these teeth. The enamel of both the incisors and cheek-teeth is finely striated.

Hab. Queensland and South Australia.

The following specimens are the types of Sceparnodon.

M. 1917. (*Fig.*) Fragment of an upper incisor, of which one extremity has been cut and polished; from the Pleistocene of King's Creek, Queensland. Figured by Owen in the 'Phil. Trans.' 1884, pl. xii. figs. 1–3 and 9.

Presented by C. H. Hartmann, Esq., 1883.

M. 1918. Cast of an upper incisor, agreeing in size with the preceding. The original was obtained from the Pleistocene near Lake Eyre, South Australia, and is preserved in the Museum at Sydney; it is figured by Owen, *op. cit.* pl. xii. figs. 4, 5.

Presented by the Trustees of the Australian Museum, 1881.

M. 1919. Cast of a larger upper incisor. The history of the original is the same as those of the preceding, and it is figured by Owen, *op. cit.* pl. xii. figs. 6–8. *Same history.*

The following specimens include those figured by Owen as Phascolonus gigas.

50075 a. An imperfect fourth upper premolar; from (? King's Creek) Queensland. The transverse diameter of this tooth is 0,019; it has the complex character of $\underline{\text{pm. 4}}$ of *Phascolomys platyrhinus*; and in the striations and reddish stains of the enamel it agrees so exactly with No. M. 1917 that it is quite probable that it may have belonged to the same individual.

Presented by G. F. Bennett, Esq., 1879.

43044. (*Fig.*) The imperfect mandible, showing on the right side the four true molars, and on the left all the cheek-teeth except $\overline{\text{m. 3}}$; from the Pleistocene of Eton Vale, Darling Downs, Queensland. Figured by Owen in the 'Phil. Trans.' 1872, pls. xxxvi.–xxxviii., and in the 'Extinct Mammals of Australia,' pls. lxii., lxiii. The length of the space occupied by the cheek-teeth is 0,108.

Presented by Sir D. Cooper, Bart. About 1871.

35977. Cast of the greater portion of the right ramus of the mandible, showing part of the incisor and all the cheek-teeth. The original was obtained from the Pleistocene of Gowrie,

Queensland, and is preserved in the Australian Museum, Sydney; it is figured by Owen in the 'Phil. Trans.' 1872, pl. xl. fig. 1, and in the 'Extinct Mammals of Australia,' pl. lxv. fig. 1. The length of the space occupied by the cheek-teeth is 1,118; in the figure the extremity of the incisor has been restored.

Presented by Sir D. Cooper, Bart., 1864.

43045. Fragment of the right ramus of the mandible of a very large individual; from the Pleistocene of the Condamine River, Queensland. Noticed by Owen in the 'Phil Trans.' 1872, p. 253. *Presented by Sir D. Cooper, Bart.*, 1871.

38606. Part of the left ramus of the mandible, containing the four
(*Fig.*) true molars; from Eton Vale. Figured by Owen, *op. cit.* pl. xl. fig. 6. *Presented by Sir D. Cooper, Bart.*, 1864.

43046. Fragment of the left ramus of the mandible, showing the
(*Fig.*) alveoli of the last three true molars; from the Pleistocene of Clifton Plains, Queensland. Figured by Owen, *op. cit.* pl. xxxviii. figs. 3, 4. *Presented by Prof. Harkness*, 1871.

M. 3634. Fragment of the right ramus of the mandible, with the four true molars ($\overline{m. 1}$ imperfect); from Queensland.

Presented by Dr. George Bennett.

43088 a. Cast of a fragment of the left ramus of the mandible, showing the last three true molars ($\overline{m. 4}$ imperfect). The original was obtained from Queensland.

Presented by the Trustees of the Australian Museum, 1871.

M. 3635. Hinder part of the right ramus of the mandible, containing the last three true molars; from Queensland.

Presented by Dr. George Bennett.

44124. The imperfect symphysial portion of the mandible, showing the basal portion of the incisors; from Darling Downs.

Presented by Dr. George Bennett, 1873.

43088 b. Cast of the anterior part of the right lower incisor. The original is from Queensland, and is preserved in the Museum at Sydney. It is figured by Owen in the 'Phil. Trans.' 1872, pl. xl. figs. 2–4, and also in the 'Extinct Mammals of Australia,' pl. lxv. figs. 2–4. The length of the abraded terminal surface is 0,034.

Presented by the Trustees of the Australian Museum, 1871.

At least the greater number of the following specimens may probably be referred to this species.

39992. The proximal half of the right ulna; from Gowrie.
Presented by Sir D. Cooper, Bart., 1865.

50084. The third left metacarpal; from Queensland.
Presented by G. F. Bennett, Esq., 1879.

46073 x. An imperfect homologous bone of the opposite side; from Queensland. *Presented by Dr. George Bennett*, 1874.

38793. The first phalangeal of the third digit of the manus; from Darling Downs. *Presented by F. N. Isaac, Esq.*, 1861.

46839. The first phalangeal of the fifth digit of the left manus; from Queensland. *Presented by Dr. George Bennett*, 1874.

40381. The distal portion of the left femur; from the Pleistocene of the Condamine River, Queensland. This bone, which agrees in relative size with the next specimen, in the prominence of the inner ridge of the patellar trochlea resembles the femur of *P. platyrhinus* rather than that of *P. wombat*. The transverse diameter across the condyles is 0,071. *Presented by Dr. F. Campbell*, 1867.

38782. The left tibia; from Darling Downs. The length is 0,180, and the width of the distal surface 0,090; the bone being relatively shorter than in existing Wombats.
Presented by F. N. Isaac, Esq., 1861.

38790. The left calcaneum; from Darling Downs. This bone, which agrees in relative size with the preceding, differs slightly in contour from the calcaneum of existing forms.
Presented by F. N. Isaac, Esq., 1861.

38793 a. The first phalangeal of the fourth digit of the pes; from Darling Downs. *Presented by F. N. Isaac, Esq.*, 1861.

43353. A terminal phalangeal of the pes; from Queensland. This bone is less flattened than in existing Wombats; it differs from the corresponding bone of *Macropus* by its greater curvature, and the absence of angulation on the dorsal aspect. *Presented by Dr. George Bennett*, 1872.

46839. A similar bone; from Queensland.
Presented by Dr. George Bennett, 1875.

46841 n. A cervical vertebra; from Queensland. The transverse diameter of the centrum is 0,062.
Presented by Dr. George Bennett, 1875.

Family NOTOTHERIIDÆ.

Dentition:—I. $\frac{3}{1}$, C. $\frac{0}{0}$, Pm. $\frac{1}{1}$, M. $\frac{4}{4}$. The cheek-teeth are rooted, and the upper true molars carry two transverse ridges, which are not connected by a longitudinal bridge; the premolar is triangular, small, and apparently had no predecessor. In the lower true molars the unworn ridges are sometimes subcrescentoid, and partially connected by an incomplete longitudinal bridge (No. 43523). The upper incisors of the two sides are widely separated and are of moderate size; while the lower incisor is procumbent and of the same relative proportions. The preorbital part of the cranium is very short; the extremities of the bones roofing the nasal cavity are expanded laterally, and the zygomatic arches are enormously wide. There are no palatal vacuities. The mandibular symphysis is anchylosed; the inferior border of the ramus highly convex, the condyle like that of *Phascolomys*, and the masseteric fossa without pit or perforation. The limbs are of equal length; the humerus [1] has an entepicondylar foramen, and closely resembles that of *Phascolomys*; the olecranon is well developed, and the other limb-bones and vertebræ are of the type of those of the latter genus. The structure of the feet is not fully known, but it apparently approximated to that of *Phascolomys*.

This family connects the *Phascolomyidæ* with the *Diprotodontidæ*; the cranium, limb-bones, and vertebræ being nearest to those of the former, the mandible showing characters common to the two, while the cheek-teeth are of the type of the latter.

It is easy to see how the structure of the cheek-teeth could pass into that of the *Phascolomyidæ*; and it is not improbable that the two families may have diverged from a common ancestor. The structure of the humerus apparently indicates fossorial habits.

Genus **NOTOTHERIUM,** Owen [2].

Since this is the only known genus, its characters are the same as those of the family.

[1] De Vis (Proc. Linn. Soc. N. S. Wales, vol. viii. p. 404) has provisionally referred another type of humerus to *Nototherium*. That bone probably, however, belonged to a small *Diprotodon*; and there is every reason for regarding Owen's determination as correct, since there is no dental evidence of the existence of any other animal to which the bones referred by Owen to the present genus could have belonged. The great difference between the skulls of *Diprotodon* and *Nototherium* would of itself indicate that an equally well-marked difference should occur in the limb-bones of the two genera.

[2] Cat. Foss. Mamm. and Aves Mus. R. Coll. Surg. p. 314 (1845).

Nototherium mitchelli, Owen[1].

Including:—*Nototherium inerme*, Owen[2].
Nototherium victoriæ, Owen[3].

The form described as *N. inerme* is somewhat smaller than the type of *N. mitchelli*, and differs slightly in the characters of the cheek-teeth; a large series of specimens show, however, such a gradation in these respects, that the writer has followed Flower in provisionally including all the specimens under one name. *N. victoriæ* is apparently founded on characters which are merely individual.

Hab. Australia.

33259. Cast of the restored skull. The original was obtained from the Pleistocene of the Condamine River, Queensland, and is preserved in the Australian Museum, Sydney. It is figured by Owen in the 'Phil. Trans.' 1872, pls. i.–iii., and also in the 'Extinct Mammals of Austraali,' pls. xxxvi.–xxxviii.; the cheek-dentition being figured in pl. xliii. fig. 4 of the same work.

Presented by the Trustees of the Australian Museum.

M. 3492. Fragments of a large cranium; from Queensland.

Presented by Dr. George Bennett.

43212. Hinder portion of the palate, showing the well-worn m. 3 and m. 4, and a fragment of m. 2; from the Pleistocene of the Albert River, Queensland.

Presented by Sir R. Daintree, 1871.

32850 x. (*Fig.*) Fragment of the right maxilla of a large individual, showing the last three molars in a half-worn condition; from the Condamine River. Figured by Owen in the 'Quart. Journ. Geol. Soc.' vol. xv. pl. ix. figs. 4, 5; in the 'Phil. Trans.' 1872, pl. ix. figs. 6, 7, and in the 'Extinct Mammals of Australia,' pl. xliii. figs. 6, 7. *Purchased*

43087. Fragment of the right maxilla of a medium-sized individual, showing m. 4, a fragment of m. 3, and the bases of the remaining cheek-teeth; from the Pleistocene of Scone, Australia. This specimen, which was referred by Owen to *N. inerme*, is intermediate in size between No. 33259 and No. 43221, but has the large pm. 4 characteristic of the former.

Presented by the Trustees of the Australian Museum, 1871.

[1] Cat. Foss. Mamm. and Aves Mus. R. Coll. Surg. p. 316 (1845).
[2] *Op. cit.* p. 314. [3] Phil. Trans. 1872, p. 61.

43087 a. Part of the right zygomatic arch, together with the associated nasal of the same side, belonging to the same individual as the preceding specimen.
Presented by the Trustees of the Australian Museum, 1871.

43221. (*Fig.*) The palate of a small (? female) individual, showing all the cheek-teeth; from the Albert River. Figured by Owen in the 'Phil. Trans.' 1872, pl. xix. fig. 5, and also in the 'Extinct Mammals of Australia,' pl. xliii. fig. 5 (as *N. inerme*); pm. 4 is of very small size.
Presented by Sir R. Daintree, 1871.

47840. Fragment of the left maxilla of a large individual, showing the well-worn pm. 4 and m. 1; from the Pleistocene of Darling Downs, Queensland.
Presented by Dr. George Bennett, 1875

39979. Part of the left maxilla of a medium-sized individual, showing m. 1, m. 2, and m. 3, the alveolus of pm. 4, and the broken base of m. 4; from Darling Downs.
Presented by Sir D. Cooper, Bart., 1866.

42304. Cast of the left maxillary region of a small individual, showing the five cheek-teeth. The original was obtained from Darling Downs; pm. 4 is of small size.
Presented by Prof. A. M. Thomson, 1870.

33260. Cast of the palatal region of a medium-sized individual. The original was obtained from the Condamine River, and is preserved in the Australian Museum, Sydney.
Presented by the Trustees of the Australian Museum.

43088. Cast of part of the right maxilla of a medium-sized individual, showing the four true molars in a half-worn condition. The original is from Queensland, and is preserved in the Australian Museum.
Presented by the Trustees of the Australian Museum, 1871.

43364. Fragment of the right maxilla, containing the slightly-worn m. 2; from Gowrie, Darling Downs.
Presented by Dr. George Bennett, 1872.

39981. (*Fig.*) Fragment of the right maxilla of a small individual, showing the slightly-worn m. 2 and m. 3; from Gowrie. Figured by Owen in the 'Phil. Trans.' 1872, pl. ix. fig. 8.
Presented by Sir D. Cooper, Bart., 1866.

43336. Fragment of the right maxilla of a somewhat larger individual, containing the half-worn m. 3 and m. 4; from Warra-Warra Station, Condamine River.
Presented by Dr. George Bennett, 1872.

39982. The germ of the third right upper true molar of a small individual; from Gowrie.
Presented by Sir D. Cooper, Bart., 1866.

M. 3643. Cast of the crown of the right upper premolar. The original is from Darling Downs, and is preserved in the Museum at Brisbane; it indicates a small-sized individual.
Presented by Dr. George Bennett, 1872.

35958. A worn specimen of the first right upper incisor; from
(*Fig.*) Gowrie. Figured by Owen in the 'Phil. Trans.' 1872, pl. ix. figs. 1, 2; and also in the 'Extinct Mammals of Australia,' pl. xliii. figs. 1, 2.
Presented by Sir D. Cooper, Bart., 1866.

43339. The crown of a first right upper incisor of a large individual; from the Condamine River.
Presented by Dr. George Bennett, 1872.

46057 a. A first upper incisor; from Queensland.
Presented by W. L. R. Gipps, Esq., 1875.

43338, 46068. Two upper (?) incisors; from Queensland.
Presented by Dr. George Bennett, 1872–74.

32850. The imperfect mandible, showing the last three true molars;
(*Fig.*) from the Condamine River. Figured by Owen in the 'Quart. Journ. Geol. Soc.' vol. xv. pl. ix. figs. 1, 2; in the 'Phil. Trans.' 1872, pl. iv., and pl. x. figs. 1, 2; and also in the 'Extinct Mammals of Australia,' pl. xxxviii., and pl. xliv. figs. 1, 2. *Purchased.*

43088. Cast of the symphysis and the greater part of the right ramus of the mandible, showing the last three true molars and the alveoli of the anterior teeth. The original was obtained from Queensland, and is preserved in the Museum at Sydney; it is figured by Owen in the 'Phil. Trans.' 1872, pl. v., and also in the 'Extinct Mammals of Australia,' pl. xxxix.
Presented by the Trustees of the Australian Museum, 1871.

43577. The slightly imperfect mandible, showing the dentition in a well-worn state; from Darling Downs. The cingulum of the molars is strongly developed.

Presented by Dr. George Bennett, 1872.

M. 3637. Cast of the slightly imperfect left ramus of the mandible, showing the four true molars ($\overline{\text{m. 1}}$ imperfect). The original was obtained from the Pleistocene near Lake Victoria, South Australia, and is preserved in the Museum at Adelaide; it is figured by Owen in the 'Phil. Trans.' 1872, pl. vii., and pl. x. figs. 4–6, as well as in the 'Extinct Mammals of Australia,' pl. xli., and pl. xliv. figs. 4–6, under the name of *N. victoriæ*, of which it is the type. One of the alleged specific characters is the posterior opening of the dental canal being situated on the level of $\overline{\text{m. 4}}$.

Presented by the Trustees of the Adelaide Museum of Natural History, 1871.

43578. Fragment of the left ramus of the mandible, containing $\overline{\text{m. 4}}$; from Darling Downs. This specimen, which was referred in MSS. by Owen to *N. victoriæ*, agrees very closely with the preceding; but the aperture of the dental canal is placed higher up, being intermediate in position between that specimen and No. 43577.

Presented by Dr. George Bennett, 1872.

46834. Part of the left ramus of the mandible of a large individual, showing the four true molars in a well-worn condition; from Australia. *Presented by Dr. George Bennett*, 1874.

39980. Part of the right ramus of the mandible, containing the first three true molars in an imperfect condition; from Darling Downs. *Presented by Sir D. Cooper, Bart.*, 1866.

39979. Fragment of the left ramus of the mandible of a smaller individual, showing $\overline{\text{m. 3}}$ and $\overline{\text{m. 4}}$; from Darling Downs.

Presented by Sir D. Cooper, Bart., 1866.

43523. (*Fig.*) Hinder part of the left ramus of the mandible, showing the last three true molars in an early stage of wear; from Queensland. Figured by Owen in the 'Phil. Trans.' 1872, pl. xi., and also in the 'Extinct Mammals of Australia,' pl. xlv. The cingulum of the molars is strongly developed.

Presented by H.R.H. The Duke of Edinburgh, K.G., 1872.

M. 3638. Cast of the hinder part of the left ramus of the mandible of a small individual, showing the last three true molars in an imperfect condition. The original was obtained from South Australia, and is preserved in the Museum at Adelaide. *Presented by the Trustees of the Adelaide Museum of Natural History*, 1871.

40377. Fragment of the left ramus of the mandible, containing the
(*Fig.*) slightly-worn $\overline{\mathrm{m.\,3}}$; from the Condamine River. Figured by Owen in the 'Phil. Trans.' 1872, pl. x. fig. 7, and in the 'Extinct Mammals of Australia,' pl. xliv. fig. 7. *Presented by Dr. F. Campbell*, 1867.

43088. Cast of a fragment of the left ramus of the mandible, showing $\overline{\mathrm{m.\,3}}$. The original was obtained from Queensland, and is preserved in the Museum at Sydney.
Presented by the Trustees of the Australian Museum, 1871.

46065. Part of the left ramus of the mandible of a small individual, containing the slightly-worn $\overline{\mathrm{m.\,3}}$ and $\overline{\mathrm{m.\,4}}$; from Australia. The aperture of the dental canal opens on the same level as in the type mandible of the so-called *N. victoriæ*. *Presented by Dr. George Bennett*, 1874.

43953. Fragment of the left ramus of the mandible of a small indi-
(*Fig.*) vidual, showing the well-worn $\overline{\mathrm{m.\,4}}$ and the hinder half of $\overline{\mathrm{m.\,3}}$; from Australia. Figured by Owen in the 'Phil. Trans.' 1872, pl. x. fig. 8, and in the 'Extinct Mammals of Australia,' pl. xliv. fig. 8.
Presented by Dr. George Bennett, 1871.

39988. Fragment of the right ramus of the mandible of a small individual, showing the much-worn $\overline{\mathrm{m.\,2}}$ and $\overline{\mathrm{m.\,3}}$, and the hinder half of $\overline{\mathrm{m.\,1}}$; from Darling Downs.
Presented by Sir D. Cooper, Bart., 1866.

43952. The greater part of the right ramus of the mandible of a
(*Fig.*) young individual; from Darling Downs. Figured by Owen in the 'Phil. Trans.' 1872, pl. vi., and also in the 'Extinct Mammals of Australia,' pl. xl. The incisor and the first three cheek-teeth are protruded, the germ of $\overline{\mathrm{m.\,3}}$ is *in alveolo*, and there is also the alveolus of the germ of $\overline{\mathrm{m.\,4}}$. The outer wall of the ramus has been cut away in order to exhibit the absence of any replacing tooth.
Presented by Prof. Sir R. Owen, K.C.B., 1872.

46066. The greater portion of the right ramus of the mandible of a (*Fig.*) slightly older but smaller individual; from Queensland. The first and second true molars are in use, and $\overline{\text{m. 3}}$ and $\overline{\text{m. 4}}$ *in alveolo*. Figured by Owen in the 'Extinct Mammals of Australia,' pl. cxxv.

Presented by Dr. George Bennett, 1874.

M. 3639. Hinder part of the left ramus of the mandible, showing the slightly worn $\overline{\text{m. 4}}$; from Australia. *Same history.*

45871. Part of the right ramus of the mandible, with the crowns of the teeth broken off; from Australia.

Presented by Dr. George Bennett, 1874.

46838. The condyle of a mandible, provisionally referred to this genus; from Queensland. This specimen closely resembles the corresponding element of *Phascolomys* and of *Diprotodon*, and is totally unlike that of *Macropus*.

Presented by Dr. George Bennett, 1874.

48424. A worn tooth, which is apparently a first lower true molar of this species; from Queensland. Labelled by Sir R. Owen *Palorchestes major*.

Presented by Dr. George Bennett, 1878.

46914. The right humerus; from the Pleistocene of the Castlereagh (*Fig.*) River, Kirban, near Mendoran, New South Wales. Figured by Owen in the 'Extinct Mammals of Australia,' pl. cxxvii. (reversed). In general structure this bone closely resembles the humerus of *Phascolomys*, but the deltoid ridge is double, and the entocondyle more flattened.

Presented by W. L. R. Gipps, Esq., 1875.

47828. The proximal extremity of the left ulna; from Australia. This specimen evidently belongs to the same species as the preceding; it is almost indistinguishable from the ulna of *Phascolomys*, but has the inner articular facette flat, instead of slightly cupped; it differs widely from the corresponding bone of *Diprotodon*, in which the latter surface is very deeply cupped and the olecranal process absent.

Presented by Dr. George Bennett, 1877.

48419 a. The right radius; from Queensland. This bone differs from the radius of *Diprotodon* and *Phalangista* by the transverse elongation of the proximal articular cup, and thereby agrees with *Phascolomys*, although presenting certain minor points of distinction.

Presented by Dr. George Bennett, 1877.

48418. The distal portion of the right femur; from Queensland.
(*Fig.*) This specimen agrees in structure with the femur of *Phascolomys* and in relative size with the humerus, No. 46914; it is figured by Owen in the 'Quart. Journ. Geol. Soc.' vol. xxxviii. pl. xvi.

Presented by Dr. George Bennett, 1877.

39983. The left tibia, imperfect proximally; from Darling Downs. This specimen differs from the corresponding bone of *Diprotodon* by the more prominent inner malleolus, by the less flattened distal surface, and the greater development of the muscular ridges on the shaft; in all of which respects it approximates to *Phascolomys*.

Presented by Sir D. Cooper, Bart., 1866.

38787. The right calcaneum; from Darling Downs. This bone, which agrees in relative size with the preceding specimens, comes very close to the calcaneum of *Phascolomys*, but makes some approximation to that of *Diprotodon*.

Presented by F. N. Isaac, Esq., 1861.

M. 3640. The atlas vertebra; from Australia. This bone agrees
(*Fig.*) closely with the atlas of *Phascolomys*, but is relatively less expanded laterally than in *Diprotodon*. Figured by Owen in the 'Extinct Mammals of Australia,' pl. cxxvi. The anterior artery passed through a foramen.

Presented by Dr. George Bennett.

39986. The right half of the atlas vertebra; from Darling Downs.
(*Fig.*) Figured by Owen in the 'Phil. Trans.' 1870, pl. xliii. fig. 2. *Presented by Sir D. Cooper, Bart.*, 1886.

44007. An imperfect sacral vertebra of a small individual; from Australia. This bone agrees very closely with the sacral vertebræ of *Phascolomys*.

Presented by the Trustees of the Adelaide Museum of Natural History, 1871.

46067. An anterior sacral vertebra of a large individual; from Australia. *Presented by Dr. George Bennett*, 1874.

38618. An imperfect and water-worn sacral vertebra; from Darling Downs. *Presented by Sir D. Cooper, Bart.*, 1864.

The following probably belong either to the present or the preceding family.

46841 l. Three imperfect metacarpals; from Queensland. The distal articular surfaces are more rounded and less laterally expanded than in existing species of *Phascolomys*.
Presented by Dr. George Bennett, 1875.

46841 m. Numerous phalangeals of the manus, associated with the preceding. The terminal phalangeals are compressed laterally, instead of vertically as in *Phascolomys*, while the distal articular trochleæ of the second phalangeals are much more developed than in that genus.
Same history.

M. 3641. Several metapodials (? metacarpals), agreeing in structure with the preceding specimens; from Queensland.
Same history.

M. 3642. Four terminal phalangeals, agreeing in structure with the corresponding bones of No. 46841 *m*; from Queensland.
Presented by Dr. George Bennett, 1874.

The following may perhaps belong either to the present or the next family.

42705. Cast of an undetermined imperfect bone. The original was probably obtained from Queensland. *No history.*

Family DIPROTODONTIDÆ.

Dentition:—I. $\frac{3}{1}$, C. $\frac{0}{0}$, Pm. $\frac{1}{1}$, M. $\frac{4}{4}$. The general structure of the cheek-teeth agrees with that of the *Nototheriidæ*, but there is no trace of a longitudinal bridge in the lower true molars. The upper incisors decrease in size from the first to the third, and those of opposite sides are in contact in the median line of the palate; the first pair are scalpriform, coated with enamel only on the anterior surface, and grow from persistent pulps. The lower incisors are very large and proclivous. The cranium (fig. 27) is elongated, and has no vacuities on the palate. The general structure of the mandible resembles that of *Nototherium*, but the inferior border of the ramus is much less elongated. The limbs are of approximately equal length; the scapula is long and falciform; the humerus elongated and slender, with a slight distal expansion, and no entepicondylar foramen; and the ulna has no distinct olecranal process. The proximal bones of the limbs were apparently placed nearly vertically, and appear to have been adapted solely for walking. The feet are only imperfectly known; but it is probable they were plantigrade and pentadactylate, that the covering of the terminal phalangeals approximated more to the nature of hoofs than claws, and that the phalangeals of the fifth digits were aborted.

The family apparently connects the *Nototheriidæ* with the *Phalangistidæ*. The absence of a pit and perforation in the masseteric fossa of the mandible is a character common to the three families; the limb-bones (especially in the fore limb) are nearest to those of the *Phalangistidæ* (the absence of an entepicondylar foramen to the humerus being a peculiar feature); while the first upper incisors approximate to those of the *Phascolomyidæ*, more especially to those referred to *Phascolonus*. The calcaneum and astragalus also approximate to the corresponding bones of the same two families, and are totally unlike those of the *Macropodidæ*.

Genus **DIPROTODON**, Owen[1].

Since this is the only known genus, its characters are the same as those of the family.

Diprotodon australis, Owen[2].

This is the type and only species. The largest individuals appear of larger bulk than *Rhinoceros unicornis*; but others were considerably smaller, although not presenting any structural differences which can be regarded as of specific value.

Hab. Australia.

32851. (*Fig.*) The nearly entire cranium; said to have been obtained from the Pleistocene at Gowrie, near Drayton, Darling Downs, Queensland. This specimen (woodcut, fig. 27) is figured by Owen in the 'Phil. Trans.' 1870, pl. xxxv. figs. 1–4, and also in the 'Extinct Mammals of Australia,' frontispiece and pl. xix. figs. 1–4. *Purchased.*

32869. Fragment of the occipital region, showing the foramen magnum and condyles; from Australia. *Purchased.*

50080. The palatal region of the cranium, showing all the teeth

[1] In Mitchell's 'Three Expeditions into Eastern Australia,' 2nd ed. vol. ii. p. 362 (1838).
[2] *Loc. cit.*

except the right i. 2 and i. 3 of both sides; from Gowrie. The cheek-teeth are in an early stage of wear.

Presented by G. F. Bennett, Esq., 1879.

Fig. 27.

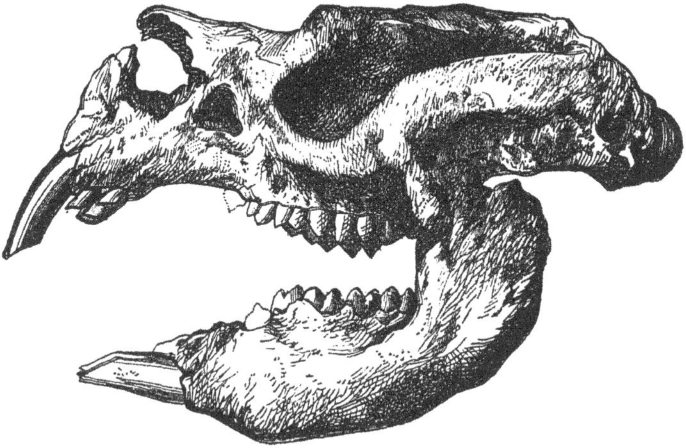

Diprotodon australis.—The skull; from the Pleistocene of Australia. $\frac{1}{10}$.

M. 1. The palatal region of the cranium, showing the four true molars in a much-worn condition, the right i. 2 and i. 1 of both sides; from the Pleistocene of Queensland.

Presented by Sir G. Verdon, 1877.

47821. Fragment of the premaxillæ, showing the three pairs of incisors; from the Pleistocene of Australia.

Presented by Dr. George Bennett, 1875.

38601 x. Part of the premaxillæ, transversely bisected, and showing sections of i. 1 and of the alveoli of i. 2 and i. 3; from Eton Vale, Darling Downs. Figured by Owen in the 'Phil. Trans.' 1870, pl. xxxvi. fig. 5, and in the 'Extinct Mammals of Australia,' pl. xx. fig. 5.

Presented by Sir D. Cooper, Bart., 1869.

46008. Part of the left premaxilla, containing the three incisors, together with a fragment of the maxilla of the same side with the last three true molars; from Queensland. These specimens apparently belong to the same individual as the mandible No. 46008 *a* (p. 175).

Presented by the Trustees of the Adelaide Museum of Natural History, 1871.

46059. The extremity of the premaxillæ, showing all the incisors (mostly broken) except the right i. 3; from Queensland.
Presented by Dr. George Bennett, 1874.

36281. Fragment of the diastemal region of the palate, showing the alveolus of pm. 4 on both sides; from the Pleistocene near Melbourne, Victoria.
Presented by F. M. Raynal, Esq., 1862.

48065. The hinder portion of the palate, showing the cheek-teeth of either side; from Queensland. The fourth true molar is not fully protruded, and only the summit of its anterior ridge is abraded. *Presented by Dr. George Bennett*, 1877.

32848. Part of the right maxilla, containing the four true molars;
(*Fig.*) from Queensland. Figured by Owen in the 'Phil. Trans.' 1870, pl. xxxviii. figs. 1, 2, and in the 'Extinct Mammals of Australia,' pl. xxi. figs. 1, 2. The last molar is but slightly worn. *Purchased*, 1857

39273. Part of the right maxilla, showing the four true molars;
(*Fig.*) from the Pleistocene of the Condamine River, Queensland. Figured by Owen in the 'Phil. Trans.' 1870, pl. xxxvii. figs. 1, 2, and in the 'Extinct Mammals of Australia,' pl. xxii. figs. 1, 2; pm. 4 appears to have been shed.
Presented by Edward Hill, Esq., 1865.

47822. Fragment of the right maxilla of a small (? female) individual; from Australia.
Presented by Dr. George Bennett, 1875.

35913. Fragment of the right maxilla, containing m. 3 and m. 4; from Gowrie. *Presented by Sir D. Cooper, Bart.*, 1861.

47823. Fragment of the right maxilla, containing m. 3 and m. 4, which are rather more worn than in the preceding specimen; from Australia.
Presented by Dr. George Bennett, 1875.

43955. Fragment of the left maxilla, containing the well-worn m. 3 and m. 4; from Australia.
Presented by Dr. George Bennett, 1871.

48415. Fragment of the left maxilla, containing m. 2, m. 3, and the first ridge of m. 4; from Australia.
Presented by Dr. George Bennett, 1877.

38609. Part of the right maxilla of a small (? female) individual, showing the four true molars in a half-worn condition; from Eton Vale. *Presented by Sir D. Cooper, Bart.*, 1864.

39274. (*Fig.*) Part of the left maxilla of a specimen agreeing in size with the preceding; from the Condamine River. This specimen, which contains the last three true molars, has been vertically and longitudinally bisected; and is figured by Owen in the 'Phil. Trans.' 1870, pl. xlii. fig. 1, and also in the 'Extinct Mammals of Australia,' pl. xxvii. fig. 1.
Presented by Edward Hill, Esq., 1865.

32845. (*Fig.*) Fragment of the left maxilla, containing the broken m. 3 and m. 4 in an early stage of wear; from Queensland. Figured by Owen in the 'Phil. Trans.' 1870, pl. xxxix. fig. 3, and also in the 'Extinct Mammals of Australia,' pl. xxiii. fig. 3. *Purchased.*

44125. Fragment of the left maxilla, containing the broken m. 2 and m. 3; from Australia.
Presented by Dr. George Bennett, 1873.

43955 a. Fragment of the right maxilla, containing the much-worn m. 4 and a fragment of m. 3; from Australia.
Presented by Dr. George Bennett, 1871.

M. 3639. Fragment of the left maxilla, containing the well-worn m. 3; from Australia.
Presented by Sir R. Owen, K.C.B.

45896. Cast of a fragment of the left maxilla, containing m. 4. The original is from Australia, and is probably in the Museum at Sydney. *Presented by Dr. George Bennett*, 1874.

32849. Fragment of the left maxilla, containing the much-worn m. 3; from Queensland. *Purchased.*

40374. Small fragment of the left maxilla, showing the well-worn m. 3; from St. Ruth's Station on the Condamine River.
Presented by Dr. Francis Campbell, 1867.

40378. Fragment of a left maxilla, probably belonging to the present species, and showing the alveoli of pm. 4 and m. 1, and a fragment of m. 2, of which the ridges have been totally worn away; from the same locality. *Same history.*

47824. Fragment of the right maxilla, showing the partially-worn m. 4; from Australia.
Presented by Dr. George Bennett, 1875.

39969. The well-worn fourth right upper true molar; from Darling
(*Fig.*) Downs. Figured by Owen in the 'Phil. Trans.' 1870, pl. xxxvii. figs. 4–6, and pl. xxxix. fig. 2, as well as in the 'Extinct Mammals of Australia,' pl. xxiii. fig. 2.
Presented by Sir D. Cooper, Bart., 1866.

39985. The slightly-worn fourth right upper true molar of a small
(*Fig.*) (? female) individual; from Darling Downs. Figured by Owen in the 'Phil. Trans.' 1870, pl. xxxviii. figs. 3, 4, and pl. xxxix. fig. 1; and also in the 'Extinct Mammals of Australia,' pl. xxii. figs. 3, 4, and pl. xxiii. fig. 1.
Presented by Sir D. Cooper, Bart., 1866.

M. 1 a. The germ of the fourth right upper true molar; from Queensland. *Presented by Sir G. Verdon*, 1882.

47824 a. An upper premolar from Queensland.
Presented by Dr. George Bennett, 1875.

M. 479. The left upper premolar, in an early condition of wear; from Queensland. *Presented by Dr. George Bennett.*

46074. The first left upper incisor, longitudinally and transversely bisected; from Australia.
Presented by Sir R. Owen, K.C.B., 1878.

36209. Two specimens of the extremity of the first left upper incisor;
(*Fig.*) from Australia. Figured by Owen in the 'Phil. Trans.' 1870, pl. xxxvi. figs. 1–4, and also in the 'Extinct Mammals of Australia,' pl. xx. figs. 1–4.
Presented by Sir D. Cooper, Bart., 1861.

35914–5. Two imperfect specimens of the first upper incisor; from
(*Fig.*) Gowrie. Figured by Owen in the 'Phil. Trans.' 1870, pl. xxxvi. figs. 1, 2, 6.
Presented by Sir D. Cooper, Bart., 1861.

36212. Cast of the first right upper incisor; the original was obtained from Queensland.
Presented by Sir D. Cooper, Bart., 1861.

43369. The first right upper incisor; from Clifton, Queensland.
Presented by Dr. George Bennett, 1872.

49656. Two specimens of the exposed portion of the first upper incisor; from the Pleistocene of Wirialpa, South Australia.
Presented by Rev. W. B. Clarke, 1878.

48067. The imperfect first left upper incisor of a smaller individual: from Queensland. *Presented by Dr. George Bennett*, 1874.

44088. The imperfect first left upper incisor of a still smaller individual; from Australia.
Presented by the Trustees of the Adelaide Museum of Natural History, 1871.

48067 a. The imperfect first left upper incisor of a young individual; from Queensland. *Presented by Dr. George Bennett*, 1874.

32911. (*Fig.*) A well-worn second upper incisor; from Australia. Figured by Owen in the 'Phil. Trans.' 1870, pl. xxxvi. fig. 7, and pl. xxxix. figs. 7, 8, as well as in the 'Extinct Mammals of Australia,' pl. xxiii. figs. 7, 8. *Purchased.*

46060. The left second upper incisor, in a less worn condition; from Australia. *Presented by Dr. George Bennett*, 1874.

50081. Four specimens of the second and third upper incisors; from Queensland. *Presented by G. F. Bennett, Esq.*, 1872.

32851. (*Fig.*) The symphysis and left ramus of the mandible; from Australia. This specimen (woodcut, fig. 27, p. 171) is figured by Owen in the 'Phil. Trans.' 1870, pl. xxxv. figs. 1–4, and also in the 'Extinct Mammals of Australia,' pl. xix. figs. 1–4. *Purchased.*

47855. The nearly complete mandible; from Darling Downs.
Presented by Dr. George Bennett, 1875.

M. 3645. The entire mandible; from Australia. The incisors are broken and $\overline{\text{pm. 4}}$ has been shed; the true molars are in a middle stage of wear. *Same history.*

46008 a. The anterior portion of the mandible; showing the greater part of the dentition; from Queensland. This specimen was apparently associated with the upper jaw No. 46008 (p. 171). *Presented by the Trustees of the Adelaide Museum of Natural History*, 1871.

39968. (*Fig.*) The greater part of the left ramus of the mandible, showing all the teeth except $\overline{\text{pm. 4}}$; from Darling Downs. Figured by Owen in the 'Phil. Trans.' 1870, pl. xl. figs. 4, 8, 11, and pl. xlii. fig. 2, as well as in the 'Extinct Mammals of Australia,' pl. xxiv. figs. 4, 8, 11, and pl. xxvii. fig. 2.
Presented by Sir D. Cooper, Bart., 1866.

43210. The hinder part of the right ramus of the mandible of a very old individual, showing the much-worn $\overline{\text{m. 4}}$; from Darling Downs. *Presented by Sir R. Daintree*, 1871.

42691. Cast of a fragment of the left ramus of the mandible of a small immature individual, showing $\overline{\text{m. 3}}$ and the alveolus of $\overline{\text{m. 4}}$. The original is from Australia.
Presented by the Trustees of the Australian Museum, 1871.

43088 d. Cast of a fragment of the right ramus of the mandible, showing $\overline{\text{m. 3}}$ and $\overline{\text{m. 4}}$. The original is from Australia.
Presented by the Trustees of the Australian Museum, 1871.

43088 e. Cast of a fragment of the right ramus of the mandible, showing the broken bases of $\overline{\text{m. 3}}$ and $\overline{\text{m. 4}}$. The original is from Australia.
Presented by the Trustees of the Australian Museum, 1871.

46056. The imperfect right ramus of the mandible; from the Pleistocene of Mundoman, New South Wales.
Presented by W. L. R. Gipps, Esq., 1874.

38745. Small fragment of the left ramus of the mandible, containing $\overline{\text{m. 4}}$ and the hinder ridge of $\overline{\text{m. 3}}$, in a very much-worn condition; from Darling Downs.
Presented by F. N. Isaac, Esq., 1861.

43954. Fragment of the right ramus of the mandible, containing the last three true molars in a middle condition of wear; from Queensland. *Presented by Dr. George Bennett*, 1871.

45874. The imperfect left half of the mandibular symphysis, containing a fragment of the incisor; from Queensland.
Presented by Dr. George Bennett, 1874.

48066. Fragment of the left ramus of the mandible of a small (? female) individual, showing the greater part of the incisor, and the first three true molars, which are in an intermediate stage of wear; from Queensland.
Presented by Dr. George Bennett, 1877.

43088 c. Cast of part of the left ramus of the mandible of an individual agreeing in size with the preceding, and showing the last three true molars and a portion of the canine. The original is from Australia.
Presented by the Trustees of the Australian Museum, 1871.

38744. Fragment of the right ramus of the mandible; showing
(*Fig.*) $\overline{\text{m. 3}}$ and $\overline{\text{m. 4}}$; from Darling Downs. Figured in Owen's 'Extinct Mammals of Australia,' pl. xxv.
Presented by F. N. Isaac, Esq., 1861.

39984. Part of the left ramus of the mandible of an immature (*Fig.*) individual; containing the first three true molars and the incisor, the latter having been longitudinally bisected; from Darling Downs. Figured by Owen in the 'Phil. Trans.' 1870, pl. xl. figs. 3, 7, 10, pl. xli. figs. 1, 2, and pl. xlii. fig. 5; as well as in the 'Extinct Mammals of Australia,' pl. xxiv. figs. 3, 7, 10, pl. xxvi. figs. 1, 2, and pl. xxvii. fig. 5.

Presented by Sir D. Cooper, Bart., 1866.

35912. Fragment of the left ramus of the mandible of a very old (*Fig.*) male, showing $\overline{\text{m. 3}}$ and $\overline{\text{m. 4}}$ in a very much-worn condition; from Darling Downs. Figured by Owen in the 'Phil. Trans.' 1870, pl. xl. figs. 17, 18, and in the 'Extinct Mammals of Australia,' pl. xxiv. figs. 17, 18.

Presented by Sir D. Cooper, Bart., 1861.

M. 475. Fragment of the right ramus of the mandible, containing $\overline{\text{m. 2}}$ and $\overline{\text{m. 3}}$; from Queensland.

Presented by Dr. George Bennett.

M. 1528. Cast of the symphysial extremity of the mandible, showing the bases of the incisors. The original was obtained from the Pleistocene of Gelgoine Station, Cannon-bar, and is preserved in the Australian Museum, Sydney.

Presented by the Trustees of the Australian Museum, 1883.

M. 1529. Cast of the symphysial extremity of a rather smaller mandible. The original was obtained from Gelgoine Station. *Same history.*

50080 a. The imperfect mandibular symphysis, showing the incisive alveoli; from Queensland.

Presented by G. F. Bennett, Esq., 1879.

46841 g. Part of the right ramus of the mandible of an immature (*Fig.*) individual, showing the first three cheek-teeth and the base of the incisor; from Darling Downs. Figured by Owen in the 'Extinct Mammals of Australia,' pl. cxxiv.

Presented by Dr. George Bennett, 1875.

43965. Part of the right ramus of the mandible of a smaller but slightly older individual, containing $\overline{\text{m. 1}}$ and $\overline{\text{m. 2}}$, and the imperfect incisor; from Clifton, Darling Downs.

Presented by F. Nicholson, Esq., 1872.

M. 474. Part of the left ramus of the mandible of a young individual, showing the first four cheek-teeth in an early stage of wear, and the incisor; from Queensland.
Presented by Dr. George Bennett.

46880. (*Fig.*) A condyle of the mandible; from Queensland. Figured by Owen in the 'Phil. Trans.' 1870, pl. xlii. figs. 3, 4, and also in the 'Extinct Mammals of Australia,' pl. xxvii. figs. 3, 4. *Presented by Dr. George Bennett,* 1875.

47844. An imperfect mandibular condyle; from Queensland.
Presented by Dr. George Bennett, 1875.

39970. (*Fig.*) The exposed portion of a lower incisor, of which the base has been transversely cut; from Gowrie. The section is figured by Owen in the 'Phil. Trans.' 1870, pl. xxxix. fig. 6. *Presented by Sir D. Cooper, Bart.,* 1866.

47843. The anterior portion of the right lower incisor; from Darling Downs. *Presented by Dr. George Bennett,* 1875.

M. 3646. Two specimens of lower incisors of small individuals; from Queensland. *Same history.*

43579. The anterior portion of a lower incisor of a small individual; from Darling Downs.
Presented by Dr. George Bennett, 1872.

43347. The anterior portion of the left lower incisor of an immature individual; from Gowrie.
Presented by Dr. George Bennett, 1872.

47827. The middle portion of a small lower incisor; from Darling Downs. *Presented by Dr. George Bennett,* 1875.

M. 3647. (*Fig.*) Fragment of a mandible with the incisor in its alveolus, which has been longitudinally bisected; from Australia. This is part of the specimen figured in the 'Phil. Trans.' 1870, pl. xl. *No history.*

47827 a. Two fragments of lower incisors; from Queensland.
Presented by Dr. George Bennett, 1875.

32861. (*Fig.*) The right scapula; from Australia. Figured by Owen in the 'Phil. Trans.' 1870, pl. xlv. figs. 1–3, and in the 'Extinct Mammals of Australia,' pl. xxx. In the narrow blade, long coracoidal, and expanded acromial process, this bone agrees very closely with the scapula of the *Phalangistidæ*, and differs widely from that of the *Macropodidæ*.
Purchased, 1854.

32860. The left scapula of a somewhat larger individual; from Australia. *Purchased*, 1854.

45876. The imperfect glenoidal two thirds of the right scapula; from Australia. *Presented by Dr. George Bennett*, 1874.

39971. The glenoidal region of the right scapula; from Gowrie. *Presented by Sir D. Cooper, Bart.*, 1866.

39275. Portion of the glenoidal and spinal region of the left scapula; from the Condamine River.
Presented by E. Hill, Esq., 1865.

43964. Fragment of the glenoidal region of the left scapula; from King's Creek, Clifton, Queensland.
Presented by Dr. George Bennett, 1872.

35916. The glenoidal cavity of a scapula; from Darling Downs. *Presented by Sir D. Cooper, Bart.*, 1861.

45877. A clavicle; from Queensland.
Presented by Dr. George Bennett, 1874.

32866. The right humerus; from Australia. Figured by Owen in
(*Fig.*) the 'Phil. Trans.' 1870, pl. xlvi. figs. 1–5, and in the 'Extinct Mammals of Australia,' pl. xxxi. figs. 1–5. Agrees in general shape with the humerus of the *Phalangistidæ*, but differs by the absence of the entepicondylar foramen. *Purchased*, 1857.

32868. The distal extremity of the left humerus; from Australia. *Purchased*, 1857.

44005. Proximal extremity of the left humerus; from Australia. This bone corresponds very closely with the humerus of *Phalangista*. *Presented by the Trustees of the Adelaide Museum of Natural History*, 1871.

M. 3504. The head of the humerus of a very large individual; from Monaba, New South Wales. *Purchased*, 1886.

47857. The left ulna, wanting the distal extremity; from Queensland. There is no distinct olecranon, and the articular facette for the entocondyle of the humerus is deeply cupped and looks nearly vertically. The ulna of the *Phalangistidæ* comes near to this type, but there is a distinct olecranon, and the entocondylar cup is directed anteriorly; the limb of the fossil was evidently less flexed during progression.
Presented by Dr. George Bennett, 1875.

42702. Cast of the proximal extremity of the left ulna. The original is from Australia.
Presented by the Trustees of the Australian Museum, 1871.

35929. The proximal half of the left ulna of a small individual; from Queensland.
Presented by Sir D. Cooper, Bart., 1861.

43088 j. Cast of the imperfect proximal extremity of the right ulna of a very large individual. The original is from Australia.
Presented by the Trustees of the Australian Museum, 1871.

47856. The proximal extremity of the right ulna; from Queensland.
Presented by Dr. George Bennett, 1875.

45897. Cast of the right ulna. The original is from Australia.
Presented by Dr. George Bennett, 1874.

35924. The imperfect proximal half of the right ulna of a small individual; from Gowrie.
Presented by Sir D. Cooper, Bart., 1861.

44207. The proximal extremity of the right ulna of a medium-sized individual; from Queensland.
Presented by Dr. George Bennett, 1872.

40036. The proximal extremity of the left ulna of a very small individual; from Gowrie.
Presented by Sir D. Cooper, Bart., 1866.

48419. The left ulna of a small immature individual, wanting part of the distal epiphysis, and with both the epiphyses still ununited to the shaft; from Queensland. This bone differs from the ulna of *Nototherium* by the absence of a distinct olecranal process.
Presented by Dr. George Bennett, 1877.

42703. Cast of the distal portion of the ulna of a very large individual. The original was obtained from Queensland, and is preserved in the Australian Museum, Sydney.
Presented by the Trustees of the Australian Museum, 1871.

48417. The distal half of a somewhat smaller ulna; from Queensland. *Presented by Dr. George Bennett*, 1877.

47858. The left radius of a full-sized individual; from Queensland. In its circular proximal extremity this bone agrees with the radius of *Phalangista*, and differs from that of *Nototherium* and *Phascolomys*, in which that surface is oval.
Presented by Dr. George Bennett, 1875.

46057. The left radius; from the Pleistocene of Mundaman, New South Wales. *Presented by W. L. R. Gipps, Esq.*, 1874.

47859. The distal half of the right radius; from Queensland. *Presented by Dr. George Bennett*, 1875.

45878. The shaft of the right radius; from Australia. *Presented by Dr. George Bennett*, 1874.

46841 o. The left scaphoid; from Queensland. Figured by Owen
(*Fig.*) in the 'Extinct Mammals of Australia,' pl. cxxii. figs. 1–3 (as the scapho-lunar[1]). This bone agrees very closely with the corresponding element in *Phascolomys* and *Phalangista*. *Presented by Dr. George Bennett*, 1875.

46841 h. The left scaphoid; from Queensland. *Presented by Dr. George Bennett*, 1875.

43376. The slightly imperfect right scaphoid; from King's Creek, Clifton. *Presented by G. F. Bennett, Esq.*, 1872.

46841 c. The outer half of the right scaphoid; from Queensland. *Presented by Dr. George Bennett*, 1875.

46841 j. The left cuneiform; from Queensland. *Presented by Dr. George Bennett*, 1875.

46062. The left magnum; from Queensland. *Presented by Dr. George Bennett*, 1874.

46062 a. The right magnum; from Australia. *Presented by Dr. George Bennett*, 1874.

46841 e. The left fifth (?) metacarpal; from Queensland. Figured
(*Fig.*) by Owen in the 'Extinct Mammals of Australia,' pl. cxxii. figs. 7–10. It is compared by Owen to *Macropus*, but its resemblance appears much closer to the corresponding bone of *Phascolomys*. *Presented by Dr. George Bennett*, 1875.

46841 i. A similar specimen, imperfect; from Queensland. *Presented by Dr. George Bennett*, 1875.

46841 f. The terminal phalangeal of (? the second digit of) the left
(*Fig.*) manus; from Queensland. Figured by Owen in the 'Extinct Mammals of Australia,' pl. cxxiii. figs. 1–5. *Presented by Dr. George Bennett*, 1875.

M. 3453 a. A larger terminal phalangeal; from Queensland. *Same history.*

[1] See Flower 'Osteology of Mammalia,' 1st ed. pp. 277–278 (1870).

M. 16. The proximal half of a similar bone; from Australia.
Presented by Sir G. Verdon, 1877.

38604. The imperfect pelvis and sacrum; from Darling Downs.
(*Fig.*) Figured by Owen in the 'Phil. Trans.' 1870, pl. xlvii., and also in the 'Extinct Mammals of Australia,' pl. xxxii., The pelvis is intermediate between that of *Phascolomys* and *Phalangista*, the lateral anterior deflection of the ilia being even greater than in the former; it is totally unlike the pelvis of *Macropus*. The specimen belongs to a small individual. *Presented by Sir D. Cooper, Bart.*, 1864.

40373. Portion of the pelvis and sacrum of a somewhat larger individual; from St. Ruth's Station, Condamine River.
Presented by Dr. F. Campbell, 1867.

44195. The left acetabular region of a very large individual; from King's Creek, Clifton, Queensland.
Presented by Dr. George Bennett, 1873.

45879. Small fragment of the left acetabular region, belonging to an individual of the same size as the last specimen; from Australia. *Presented by Dr. George Bennett*, 1874.

M. 11. Part of a left innominate agreeing precisely in structure with the preceding, but of much smaller size; from Queensland. *Presented by Sir G. Verdon*, 1877.

32864. The right femur; from Darling Downs. Figured by Owen
(*Fig.*) in the 'Phil. Trans.' 1870, pls. xlviii. and xlix., and also in the 'Extinct Mammals of Australia,' pl. xxxiii. and pl. xxxiv. fig. 1. This bone comes nearest to the femur of *Phascolomys* and *Phalangista*, but differs in the excessive prominence of the inner border of the trochlea for the patella, and in the head being the highest point of the whole bone. The femur of *Macropus* differs widely in the form of the distal extremity, as well as in the less perfectly globular head. *Purchased.*

M. 3649. The imperfect right femur; from Australia.
Presented by Surgeon B. Innis, R.N., 1873.

32864*. The imperfect left femur, transversely bisected; from
(*Fig.*) Australia. Figured by Owen in the 'Phil. Trans.' 1870, pl. xlviii. fig. 3, and in the 'Extinct Mammals of Australia,' pl. xxxiii. fig. 3. *Purchased*, 1857.

M. 3650. The left femur, wanting the head; from Australia. *No history.*

44001. The imperfect left femur; from Australia. *Presented by the Trustees of the Adelaide Museum of Natural History*, 1871.

44192. The left femur of an immature individual; from King's Creek, Clifton. *Presented by Dr. George Bennett*, 1873.

45872. The imperfect left femur of an immature individual; from Australia. *Presented by Dr. George Bennett*, 1874.

44002. The distal extremity of the right femur of a very large individual; from Australia. *Presented by the Trustees of the Adelaide Museum of Natural History*, 1871.

42699. Cast of the distal extremity of the left femur of an individual nearly equal in size to the preceding. *No history.*

38605. (*Fig.*) The right tibia; from Darling Downs. Figured by Owen in the 'Phil. Trans.' 1870, pl. xlix. figs. 2–5, and also in the 'Extinct Mammals of Australia,' pl. xxxiv. figs. 2–5. This bone comes nearest to the femur of *Phascolomys* and *Phalangista*; agreeing with the former in its shortness and stoutness, but differing in the deep concavity of the facette for the inner condyle of the femur, and in the absence of a prominent internal malleolus; in *Phalangista* there is a slight cupping of the inner femoral facette, and the internal malleolus is less prominent than in *Phascolomys*. *Presented by Sir D. Cooper, Bart.*, 1864.

M. 13. The right tibia; from Queensland. *Presented by Sir G. Verdon*, 1877.

44003. The proximal extremity of the right tibia; from Australia. *Presented by the Trustees of the Adelaide Museum of Natural History*, 1871.

42700. Cast of the shaft of the right tibia. *Same history.*

39974. The proximal extremity of the left tibia; from Gowrie. *Presented by Sir D. Cooper, Bart.*, 1866.

38748 a. The distal extremity of the right tibia; from Darling Downs. *Presented by F. N. Isaac, Esq.*, 1861.

43217. The distal extremity of the right tibia; from Australia. *Presented by Sir R. Daintree*, 1871.

38748. The distal portion of the right tibia of an immature individual; from Darling Downs.

Presented by F. N. Isaac, Esq., 1861.

39977. The proximal half of the tibia or fibula of an immature individual; from Gowrie. This bone is almost indistinguishable from the fibula of *Phalangista*, and is widely different from that of *Macropus*.

Presented by Sir D. Cooper, Bart., 1866.

39975. The distal portion of a similar bone; from Australia.

Presented by Sir D. Cooper, Bart., 1866.

M. 14. The distal half of another similar bone; from Australia.

Presented by Sir G. Verdon, 1877.

47860. The left calcaneum; from Queensland. In its general plan of structure this bone agrees with the calcaneum of *Phascolomys* and *Phalangista*, and differs widely from that of *Macropus*; the facette for the astragalus is, however, more concave, and that for the unciform relatively smaller.

Presented by Dr. George Bennett, 1876.

36215. Cast of the articular half of the right calcaneum of a larger individual. The original is from Australia, and is preserved in the Adelaide Museum.

Presented by the Trustees of the Adelaide Museum of Natural History, 1871.

47841. The middle portion of the left calcaneum of a young individual; from Queensland.

Presented by Dr. George Bennett, 1876.

M. 15. The left astragalus; from Australia. This bone comes nearest to the astragalus of *Phalangista*, and is of a totally different type from that of *Macropus*.

Presented by Sir G. Verdon, 1877.

M. 3648. The imperfect left astragalus; from Australia.

No history.

50083. The left astragalus of a small individual; from Queensland.

Presented by G. F. Bennett, Esq., 1879.

42711. Cast of the right astragalus of a medium-sized individual. The original is from Australia, and is preserved in the Adelaide Museum.

Presented by the Trustees of the Adelaide Museum of Natural History, 1871.

46841 k. The right astragalus, agreeing in size with the preceding; from Queensland.
Presented by Dr. George Bennett, 1875.

42708. Cast of the right astragalus. The original is from Australia, and is preserved in the Adelaide Museum.
Presented by the Trustees of the Adelaide Museum of Natural History, 1871.

39987. The right astragalus of a small individual; from Darling Downs. *Presented by Sir D. Cooper, Bart.*, 1866.

47842. The right astragalus of a small individual; from Darling Downs. *Presented by Dr. George Bennett*, 1875.

42709. Cast of a navicular. The original is from Australia, and is preserved in the Adelaide Museum.
Presented by the Trustees of the Adelaide Museum of Natural History, 1871.

46837. The fifth left metatarsal; from Queensland. This bone agrees in general characters with the corresponding metatarsal of *Phascolomys* and *Phalangista*, but in its small distal condyle shows indications that the phalangeals (as in the corresponding digit of the manus) were aborted.
Presented by Dr. George Bennett, 1875.

48416. The slightly imperfect atlas vertebra; from Queensland. In the great relative size of the vertical diameter this bone comes nearer to *Phalangista* than to *Phascolomys* or *Macropus*; it differs from the latter in the greater thickness of the transverse processes, and from that of *Phascolomys* and *Nototherium* in the near approximation to the middle line of the two portions of the inferior arch; in *Phalangista* these two portions unite.
Presented by Dr. George Bennett, 1874.

48416 a. The slightly imperfect atlas vertebra; from King's Creek, Clifton. *Presented by Dr. George Bennett*, 1873.

M. 3. The complete atlas vertebra of a very large individual; from Australia. *Presented by Sir G. Verdon*, 1877.

32870. The axis vertebra; from Australia. Figured by Owen in (*Fig.*) the 'Phil. Trans.' 1870, pl. xliii. fig. 1, and pl. xliv. figs. 1–3, as well as in the 'Extinct Mammals of Australia, pl. xxviii. fig. 1, and pl. xxix. figs. 1–3. The relative proportions of this bone agree very closely with those of the axis of *Phalangista*, in which the relative length of the centrum is intermediate between that of *Phascolomys* on the one hand and *Macropus* on the other. *Purchased.*

46061. The centrum of the axis vertebra; from Queensland. The posterior epiphysis, which is absent in the preceding specimen, is present.
Presented by Dr. George Bennett, 1874.

M. 4. The imperfect centrum of the axis vertebra; from Australia.
Presented by Sir G. Verdon, 1877.

32852–4. (*Fig.*) The associated third, fourth, and fifth cervical vertebræ; from Queensland. The third vertebra is figured by Owen in the 'Phil. Trans.' 1870, pl. xliv. fig. 4, and also in the 'Extinct Mammals of Australia,' pl. xxix. fig. 4.
Purchased.

M. 6. The third cervical vertebra; from Australia.
Presented by Sir G. Verdon, 1877.

45883. The fourth cervical vertebra; from Queensland.
Presented by Dr. George Bennett, 1874.

45884. The centrum of the fifth (?) cervical vertebra; from Queensland. *Presented by Dr. George Bennett*, 1874.

M. 5. The imperfect centrum of the sixth cervical vertebra; from Australia. *Presented by Sir G. Verdon*, 1877.

32856. (*Fig.*) An early dorsal vertebra; from Australia. Figured by Owen in the 'Phil. Trans.' 1870, pl. xliv. figs. 7, 8, and also in the 'Extinct Mammals of Australia,' pl. xxix. figs. 7, 8. *Purchased*, 1857.

32857. (*Fig.*) An early dorsal vertebra; from Australia. Figured in pl. xliv. figs. 5, 6, and pl. xxix. figs. 5, 6 of the above-cited works. *Purchased.*

44196. An imperfect late dorsal vertebra; from King's Creek, Clifton. *Presented by Dr. George Bennett*, 1873.

43964. An imperfect late dorsal vertebra; from King's Creek, Clifton. *Presented by Dr. George Bennett*, 1873.

43214. An imperfect late dorsal vertebra; from Darling Downs.
Presented by Sir R. Daintree, 1871.

32858. An imperfect lumbar vertebra; from Australia. Figured
(*Fig.*) in pl. xliv. figs. 9–11 and pl. xxix. figs. 9–11 of the works cited. This and the following specimens are characterized by the extreme shortness of their centra.
Purchased, 1857.

32859. Two associated imperfect lumbar vertebræ; from Australia.
Purchased, 1857.

45881. Six imperfect dorso-lumbar vertebræ; from Queensland.
Presented by Dr. George Bennett, 1874.

45882. An imperfect lumbar vertebra; from Australia.
Presented by Dr. George Bennett, 1874.

38603. An imperfect lumbar vertebra; from Darling Downs.
Presented by Sir D. Cooper, Bart., 1864.

40375. An imperfect lumbar vertebra; from the Condamine River.
Presented by Dr. F. Campbell, 1867.

M 7. A complete late lumbar vertebra; from Australia. The corresponding vertebra of *Phalangista* makes the nearest approach to this specimen of any living Marsupial, but differs by the longer centrum, and the deflection of the transverse process, which is not carried entirely by the arch. *Presented by Sir G. Verdon*, 1877.

43375. Part of the neural arch of a late sacral vertebra of a very large individual; from King's Creek, Clifton.
Presented by Dr. George Bennett, 1872.

32855. Two specimens of the anchylosed epiphyses of two adjacent
(*Fig.*) vertebræ; from Australia. Figured by Owen in the 'Phil. Trans.' 1870, pl. xliv. figs. 12, 13, and also in the 'Extinct Mammals of Australia,' pl. xxix. figs. 12, 13.
Purchased, 1857.

M. 480. Two segments of the sternum; from Queensland.
Presented by Dr. George Bennett.

Family PHALANGISTIDÆ.

Dentition (except in *Tarsipes*):—I. $\frac{3}{1}$, C. $\frac{1}{(0-1)}$, Pm. $\frac{(2-3)}{(1-3)}$, M. $\frac{(1-4)}{(2-4)}$. The first upper incisor is strong, curved, and cutting, the other two being usually smaller; the lower incisor is cutting and generally more or less proclivous. The upper canine is small and rounded, and the lower either rudimentary or absent. The premolars are very variable, and may either resemble the true molars or be trenchant. The true molars may be either simply quadritubercular, or their tubercles may either coalesce into incomplete transverse ridges, with subcrescentoid summits when unworn (*Phalangista*), or may be laterally compressed, and with a subselenodont structure (*Phascolarctos*), or may be simply conical (*Thylacoleo*). The cranium is always short, and large in proportion to the body. The mandible has the masseteric fossa without pit or large perforation, and its condyle convex and transversely elongated, but not separated by a notch from the coronoid process. The limbs are of equal length; the scapula and humerus being elongated, and the latter furnished with an entepicondylar foramen. The calcaneum and astragalus approximate to those of the *Phascolomyidæ*, and differ widely from those of the *Macropodidæ*. In existing forms the manus has five subequal digits, which are all furnished with claws; while in the pes the second and third digits are very slender and partly united by integument.

Thylacoleo agrees so essentially with the existing *Phalangistinæ*, that there seems every reason for following the lead of Krefft[1] in regarding it as a highly specialized member of the same family.

Subfamily PHALANGISTINÆ.

The lower incisor is markedly proclivous, a tooth generally regarded as a rudimental lower canine is usually present, and the true molars are larger than the premolars; pm. 4 is preceded by a milk-tooth.

Genus **PSEUDOCHIRUS,** Ogilby[2].

Dentition:—I. $\frac{3}{1}$, C. $\frac{1}{1}$, Pm. $\frac{3}{3}$, M. $\frac{4}{4}$. The crowns of the upper true molars are low and broad, and carry six cusps, which are laterally compressed and have an imperfect selenodont structure; the upper row of incisors is angulated; $\underline{\text{i. 1}}$ is only slightly larger than $\underline{\text{i. 2}}$ or $\underline{\text{i. 3}}$; and the auditory bulla is inflated.

[1] 'Mammals of Australia,' Introd., p. 3 (1871).—*Diprotodon* and *Nototherium* are also included.

[2] Proc. Zool. Soc. 1836, p. 26.

Pseudochirus caudivolvulus (Kerr[1]).

Syn. *Didelphys caudivolvula*, Kerr[2].

Hab. New South Wales and Queensland.

M. 3651. Fragment of the right maxilla, containing pm. 4 and m. 1; from a cave in the Wellington Valley, New South Wales.
Presented by the Trustees of the Australian Museum.

Subfamily THYLACOLEONTINÆ.

The lower incisor is but slightly proclivous, there is no lower canine, the true molars and early premolars are small and almost or quite functionless, while the last premolar is of enormous size, laterally compressed, and trenchant; it is not known whether this tooth was preceded by a milk-molar.

Genus **THYLACOLEO**, Owen[3].

Dentition :—I. $\frac{3}{1}$, C. $\frac{1}{0}$, Pm. $\frac{3}{3}$, M. $\frac{1}{2}$. This is the only genus of the subfamily. The structure of the foot is unknown. Some of the functionless teeth were deciduous, and perhaps sometimes altogether wanting.

Fig. 28.

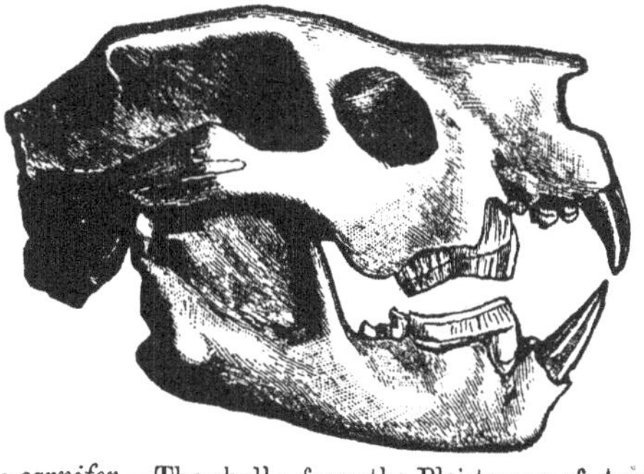

Thylacoleo carnifex.—The skull; from the Pleistocene of Australia. $\frac{1}{5}$.

Thylacoleo carnifex, Owen[4].

This is the type and only species, and is much larger than any existing member of the family.

Hab. Australia.

[1] Linn. Anim. Kingd. p. 196 (1792).—*Didelphys.* [2] *Loc. cit.*

[3] In Gervais's 'Zool. et Pal. Françaises,' 1st ed. pt. i. p. 192 (1849–52).

[4] Phil. Trans. 1859, p. 309.

M. 3652. Cast of the imperfect cranium, showing pm. 4 and m. 1 of the right side. The original, which is the type, was obtained from the Pleistocene near Melbourne, Victoria, and is preserved in the Museum of the Royal College of Surgeons (No. 3853); it is described and figured by Owen in the 'Phil. Trans.' 1859, p. 309, pls. xi., xiii., xiv., xv., and also in the 'Extinct Mammals of Australia,' pls. xi., xiii., xiv., xv.

Presented by the Council of the Royal College of Surgeons.

39271. (*Fig.*) The slightly imperfect cranium, showing pm. 4 of both sides and the right canine; from the Pleistocene of the Condamine River, Queensland. Figured by Owen in the 'Extinct Mammals of Australia,' pls. xvi.–xviii.

Presented by Edward Hill, Esq., 1865.

M. 1514. Cast of the cranium, showing nearly all the dentition. The original, which was obtained from the Pleistocene of King's Creek, Darling Downs, Queensland, is described and figured by Owen in the 'Geol. Mag.' dec. ii. vol. x. p. 289, pl. vii. (1883). *Presented by C. H. Hartmann, Esq.*, 1884.

43209. The anterior portion of the cranium, in a broken and crushed condition; from Darling Downs. Portions of the four anterior teeth remain on both sides, while on the left side is shown the entire pm. 4 and the alveolus of m. 1.

Presented by Sir R. Daintree, 1871.

M. 3654. (*Fig.*) Fragment of the left maxillary region, containing m. 1 and the much-worn pm. 4; from Queensland. Figured by Owen in the 'Extinct Mammals of Australia,' pl. vi. figs. 9, 10. *Presented by Prof. Sir R. Owen, K.C.B.*

M. 3653. (*Fig.*) Fragment of the right maxilla of the same individual, containing the corresponding teeth. Figured by Owen, *op. cit.* pl. vi. figs. 1, 3 *bis*. *Same history.*

39994. (*Fig.*) Part of the right half of the anterior portion of the cranium; from the Pleistocene of Gowrie, Queensland. Figured by Owen in the 'Phil. Trans.' 1871, pl. xi. figs. 1–5, and also in the 'Extinct Mammals of Australia,' pl. vii. figs. 1–5. The first incisor, canine, and the three premolars are *in situ*, and the alveoli of i. 2 and i. 3 also remain.

Presented by Sir D. Cooper, Bart., 1866.

42299. Fragment of the right maxilla, containing pm. 4 and m. 1; from a cave in the Wellington Valley, New South Wales.

Presented by Prof. A. M. Thomson, 1870.

M. 1952. Cast of a fragment of the right half of the anterior region of the cranium, showing the canine and the well-worn $\underline{\text{pm. 4}}$. The original is from a cave in the Wellington Valley, and is preserved in the Australian Museum, Sydney.
Presented by the Trustees of the Australian Museum, 1884.

M. 1953. Cast of a fragment of the left maxilla, with $\underline{\text{pm. 4}}$ and $\underline{\text{m. 1}}$. The history of the original is the same as that of the preceding specimen.
Presented by the Trustees of the Australian Museum, 1884.

M. 1954. Cast of a fragment of the left maxilla, showing $\underline{\text{pm. 4}}$ and $\underline{\text{m. 1}}$. History the same as that of the preceding.
Presented by the Trustees of the Australian Museum, 1884.

M. 1955. Cast of a fragment of the right maxilla, with $\underline{\text{pm. 4}}$. History the same as that of the preceding.
Presented by the Trustees of the Australian Museum, 1884.

M. 1956. Cast of a fragment of the left premaxilla, showing the first incisor. History the same as that of the preceding.
Presented by the Trustees of the Australian Museum, 1884.

42526. The left first upper incisor; from a cave in the Wellington
(*Fig.*) Valley. Figured by Owen in the 'Phil. Trans.' 1871, pl. xi. figs. 6, 7, and also in the 'Extinct Mammals of Australia,' pl. vii. figs. 6, 7.
Presented by the Trustees of the Australian Museum, 1871.

42518–19–22. Three specimens of second and third upper incisors;
(*Fig.*) from the caverns of the Wellington Valley. Figured by Owen in pl. xi. figs. 9, 10, 12, and pl. vii. figs. 9, 10, 12, of the two memoirs cited above.
Presented by the Trustees of the Australian Museum, 1871.

M. 1817. Cast of the nearly entire mandible, showing all the teeth except $\overline{\text{m. 2}}$. The original was obtained from a cave in the Wellington Valley, and is preserved in the Australian Museum.
Presented by the Trustees of the Australian Museum, 1884.

42305. Cast of a partially restored mandible, showing the incisor, $\overline{\text{pm. 4}}$, and $\overline{\text{m. 1}}$ of both sides.
Presented by Prof. A. M. Thomson, 1870.

43088. Cast of the middle portion of the mandible, showing $\overline{\text{pm. 4}}$ and $\overline{\text{m. 1}}$ of both sides. History the same as that of No. M. 1817.
Presented by the Trustees of the Australian Museum, 1871.

46835. The nearly entire right ramus of the mandible, showing the incisor, $\overline{\text{pm. 4}}$, and $\overline{\text{m. 1}}$; from Australia. There is no trace of the alveolus of $\overline{\text{m. 2}}$.

Presented by Dr. George Bennett, 1875.

M. 3570. The greater portion of the right ramus of the mandible, with $\overline{\text{pm. 4}}$ and $\overline{\text{m. 1}}$; from King's Creek, Clifton, Queensland. Noticed by Owen in the 'Phil. Trans.' 1887, p. 3.

Presented by G. F. Bennett, Esq., 1887.

M. 1957. Cast of the right ramus of the mandible, showing the incisor, $\overline{\text{pm. 2}}$, $\overline{\text{pm. 4}}$, and $\overline{\text{m. 1}}$, and the alveoli of $\overline{\text{pm. 3}}$ and $\overline{\text{m. 2}}$. The original was obtained from a cave in the Wellington Valley, and is preserved in the Australian Museum; it is figured by Owen in the 'Phil. Trans.'[1] 1887, pl. i.

Presented by the Trustees of the Australian Museum, 1884.

M. 1958. Cast of the right ramus of the mandible, wanting part of the coronoid process, and showing $\overline{\text{i. 1}}$, $\overline{\text{pm. 4}}$, and $\overline{\text{m. 1}}$, and the alveoli of the other three teeth. History the same as that of the preceding.

Presented by the Trustees of the Australian Museum, 1884.

M. 1959. Cast of the imperfect left ramus of the mandible, showing $\overline{\text{pm. 4}}$ and $\overline{\text{m. 1}}$. History the same as that of the preceding.

Presented by the Trustees of the Australian Museum, 1884.

M. 1960. Fragment of the left ramus of the mandible, showing $\overline{\text{i. 1}}$, $\overline{\text{pm. 4}}$, and $\overline{\text{m. 1}}$. History the same as that of the preceding.

Presented by the Trustees of the Australian Museum, 1884.

M. 1518. Casts of the two associated mandibular rami, in a somewhat crushed condition. History the same as that of the preceding.

Presented by the Trustees of the Australian Museum, 1883.

43088. Cast of the imperfect right ramus of the mandible, showing the incisor, $\overline{\text{pm. 4}}$, and $\overline{\text{m. 1}}$. History the same as that of the preceding. Figured in the 'Phil. Trans.' 1871, pl. xiii. fig. 1.

Presented by the Trustees of the Australian Museum, 1871.

M. 4. Cast of a fragment of the right ramus of the mandible. The original is from Hodgson's Creek, Darling Downs, and is figured by Owen in the 'Phil. Trans.' 1859, pl. xi. fig. 3,

[1] Erroneously stated in the text to have been obtained in 1886.

and pl. xiii. figs. 4, 5; and also in the 'Extinct Mammals of Australia,' pl. xi. fig. 3, and pl. xiii. figs. 4, 5.
Made in the Museum.

39995. (*Fig.*) The greater part of the left ramus of the mandible, containing $\overline{\text{pm. 4}}$ and the root of the incisor; from Gowrie, Queensland. Figured by Owen in the 'Phil. Trans.' 1871, pl. xii. figs. 1–5, and also in the 'Extinct Mammals of Australia,' pl. viii. figs. 1–5.
Presented by Sir D. Cooper, Bart., 1866.

M. 39. Part of the right ramus of the mandible, containing $\overline{\text{pm. 4}}$, and the alveoli of $\overline{\text{pm. 2}}$, $\overline{\text{pm. 3}}$, $\overline{\text{m. 1}}$, and $\overline{\text{m. 2}}$; from Queensland. *Presented by G. F. Bennett, Esq.*

M. 3664. Fragment of the right ramus of the mandible, showing the hinder half of $\overline{\text{pm. 4}}$ and the alveoli of most of the other teeth; from Australia. *Same history.*

M. 2572. The imperfect right ramus of the mandible, showing $\overline{\text{pm. 4}}$ and $\overline{\text{m. 1}}$, and the alveoli of the other teeth; from Eton Vale, Darling Downs.
Presented by G. F. Bennett, Esq., 1885.

43088. Cast of the anterior part of the right ramus of the mandible, showing $\overline{\text{i. 1}}$, $\overline{\text{pm. 4}}$, and $\overline{\text{m. 1}}$. The original, which was obtained from a cave in the Wellington Valley, and is preserved in the Museum at Sydney, is figured by Owen in the 'Phil. Trans.' 1871, pl. xiii. fig. 2.
Presented by the Trustees of the Australian Museum, 1871.

40060. Cast of a fragment of the left ramus of the mandible, showing the worn $\overline{\text{pm. 4}}$. History the same as that of the preceding.
Presented by the Trustees of the Australian Museum, 1866.

50066. Small fragment of the right half of the mandibular symphysis, showing the broken bases of the anterior teeth; from Gowrie. *Presented by G. F. Bennett, Esq.*, 1879.

M. 3665. Part of the right ramus of the mandible of a young individual, containing the broken and unworn $\overline{\text{pm. 4}}$; from Australia. *Same history.*

42516. (*Fig.*) The fourth right lower premolar; from a cavern in the Wellington Valley. Figured by Owen in the 'Phil. Trans.' 1871, pl. xii. fig. 6, and also in the 'Extinct Mammals of Australia,' pl. viii. fig. 6.
Presented by the Trustees of the Australian Museum, 1870.

50067. The well-worn fourth left lower premolar belonging to an unusually large individual; from Gowrie.
Presented by G. F. Bennett, Esq., 1879.

47012 c. The right lower incisor; from Darling Downs.
Presented by Dr. George Bennett, 1876.

42536. The left lower incisor; from a cavern in the Wellington Valley. The peculiar distribution of the enamel cap characteristic of the genus is well displayed.
Presented by the Trustees of the Australian Museum, 1870.

50074. Two specimens of the left lower incisor; from Gowrie.
Presented by G. F. Bennett, Esq., 1879.

M. 3663. Cast of the brain-cavity; taken from the cranium No. 39271. *Made in the Museum.*

The following specimens probably belong to this species.

42617. The proximal half of the left humerus; from a cave in the Wellington Valley. This bone comes nearest in contour to the humerus of *Phalangista*, but has the head elongated transversely.
Presented by the Trustees of the Australian Museum, 1870.

42300. The proximal portion of a similar right humerus; from the same locality. *Presented by Prof. A. M. Thomson*, 1870.

M. 1526. Cast of the imperfect right ulna. The original is from the same locality as the last, and is preserved in the Museum at Sydney; it is figured by Owen in the 'Phil. Trans.' 1883, pl. xl. fig. 4. It differs from the ulna of *Phascolomys* by the superior ridge of the greater sigmoid cavity extending more completely over the inner surface of the humeral condyle, in which respect it agrees with *Phalangista*.
Presented by the Trustees of the Australian Museum, 1883.

M. 1925. Cast of the proximal extremity of the left ulna. The original, of which the history is the same as that of the last specimen, is figured by Owen, *op. cit.* pl. xl. fig. 5.
Presented by the Trustees of the Australian Museum, 1883.

35933. The proximal half of a similar left ulna; from Gowrie.
Presented by Sir D. Cooper, Bart., 1866.

M. 1525. Cast of the right radius. The original, of which the history is the same as that of No. M. 1526, is figured by Owen, *op. cit.* pl. xl. figs. 1–3.

Presented by the Trustees of the Australian Museum, 1883.

M. 1523. Cast of the left innominate. The original, of which the history is the same as the last, is figured by Owen, *op. cit.* pl. xlvi. fig. 1; it closely resembles the corresponding bone of *Phalangista*.

Presented by the Trustees of the Australian Museum, 1883.

M. 1526. Cast of an ungual phalangeal. The history of the original is the same as that of the last. The bone was evidently covered by a horny claw, like that of *Phalangista*.

Presented by the Trustees of the Australian Museum, 1883.

M. 3666. A right calcaneum, perhaps belonging to a young individual of this species; from a cave in the Wellington Valley.

Presented by the Trustees of the Australian Museum.

Family PLAGIAULACIDÆ.

This and the three following families, forming the suborder Multituberculata of Cope [1], are regarded as primitive Diprotodonts, presenting, however, some very specialized characters, and are provisionally placed here. They are characterized by the true molars carrying longitudinal rows of tubercles, separated by one or more grooves; there being either two or three such rows in the upper molars of those forms in which these teeth are known, while there

Fig. 29.

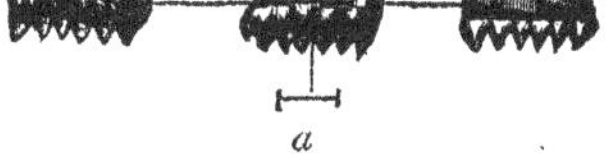

Neoplagiaulax eocænus, Lemoine.—Oral and lateral views of an upper true molar; from the Lower Eocene of Rheims. *a*. Nat. size. (From the 'Quart. Journ. Geol. Soc.')

are, at least usually, only two in those of the lower jaw. The external surface of the masseteric fossa of the mandible has neither pit nor perforation, as in the *Phalangistidæ*. In the present family, the number of upper incisors was probably two or three [2]; the lower premolars have a highly convex cutting-edge, and are always large, compressed, and usually either serrated, or obliquely grooved and

[1] Amer. Nat. vol. xviii. p. 687 (1884).

[2] Lemoine describes two incisors and a canine in *Neoplagiaulax*.

ridged, although in the genus *Liotomus*, Cope[1], they are smooth; their number varies from one (*Neoplagiaulax*) to four (*Plagiaulax minor* and *Ctenacodon*); the lower true molars are two in number, of small size, and marked by one longitudinal groove. The upper true molars of the European and North-American genus *Neopla-*

Fig. 30.

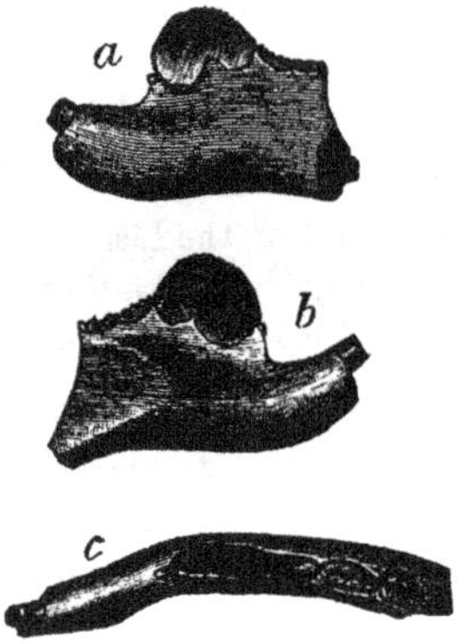

Ptilodus mediævus, Cope.—The left ramus of the mandible, from the outer (*a*), inner (*b*), and oral (*c*) aspect; from the Puerco Eocene of New Mexico. $\frac{1}{1}$. (From the 'Amer. Nat.')

giaulax have three ridges[2] (fig. 29). The mandible of the closely allied North-American *Ptilodus* is represented in fig. 30; in that genus a minute $\overline{\text{pm. 3}}$ is present, and the other slight characters by which it is distinguished from *Neoplagiaulax* are mentioned by Lemoine[3].

Genus **PLAGIAULAX**, Falconer[4].

Including *Plioprion*, Cope[5].

Dentition:—I. $\frac{?\,(2\text{ or }3)}{1}$, C. $\frac{?\,1}{0}$, Pm. $\frac{?}{(3\text{–}4)}$, M. $\frac{?}{2}$. The lower premolars are obliquely grooved and ridged, and increase gradually in size from the first to the last (woodcut, fig. 31). The condyle of the mandible (as in *Thylacoleo*) is placed below the level of the alveolar line of the cheek-teeth.

[1] Amer. Nat. vol. xviii. p. 691 (1884), = *Neoplagiaulax marshi*, Lemoine.

[2] From its resemblance to *Bolodon*, and the difference of its premolars from those referred by Lemoine to the upper jaw of *Neoplagiaulax*, the writer regards the maxilla figured by Marsh (Amer. Journ. 1887, pl. viii. figs. 2, 3) as *Ctenacodon* as probably belonging to the *Bolodontidæ*.

[3] Bull. Soc. Géol. France, sér. 3, vol. xi. p. 271 (1883).

[4] Quart. Journ. Geol. Soc. vol. xiii. p. 261 (1857).

[5] Amer. Nat. vol. xviii. p. 691 (1884).

Plagiaulax becklesi, Falconer [1].

This is one of the two type species, and has three lower premolars; it is of relatively large size, and characterized by the great depth of the mandibular ramus, and the stoutness of the incisor.

Hab. England.

Fig. 31.

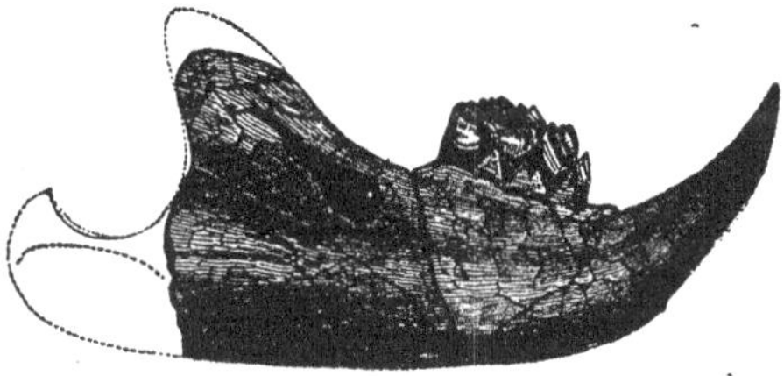

Plagiaulax becklesi.—The right ramus of the mandible; from the Middle Purbeck of Swanage. $\frac{2}{1}$.

47731. (*Fig.*) The right ramus of the mandible; from the Middle Purbeck of Durdlestone Bay, Swanage, Dorsetshire. This specimen (woodcut fig. 31) is the type, and is described and figured by Falconer in the 'Quart. Journ. Geol. Soc.' vol. xiii. p. 278, figs. 1–5, and by Owen in the 'British Mesozoic Mammalia' (Mon. Pal. Soc.), pl. iv. figs. 10, 10 *a*; it is also figured in 'Falconer's Palæontological Memoirs,' vol. ii. pl. xxxviii. figs. 1–5. *Beckles Collection. Purchased*, 1876.

47732. (*Fig.*) Fragment of the right ramus of the mandible, showing the incisor and the three premolars ($\overline{\text{pm. 4}}$ imperfect); from Durdlestone Bay. Figured by Falconer, *op. cit.* p. 279, figs. 11–13, and by Owen, *op. cit.* pl. iv. figs. 13, 14. *Same history.*

47733. (*Fig.*) Fragment of a mandibular ramus, showing the two true molars; from Durdlestone Bay. Figured by Falconer, *op. cit.* p. 279, figs. 7–10, and by Owen, *op. cit.* pl. iv. fig. 12. *Same history.*

47734. (*Fig.*) Fragment of the left ramus of the mandible, showing th incisor and premolars; from Durdlestone Bay. Figured by Owen, *op. cit.* pl. iv. fig. 15. *Same history.*

[1] Quart. Journ. Geol. Soc. vol. xiii. p. 262 (1857).

Plagiaulax medius, Owen [1].

This form is considerably smaller than the preceding, from which it apparently differs by the contour of the coronoid process of the mandible.

Hab. England.

47728. (*Fig.*) The left ramus of the mandible, showing the incisor and premolars; from the Middle Purbeck of Durdlestone Bay, Swanage, Dorsetshire. This specimen, which is the type, is described and figured by Owen in the 'British Mesozoic Mammalia' (Mon. Pal. Soc.), p. 85, fig. 7, and p. 90, fig. 12; it is also figured by Falconer in the 'Quart. Journ. Geol. Soc.' vol. xiii. p. 280, fig. 14, under the name of *P. becklesi*. *Beckles Collection. Purchased*, 1876.

Plagiaulax falconeri, Owen [2].

Approximately equal in size to *P. becklesi*, but distinguished by the smaller depth of the mandibular ramus, the more slender incisor, and the larger $\overline{\text{pm. 2}}$.

Hab. England.

47730. (*Fig.*) The imperfect right ramus of the mandible, showing the premolars and incisor; from the Middle Purbeck near Swanage, Dorsetshire. This specimen is the type, and is figured by Owen in the 'British Mesozoic Mammalia' (Mon. Pal. Soc.), pl. iv. fig. 6.

Beckles Collection. Purchased, 1876.

Plagiaulax minor, Falconer [3].

Syn. *Plioprion minor*, Cope [4].

This is the smallest species, and is distinguished by the presence of four lower premolars, on which account it has been made the type of a distinct genus by Cope.

Hab. England.

47729. (*Fig.*) The imperfect right ramus of the mandible, showing the complete dentition; from the Middle Purbeck near Swanage, Dorsetshire. This specimen is the type, and is figured by Falconer in the 'Quart. Journ. Geol. Soc.' vol. xiii. p. 28, fig. 15, and by Owen in the 'British Mesozoic Mammalia' (Mon. Pal. Soc.), pl. iv. fig. 9.

Beckles Collection. Purchased, 1876.

[1] 'British Mesozoic Mammalia' (Mon. Pal. Soc.), p. 85 (1871).

[2] *Op. cit.* p. 84.

[3] Quart. Journ. Geol. Soc. vol. xiii. p. 264 (1857).

[4] Amer. Nat. vol. xviii. p. 691 (1884).

The following specimens are not specifically determined.

48399. Fragment of the left ramus of the mandible, showing the three premolars, and a detached fourth lower premolar; from the Middle Purbeck of Durdlestone Bay.

Beckles Collection. Purchased, 1876.

Family POLYMASTODONTIDÆ.

In the one known genus of this family the dentition is I. $\frac{?}{1}$, C. $\frac{?}{0}$, Pm. $\frac{1}{1}$, M. $\frac{2}{2}$. The premolar is tubercular, and agrees in structure with the true molars, although somewhat simpler; the true molars are elongated antero-posteriorly, those of the upper jaw having three ridges, $\overline{\text{m. 1}}$ being longer than $\overline{\text{m. 2}}$ and having six tubercles on each side. The angle of the mandible is inflected; the humerus has no entepicondylar foramen, and the astragalus is without a trochlea; the caudal vertebræ indicate a large tail.

Fig. 32.

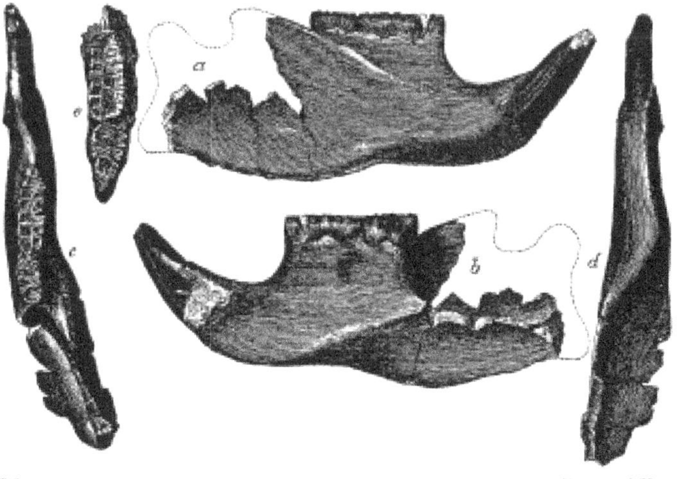

Polymastodon taoensis.—*a, b, c, d.* Right mandibular ramus; from the Puerco Eocene of New Mexico. *e.* Fragment of a maxilla. ½. (From the 'Amer. Nat.')

Genus **POLYMASTODON,** Cope[1].

Syn. *Catopsalis*, Cope[2].
Tæniolabis, Cope[3].

As this is the only genus at present known, its characters are those of the family; five species have been named.

Polymastodon taöensis, Cope[4].

Syn. *Catopsalis pollux*, Cope[5].
Tæniolabis sulcatus, Cope[6].

This is the type and largest species, its limb-bones being described as equal in size to those of *Macropus giganteus*. The mandible and dentition are represented in the accompanying woodcut (fig. 32).

Hab. North America.

M. 2568. Part of the left ramus of the mandible, showing the three cheek-teeth in an almost unworn condition; from the Puerco Eocene of New Mexico, U.S.A.

Presented by R. Lydekker, Esq., 1885.

Family TRITYLODONTIDÆ.

The upper true molars have three antero-posterior ridges and are elongated transversely; while the premolars in the type genus are of the same general structure, although somewhat smaller and simpler. It seems probable that *Stereognathus* should be included in this family[7]; but the close resemblance of the molars of the type specimen of that genus to those of *Tritylodon* suggests that the specimen in question belongs to the upper rather than to the lower jaw.

Cope[8] suggests that from the type genus may have been derived both the *Polymastodontidæ* and *Plagiaulacidæ*; while *Thylacoleo* is directly descended from the latter, and the *Macropodidæ* and *Phascolomyidæ* are lateral branches from the same stock. It is, however,

[1] Amer. Nat. vol. xvi. p. 684 (1882).

[2] *Ibid.* p. 416. This name has the priority, but has apparently been withdrawn by its author in favour of *Polymastodon*. See Amer. Nat. vol. xviii. p. 688 (1884).

[3] *Ibid.* p. 604. Founded on an incisor; see Amer. Nat. vol. xix. p. 493 (1885)

[4] *Ibid.* p. 684. [5] *Ibid.* p. 685.

[6] *Ibid.* p. 604. Subsequently quoted as *T. scalper*.

[7] See Cope, Amer. Nat. vol. xviii. p. 688 (1884). *Tritylodon*, although the latest in date, is taken as the type of the family on account of the more satisfactory evidence afforded by the type specimen.

[8] *Op. cit.* p. 695.

pretty evident that *Thylacoleo*, with its three upper incisors, could not have been derived from *Tritylodon*, in which there are only two of these teeth; the same remark being also applicable in the case of the *Macropodidæ*.

Genus **TRITYLODON**, Owen[1].

Syn. *Triglyphus*, Fraas[2].

Dentition:—I. $\frac{2}{?}$, C. $\frac{0}{?}$, Pm. $\frac{2}{?}$, M. $\frac{4}{?}$. The diastema is of great length, the first upper incisor is large and scalpriform, and the second small and functionless; each ridge of the upper cheek-teeth usually carries three subconical tubercles.

Fig. 33.

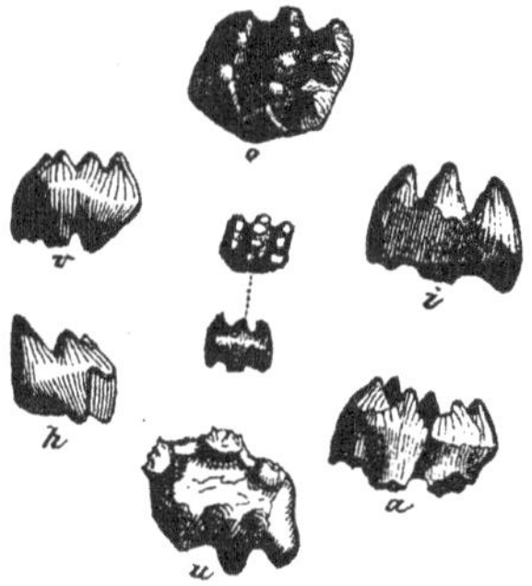

Tritylodon fraasi, n. sp. Lyd.—Upper molar; from the Triassic bone-bed near Strasbourg. The central figures are of the natural size, the others $\frac{2}{1}$. (From the 'Neues Jahrb.')

The upper molar from the Triassic bone-bed near Strasbourg (fig. 33), figured by Fraas under the name of *Triglyphus*, presents no characters by which it can be generically distinguished[3] from *Tritylodon*, and may therefore be at least provisionally referred to that genus; but as its smaller size[4] indicates its specific distiuction, it may be named *T. fraasi*.

Tritylodon longævus, Owen[5].

This is the type species, and appears to have been somewhat smaller than a Badger.

Hab. South Africa.

[1] Quart. Journ. Geol. Soc. vol. xl. p. 146 (1884).

[2] 'Vor der Sündfluth,' p. 215 (Stuttgart, 1866). The name is preoccupied for a genus of Diptera.

[3] The character mentioned by Neumayr, Neues Jahrb. 1884, vol. i. p. 279, cannot be regarded as of more than specific value.

[4] It is of too large dimensions to have been the upper molar of *Microlestes*, which was probably two-ridged.

[5] *Loc. cit.*

M. 1951. (*Fig.*) The imperfect cranium; from the Karoo series of Basuto-land, South Africa. This is the type and only known specimen, and is described and figured by Owen in the 'Quart. Journ. Geol. Soc.' vol. xl. p. 146, pl. vi. figs. 1–7, and also by Cope in the 'Amer. Nat.' vol. xviii. p. 688, fig. 2; the latter figure being reproduced in the accompanying woodcut.

By exchange with the Bloemfontein Museum, South Africa, 1884.

Fig. 34.

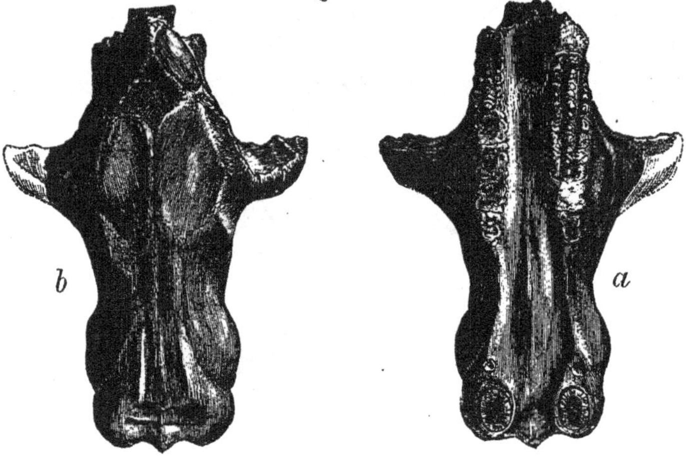

Tritylodon longævus.—Oral (*a*) and frontal (*b*), aspects of the cranium; from the Karoo series of of Basuto-land. ⅔. (From the 'Amer. Nat.')

Family BOLODONTIDÆ.

Upper true molars numerous and antero-posteriorly elongated, with a deep median groove, bordered by two antero-posterior ridges surmounted by blunt tubercles; the upper premolars with tricuspid non-secant crowns. The serial position of the family is provisional, but it is probably allied to the three preceding ones [1].

Genus **MICROLESTES**, Plieninger [2].

Syn. *Hypsiprymnopsis*, Dawkins [3].

Owen recognizes three equally minute species, viz. *M. antiquus*, Plieninger, from the Upper Trias (Keuper) of Stuttgart; *M. rhæticus*,

[1] Marsh (Amer. Journ. 1887, p. 329) places *Bolodon* in the *Plagiaulacidæ*, in which Osborn (MS.) includes *Microlestes*. The former, as the best-known genus, is made the type of the family.

[2] Jahresh. Ver. Nat. Württ. vol. iii. p. 164 (1847).

[3] Quart. Journ. Geol. Soc. vol. xx. p. 406 (1864).

Dawkins, from the Upper Trias (Rhætic) of Somerset; and the following form. There are, however, no characters by which the two latter can be distinguished from the former, and their claim to specific separation must therefore be regarded as provisional. The structure of the teeth so closely resembles that of *Bolodon*, that there can be but little hesitation in placing both in the same family.

Microlestes moorei, Owen[1].

Hab. England.

M. 2401. Three teeth; from the Upper Triassic (Rhætic) bone-bed of Frome, Somersetshire. These specimens agree with those figured by Owen in the memoir cited; but the reference of the latter to upper and lower jaws is doubtful.

By exchange with the Bath Institute, 1884.

Genus **BOLODON,** Owen[2].

Dentition:—I. $\frac{3}{?\,1}$, C. $\frac{0}{0}$, Pm. $\frac{3}{?}$, M. $\frac{4}{?}$. In the allied or identical American *Allodon*, Marsh[3] regards the check-formula as Pm. $\frac{5}{?}$, M. $\frac{2}{?}$; but from No. 47735 *a* and the analogy of *Tritylodon* the above formula is provisionally adopted. The upper premolars are conical and have three cusps on their summit.

Bolodon crassidens, Owen[4].

This is the type species.

Hab. England.

The following speçimens are from the Middle Purbeck group of Durdlestone Bay, Swanage, Dorsetshire; and belong to the Beckles Collection. Purchased, 1876.

47735. (*Fig.*) Part of the right maxilla, showing the second and third incisors, premolars, and one true molar. This specimen is the type, and is figured by Owen in his 'British Mesozoic Mammalia' (Mon. Pal. Soc.), pl. iii. fig. 5. There is a crack behind the second tooth which appears coincident with the suture.

47735 a. Part of the right maxilla, showing the three premolars and four true molars. This specimen, which was not seen by Owen, has afforded important evidence in determining the homology of the teeth.

47736. (*Fig.*) Fragment of the palate, showing two cheek-teeth on either side. Figured by Owen, *op. cit.* pl. iii. fig. 6.

[1] 'British Mesozoic Mammalia' (Mon. Pal. Soc.), p. 6 (1871).

[2] *Ibid.* p. 54 (1871).

[3] Amer. Journ. ser. 3, vol. xxxiii. p. 329 (1887). [4] *Loc. cit.*

Family MACROPODIDÆ.

Dentition:—I. $\frac{3}{1}$, C. $\frac{(0-1)}{0}$, Pm. $\frac{2}{2}$, M. $\frac{4}{4}$. The upper incisors are secant, and arranged in a continuous arched series, distinctly separated in the median line; the first being often larger than either of the others; the lower incisor is secant and proclivous, and the teeth of opposite sides frequently bite against one another. The upper canine, when present, is often deciduous, and generally small. The diastema is long. The premolars are either secant or triangular, and in the former case are generally more or less distinctly marked by nearly vertical grooves and ridges; pm. 3 is always, and pm. 4 occasionally deciduous, the latter tooth being preceded by a milk-molar. The true molars are either quadritubercular or transversely ridged; and m. 1 is occasionally deciduous. The mandible has a deep pit and perforation in the masseteric fossa, and its condyle is flat, transversely shortened, and not separated by a deep notch from the coronoid process. The fore limbs are always shorter than the hind ones, although their relative proportion varies much in the different forms; the scapula and humerus are frequently broad, and

Fig. 35.

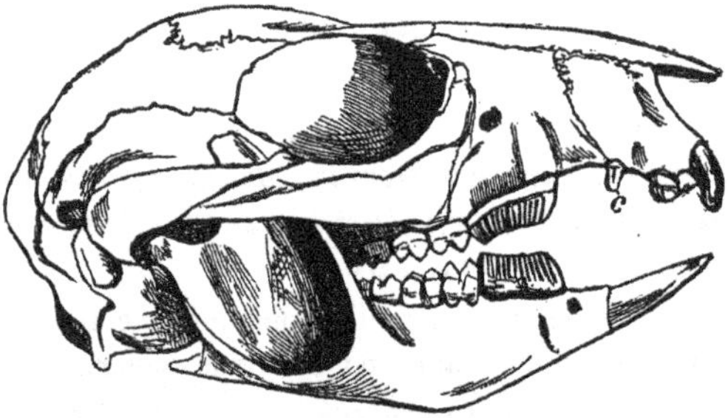

Bettongia grayi (Gould). Lateral view of skull; from Australia. $\frac{1}{1}$.
(From the 'Quart. Journ. Geol. Soc.')

the latter nearly always has an entepicondylar foramen. The olecranon is well developed; the calcaneum and astragalus are of a peculiar type, the former being characterized by the truncation of its distal articular surface. The manus has five subequal digits; but the hind foot has the second and third digits much smaller than the others, and enclosed in a common integument, the hallux being absent in all recent forms except *Hypsiprymnodon*, and the fourth digit always larger than either of the others. The femur is characterized by a projection for muscular attachments at the outer side of the external distal condyle.

Subfamily Potoroinæ.

The first upper incisor is narrow, curved, and much longer than either of the two following teeth (fig. 35); an upper canine is present; the fourth premolar is secant and much elongated, with a nearly straight or slightly concave cutting-edge, usually with the vertical grooves and ridges very distinct, and that of the upper jaw without any inner basal ridge; the true molars are quadri-tubercular, the fourth being always smaller than the third.

Genus **ÆPYPRYMNUS**, Garrod [1].

The ridges on the fourth premolar are few in number and nearly vertical; the auditory bulla is not inflated; and there are no palatal vacuities. The entepicondylar foramen of the humerus is frequently absent in both the recent and fossil races of the one species.

Aepyprymnus rufescens (Gray [2]).

Syn. *Bettongia rufescens*, Gray [3].
? *Bettongia cuneata*, Owen [4].

The fossil mandible figured under the name of *B. cuneata* apparently agrees precisely with the following specimens.

Hab. New South Wales.

42618-27. Several imperfect specimens of the cranium and mandible; from the caves of the Wellington Valley, New South Wales. Several of these and the following specimens belong to immature individuals, and show the fourth premolar in its alveolus.

Presented by the Trustees of the Australian Museum, 1870.

43868-70. Numerous fragments of the maxilla, imperfect mandibular rami, and detached teeth; from the same locality.

Same history.

32316. The anterior portion of the right ramus of the mandible of an immature individual, showing the incisor, $\overline{\text{pm. 3}}$, $\overline{\text{mm. 4}}$, $\overline{\text{m. 1}}$, $\overline{\text{m. 2}}$, and $\overline{\text{pm. 4}}$ *in alveolo*; from the same locality.

Same history.

[1] Proc. Zool. Soc. 1875, p. 59.
[2] Charlesworth's Mag. Nat. Hist. vol. i. p. 584 (1837).—*Bettongia.*
[3] *Loc. cit.*
[4] 'Extinct Mammals of Australia,' p. 107, pl. v. fig. 15 (1877).

42632–3. Two specimens of the right humerus, in one of which the entepicondylar foramen is wanting; from the same locality. *Same history.*

42668. Three specimens of the humerus; from the same locality. One specimen has no entepicondylar foramen. *Same history.*

42669. Three specimens of the ulna (two imperfect); from the same locality. *Same history.*

42630–1. Two imperfect specimens of the femur; from the same locality. *Same history.*

43928, 43939. Numerous specimens of the calcaneum; from the same locality. *Same history.*

42662. The right fourth metatarsal; from the same locality. *Same history.*

43920. Two specimens of the right fourth metatarsal; from the same locality. *Same history.*

43927. Numerous specimens of the fourth metatarsal; from the same locality. *Same history.*

43829. Numerous specimens of the phalangeals of the fourth digit of the pes; from the same locality. *Same history.*

Subfamily MACROPODINÆ.

The cutting-edges of the upper incisors are situated in nearly the same horizontal plane; the canines are generally absent or rudimentary; the true molars are bilophodont, the fourth being larger than either of the others; and the fourth upper premolar has either an inner basal ridge, or a complete inner lobe.

Genus **MACROPUS,** Shaw [1].

Syn. *Kangurus*, Lacépède [2].
Leptosiagon, Owen [3].
Pachysiagon, Owen [4].

---

[1] Naturalists' Miscellany, vol. i. pl. xxxiii. (1790).
[2] In Desmarest's Mammalogie, p. 271 (1820).
[3] Extinct Mammals of Australia, p. 450 (1877).
[4] Phil. Trans. 1874, p. 784.

Including:—*Halmaturus*, Illiger [1].
Petrogale, Gray [2].
Lagorchestes, Gould [3].
Osphranter, Gould [4].
Onychogale, Gray [5].
Phascolagus, Owen [6].
Protemnodon, Owen [7] (*in parte*).
Sthenurus, Owen [8] (*in parte*).

The mandibular symphysis is not anchylosed, the diastema is long, and the ramus not excessively deep. The fourth upper premolar has a secant edge, and an inner basal ridge or tubercle, the corresponding lower tooth being simply secant; and these teeth may be either longer or shorter than the first true molar. The true molars do not show a number of vertical ridges. The lower incisors are spatulate, and those of opposite sides can frequently be moved against one another. Palatal vacuities are present in some of the smaller forms. The fore limbs are much shorter than the hinder ones.

The writer follows Flower in including all the above-mentioned living forms in the type genus. Other writers on recent zoology retain, however, *Petrogale*, *Onychogale*, and *Lagorchestes* as generic terms, but such a classification is often inapplicable to the fossil forms.

It will be shown in the sequel, that all the upper jaws described by Owen under the name of *Sthenurus*, and the lower ones as *Protemnodon*, indicate large Wallabies belonging to the present genus; although the genus *Sthenurus*, as defined below, is a valid one.

Macropus brehus (Owen) [9].

Syn. *Sthenurus brehus*, Owen [10].
Protemnodon mimas, Owen [11].
Macropus mimas, Flower [12].

Owing to the confusion with regard to the upper dentition of *Sthenurus atlas* (*q.v.*), the upper jaws of the present form were referred by Owen to that genus, while the mandibles were made the types of the so-called *Protemnodon mimas*. A comparison of the mandible

[1] Prodr. Syst. Mamm. et Avium, p. 80 (1811).
[2] Charlesworth's Mag. Nat. Hist. vol. i. p. 583 (1837).
[3] Monograph of Macropodidæ, pt. i. (1841).
[4] *Op. cit.*
[5] In Grey's Australia, vol. ii. App. p. 402 (1841).
[6] Phil. Trans. 1874, p. 261.
[7] *Ibid.* p. 274.
[8] *Ibid.* p. 264.
[9] *Ibid.* p. 272.—*Sthenurus.*
[10] *Loc. cit.*
[11] *Ibid.* p. 278.
[12] Cat. Vert. Anim. Mus. R. Coll. Surg. pt. 2, p. 720 (1884).

associated with the cranium No. 47832 shows, however, an absolute identity with those described under the latter name [1].

The estimated length of the entire cranium is about one foot; the length of the upper series of cheek-teeth usually varying from 0,083 to 0,087, and that of the lower jaw from 0,080 to 0,083; but a larger series of specimens would probably show a more extensive variation in this respect. In this and the following three species, the upper true molars have no bridge connecting their anterior talon with the first ridge; the molars of both jaws are comparatively wide, and have a low longitudinal bridge between the ridges. In the present and largest species, the length of the last premolar considerably exceeds that of the first true molar; the lower true molars of the type form have a distinct hind talon; and the first upper incisor is much larger than either of the others.

There is considerable variation in the length of pm. 4 in the specimens referred by Owen [2] to this species, and it is probable that some of the examples in which that tooth is unusually elongated are identical with the next form.

The species appears allied to existing Wallabies like *M. ualabatus*, in which the last premolar is relatively long. The present form and its allies were regarded by Garrod [3] as showing affinity to *Dorcopsis*; but they differ by the shorter pm. 4, the distinct longitudinal bridge between the ridges of the true molars, the large i. 1, and the absence of the upper canine.

Hab. New South Wales and Queensland.

44121. (*Fig.*) The imperfect palatal region of the cranium, showing all the dentition except the last three true molars of the left side; from the Pleistocene of Clifton, Queensland. Figured by Owen in the 'Phil. Trans.' 1876, pl. xxviii. figs. 1–3, and also in the 'Extinct Mammals of Australia,' pl. cviii. figs. 1–3. *Presented by G. F. Bennett, Esq.*, 1873.

47832. (*Fig.*) The imperfect palatal region of the cranium and the associated anterior extremity of the mandible; from Clifton. These specimens, which are in a somewhat crushed condition, and belonged to an old and probably male individual,

[1] The imperfect cranium figured from a photograph by Owen in the 'Extinct Mammals of Australia,' pl. lxxxvii. figs. 1–4, under the name of *Protemnodon mimas* presents no characters by which it can be distinguished from that of *M. brehus*; the alleged simpler structure of pm. 4 being due either to the imperfection of the specimen or to want of clearness in the photograph.

[2] 'Extinct Mammals of Australia,' pl. cix.

[3] Proc. Zool. Soc. 1875, p. 275.

are figured by Owen in the 'Extinct Mammals of Australia,' pl. cix. figs. 4–8. In the palate the whole of the dentition is shown, while the mandible exhibits the incisors and the first three cheek-teeth.

Presented by G. F. Bennett, Esq., 1873.

47833. (*Fig.*) The anterior part of the palate of an old individual, showing the incisors and pm. 4 and m. 1 of the right side; from Clifton. Figured by Owen, *op. cit.* pl. cix. figs. 1, 9, 10.

Presented by G. F. Bennett, Esq., 1873.

43303 a. (*Fig.*) The middle portion of the palate, showing the true molars and the broken base of the left pm. 4; from a cave in the Wellington Valley, New South Wales. This specimen is the type, and is figured by Owen in the 'Phil. Trans.' 1874, pl. xxvii. figs. 5, 6, and also in the 'Extinct Mammals of Australia,' pl. lxxxvii. figs. 5, 6.

Presented by the Trustees of the Australian Museum, 1870.

M. 2259. (*Fig.*) Part of the right maxilla, showing all the cheek-teeth except m. 4; from the Pleistocene of Queensland. Figured by Owen in the 'Extinct Mammals of Australia,' pl. lxxxviii. figs. 5–7. *Presented by Dr. George Bennett.*

M. 3667. Fragment of the left maxilla of an immature individual, showing the last three true molars; from the Pleistocene of King's Creek, Queensland.

Presented by Dr. George Bennett.

M. 2264. Fragment of the right maxilla of an old individual, showing the last three true molars; from Queensland.

Same history.

M. 2265. A similar fragment of the right maxilla of a younger individual; from Queensland. *Same history.*

50064 c. Fragment of the left maxilla showing the last three true molars in an early stage of wear; from Queensland.

Presented by G. F. Bennett, Esq., 1879.

43580. Fragment of the left maxilla, showing the four true molars; from the Pleistocene of the Condamine River, Queensland.

Presented by Dr. George Bennett, 1872.

38614. Fragment of the left maxilla, showing the last three true molars, in an imperfect condition; from the Pleistocene of Darling Downs, Queensland.
Presented by Sir D. Cooper, Bart., 1864

32894. Fragment of the left maxilla, showing the first three true molars, in an imperfect condition; from the same locality.
Purchased, 1857.

M. 40. Fragment of the left maxilla, containing pm. 4, m. 1, and m. 2, in a well-worn condition; from Queensland.
Presented by G. F. Bennett, Esq., 1880.

50058. Fragment of the left maxilla, containing pm. 4 and m. 1; from Queensland. The premolar is shorter than in No. 43853. *Presented by G. F. Bennett, Esq.*, 1879.

M. 3668. Cast of part of the right maxilla, showing the five cheek-teeth; the exact locality whence the original was obtained is not known. *Presented by Sir R. Owen, K.C.B.*

M. 45. Fragment of the right maxilla with the last three true molars; from Queensland.
Presented by G. F. Bennett, Esq., 1880.

32301. A right upper true molar; from a cave in the Wellington Valley. *Purchased*, 1857.

43853. Fragment of the left maxilla, containing pm. 4 and m. 1;
(*Fig.*) from the same locality. Figured by Owen in the 'Phil. Trans.' 1874, pl. xxvii. figs. 7–9, and also in the 'Extinct Mammals of Australia,' pl. lxxxvii. figs. 7–9.
Presented by the Commissioners for New South Wales of the Paris Exhibition of 1867.

47834. The left lower incisor in a fragment of the mandible; from
(*Fig.*) the Pleistocene of Clifton, Queensland. Figured by Owen in the 'Phil. Trans.' 1876, pl. xxviii. fig. 5.
Presented by G. F. Bennett, Esq., 1873.

M. 35. The left half of the mandibular symphysis, containing the incisor; from Queensland.
Presented by G. F. Bennett, Esq., 1880.

38796. Crown of the right lower incisor; from Darling Downs.
Presented by F. N. Isaac, Esq., 1861.

The following specimens include the types of Protemnodon mimas *of Owen, under which name the figured specimens are described by him.*

M. 2256. The imperfect left ramus of the mandible of an old male; from Queensland. Described by Owen in the 'Extinct Mammals of Australia,' p. 447. The five cheek-teeth and part of the symphysis are shown; and the specimen presents no characters by which it can be distinguished from the mandible of No. 47832.

Presented by Dr. George Bennett.

48068. The imperfect mandible, showing the cheek-teeth of both sides; from Darling Downs. This specimen, although belonging to a younger individual, agrees precisely in characters with the preceding.

Presented by Dr. George Bennett, 1877.

43351. (*Fig.*) The left ramus of the mandible, showing the five cheek-teeth in an early stage of wear; from Gowrie, Queensland. Figured by Owen in the 'Phil. Trans.' 1874, pl. xxvi. figs. 1–3, and also in the 'Extinct Mammals of Australia,' pl. lxxxvi. figs. 1–3. Allowing for the difference in age, no dissimilarity can be detected between this specimen and the mandible of No. 47832.

Presented by Dr. George Bennett, 1872.

M. 2257. The imperfect right ramus of the mandible, showing the four true molars; from Queensland. Described by Owen in the 'Extinct Mammals of Australia,' p. 447.

Presented by Dr. George Bennett.

48423. The imperfect right mandibular ramus of a small immature individual, referred by Owen to the present form; from Queensland. The whole of the dentition is shown, but the crown of $\overline{\text{pm. 4}}$ is broken off.

Presented by Dr. George Bennett, 1877.

32882. (*Fig.*) The imperfect left ramus of the mandible of a young individual, showing $\overline{\text{mm. 4}}$, $\overline{\text{m. 1}}$, and $\overline{\text{m. 2}}$, in use, and $\overline{\text{pm. 4}}$ *in alveolo*; from the Condamine River. Figured by Owen in the 'Phil. Trans.' 1874, pl. xxvi. figs. 1–4, and also in the 'Extinct Mammals of Australia,' pl. lxxxvi. figs. 4–6.

Purchased, 1857.

40007. Part of the left ramus of the mandible of a slightly younger
(*Fig.*) individual, showing the homologous teeth (except $\overline{\text{mm. 4}}$). Figured in the 'Phil. Trans.' 1874, pl. xxiv. figs. 13–16, and also in the 'Extinct Mammals of Australia,' pl. lxxxiv. figs. 13–16. *Presented by Sir D. Cooper, Bart.*, 1866.

35962. Part of the right ramus of the mandible, showing all the cheek-teeth ; from Gowrie. *Presented by Sir D. Cooper, Bart.*, 1861.

38758. Part of the right ramus of the mandible, showing the last three true molars ; from the Pleistocene of Darling Downs, Queensland. *Presented by F. N. Isaac, Esq.*, 1861.

M. 51. The imperfect hinder part of the right ramus of the mandible, showing the four true molars and the hinder half of $\overline{\text{pm. 4}}$; from Queensland. *Presented by G. F. Bennett, Esq.*, 1880.

Macropus ræchus (Owen [1]).

Syn. *Protemnodon ræchus*, Owen [2].
Protemnodon antæus, Owen [3].

The difference in the length of the lower dental series of the two forms to which the above names have been applied not being relatively greater than that occurring in *M. giganteus*, both are regarded as specifically identical.

The mandible of this form agrees in its limits of size with that of *M. brehus*, but is distinguished by the longer $\overline{\text{pm. 4}}$ and the absence of a distinct hind talon to $\overline{\text{m. 4}}$. Since, however, as already mentioned, a similar variation occurs in $\underline{\text{pm. 4}}$ of the upper jaws included under *M. brehus*, and the case of *M. anak* shows that a talon may be present or absent in the lower molars of a single species, there is every probability that the present form is merely a variety of the former species.

Hab. Queensland.

35968. The anterior portion of the left ramus of the mandible,
(*Fig.*) containing the first four cheek-teeth ; from the Pleistocene of Gowrie, Queensland. This specimen is the type, and is

[1] Phil. Trans. 1874, p. 281.—*Protemnodon.* [2] *Loc. cit.*
[3] Extinct Mammals of Australia, p. 448 (1877).

figured by Owen in the 'Phil. Trans.' 1874, pl. xxvii. figs. 10–13, and also in the 'Extinct Mammals of Australia,' pl. lxxxvii. figs. 10–13. The length of the space occupied by the four teeth is 0,071.

Presented by Sir D. Cooper, Bart., 1866.

39999. The imperfect hinder half of the left ramus of the mandible, containing the last three true molars; from Queensland.

Same history.

38613. Hinder part of the right ramus of the mandible, containing the last two true molars; from the Pleistocene of Eton Vale, Queensland.

Presented by Sir D. Cooper, Bart., 1864.

40002. Hinder part of the right ramus of the mandible of an immature individual, showing $\overline{\text{m. 2}}$ and $\overline{\text{m. 3}}$; from Queensland. *Presented by Sir D. Cooper, Bart.*, 1866.

The following specimens include the types of the smaller form described as Protemnodon antæus.

M. 2258. The slightly imperfect left ramus of the mandible, showing
(*Fig.*) all the cheek-teeth and the alveolus of the incisor; from Queensland. Figured by Owen in the 'Extinct Mammals of Australia,' pl. cx. figs. 1–3. The length of the first four teeth is 0,061, and that of the entire series 0,080.

Presented by Dr. George Bennett.

43581. The imperfect right ramus of the mandible, showing the
(*Fig.*) first three true molars in a broken condition; from the Pleistocene of the Condamine River, Queensland. Figured by Owen, *op. cit.* pl. cx. figs. 6–9.

Presented by Dr. George Bennett, 1872.

43379. The hinder half of the right ramus of the mandible, containing the four true molars; from the Pleistocene of Clifton, Queensland. Noticed by Owen, *op. cit.* p. 449.

Presented by Dr. George Bennett, 1872.

38757. Fragment of the right ramus of the mandible, containing the last two true molars; from the same locality.

Presented by F. N. Isaac, Esq., 1861.

48422. The slightly imperfect right ramus of the mandible, showing the four true molars and the alveolus of the incisor; from the Pleistocene of Gowrie, Queensland.

Presented by Dr. George Bennett, 1877.

50051. The greater part of the right ramus of the mandible, showing the four true molars; from the same locality.

Presented by G. F. Bennett, Esq., 1879.

Macropus anak, Owen [1].

Syn. *Protemnodon anak,* Owen [2] (*in parte*).
Protemnodon og, Owen [3].
Sthenurus atlas, Owen [4] (*in parte*).

This species was founded upon a mandibular ramus, agreeing very closely in structure with the mandible of *M. brehus,* but of inferior size, and having no hind talon to $\overline{\text{m. 4}}$. Other specimens have, however, a small talon, and thus connect the typical form with the slightly larger one described by Owen as *Protemnodon og,* which may apparently be regarded merely as a variety. The confusion noticed under the head of *Sthenurus atlas* in regard to the proper upper dentition of that species led Owen to figure the upper jaws of the present species under that name, while the proper upper jaws of *S. atlas* were referred to the present species ([5]).

In the two complete maxillæ the length of the cheek-series is 0,070 and 0,072; the length of the last three molars being 0,045; but since other specimens show that the latter dimension varies from 0,043 to 0,046, it is evident that a larger variation would be found in the length of the complete series, if we had more complete specimens. In the mandible the length of the cheek-series varies from 0,057 to 0,071.

This species is apparently very closely allied to *M. brehus,* there being only a slight difference in size between the largest individuals of the former and the smallest of the latter.

Hab. Queensland, South Australia, and New South Wales.

The following specimens include those referred by Owen to Sthenurus atlas.

45934. Part of the right maxillary region, showing the five cheek-
(*Fig.*) teeth; from the Pleistocene of South Australia. Figured

[1] Proc. Geol. Soc. vol. xv. p. 185 (1859). [2] Phil. Trans. 1874, p. 275.
[3] *Ibid.* p. 277. [4] *Ibid.* p. 265.
[5] Apart from all other evidence it is quite certain that the upper premolars of the under-mentioned specimens could not have been opposed to $\overline{\text{pm. 4}}$ of *Sthenurus atlas,* while they agree precisely with the corresponding tooth of the so-called *Protemnodon anak.*

by Owen in the 'Phil. Trans.' 1874, pl. xxiv. figs. 4–6, and also in the 'Extinct Mammals of Australia,' pl. lxxxiv. figs. 4–6. The length of the cheek-series is 0,072, and that of the three true molars 0,045.

Presented by Prof. Sir R. Owen, K.C.B., 1874.

46070. The left maxillary region, showing all the cheek-teeth except m. 4; from the Pleistocene of Queensland.

Presented by Dr. George Bennett, 1874.

M. 43. (*Fig.*) Part of the right maxilla of a somewhat smaller individual, containing all the cheek-teeth; from the Pleistocene of

Fig. 36.

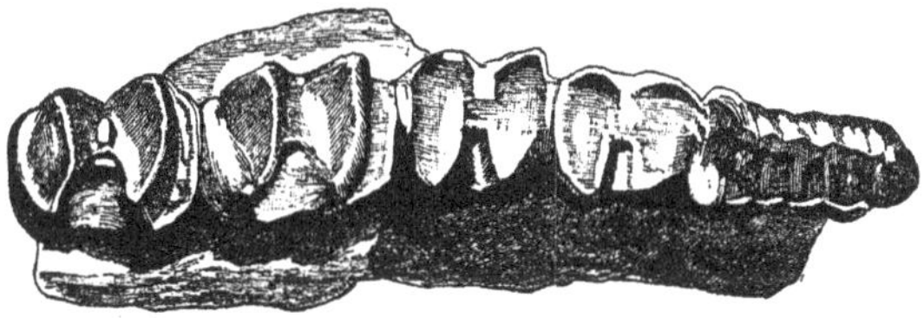

Macropus anak.—The right upper cheek-dentition; from the Pleistocene of Queensland. 1/1.

Queensland. In this specimen (woodcut fig. 36) the length of the cheek-series is 0,070, and that of the last three true molars 0,043. *Presented by G. F. Bennett, Esq.*

M. 2262. (*Fig.*) The imperfect anterior portion of the palate, showing the left pm. 4 and m. 1, and the incisive alveoli of both sides; from Queensland. Figured by Owen in the 'Extinct Mammals of Australia,' pl. lxxxviii. figs. 1–4.

Presented by Dr. George Bennett.

47838. (*Fig.*) The crushed anterior portion of the cranium of an immature individual, showing i. 1, i. 2, pm. 3, mm. 4, m. 1, and m. 2, in use, and pm. 4 *in alveolo*; from Queensland. Figured by Owen in the 'Phil. Trans.' 1876, pl. xxv. fig. 2, and pl. xxvi. fig. 4. *Presented by Dr. George Bennett*, 1876.

38751. Fragment of the left maxilla of a large individual, showing the last three true molars; from the Pleistocene of Darling Downs, Queensland. The length of the three teeth is 0,046. *Presented by F. N. Isaac, Esq.*, 1861.

50064 b. Two fragments of the maxillæ of opposite sides, each containing the last three true molars; from Queensland.
Presented by G. F. Bennett, Esq., 1879

M. 2263. Fragment of the right maxilla, containing the last three true molars; from Queensland.
Presented by Dr. George Bennett.

M. 2261. Fragment of the right maxilla, containing $\underline{\text{pm. 4}}$ and $\underline{\text{m. 1}}$;
(*Fig.*) from Queensland. Figured by Owen in the 'Extinct Mammals of Australia,' pl. lxxxviii. figs. 2, 3.
Presented by Dr. George Bennett.

50064 a. Fragment of the right maxilla of a small individual containing the four true molars, of which the first is much worn; from Queensland.
Presented by G. F. Bennett, Esq., 1879.

The following specimens include those referred by Owen to Protemnodon anak.

38753. The left ramus of the mandible, imperfect posteriorly, but
(*Fig.*) showing the entire dentition; from Queensland. Figured by Owen in the 'Phil. Trans.' 1874, pl. xxv. figs. 7–10, and also in the 'Extinct Mammals of Australia,' pl. lxxxv. figs. 7–10. The length of the cheek-series is 0,066, and that of the last three true molars 0,0425. There is no distinct hind talon to $\overline{\text{m. 4}}$.
Presented by F. N. Isaac, Esq., 1861.

35964. The greater part of the left ramus of the mandible, showing the four true molars and the hinder half of $\overline{\text{pm. 4}}$; from Gowrie. The true molars are more worn than those of the last specimen. Figured by Owen in the 'Phil. Trans.' 1874, pl. xxv. figs. 3, 4, and also in the 'Extinct Mammals of Australia,' pl. lxxxv. figs. 3–5.
Presented by Sir D. Cooper, Bart., 1861.

50060. The greater part of the left ramus of the mandible, showing the five cheek-teeth; from Gowrie. This specimen agrees almost precisely with No. 38753, with the exception that $\overline{\text{m. 4}}$ has a distinct hind talon. It also agrees so closely with the maxilla No. 46078, that it might well have belonged to the same individual.
Presented by G. F. Bennett, Esq., 1879.

50059. Fragment of the right ramus of the mandible, containing the four true molars and the base of $\overline{\text{pm. 4}}$; from Gowrie.
Presented by G. F. Bennett, Esq., 1879.

M. 1895. Part of the left ramus of the mandible, showing the five (*Fig.*) cheek-teeth; from Darling Downs. Figured by Owen in the 'Phil. Trans.' 1874, pl. xxv. figs. 1, 2. There is a slight talon to $\overline{\text{m. 4}}$.
Presented by Sir R. Owen, K.C.B., 1884.

50061. Part of the right ramus of the mandible, containing the four true molars and the hinder half of $\overline{\text{pm. 4}}$; from Gowrie. There is a small talon to $\overline{\text{m. 4}}$.
Presented by G. F. Bennett, Esq., 1879.

50054. The greater part of the left ramus of the mandible of an old individual, showing the last three true molars, and the alveolus of $\overline{\text{pm. 4}}$; from Gowrie.
Presented by G. F. Bennett, Esq., 1879.

35965. Part of the left ramus of the mandible, showing $\overline{\text{pm. 4}}$ and $\overline{\text{m. 3}}$; from Gowrie.
Presented by Sir D. Cooper, Bart., 1861.

43349 x. A detached fourth lower premolar; from Gowrie.
Presented by G. F. Bennett, Esq., 1872.

38615. Fragment of the left ramus of the mandible of a young individual, provisionally referred to this species; from Queensland. The last milk- and first true molars are in use, and $\overline{\text{pm. 4}}$ is seen *in alveolo*.
Presented by Sir D. Cooper, Bart., 1864.

The types of Protemnodon og *are included among the following.*

35963. The imperfect left ramus of the mandible, showing all the (*Fig.*) cheek-teeth; from Gowrie. The first true molar is considerably worn; the length of the series of cheek-teeth is 0,071, and that of the last three true molars 0,046; there is a distinct hind talon to $\overline{\text{m. 4}}$. Figured by Owen in the 'Phil. Trans.' 1874, pl. xxv. figs. 5, 6, and also in the 'Extinct Mammals of Australia,' pl. lxxxv. figs. 5, 6.
Presented by Sir D. Cooper, Bart., 1866.

M. 3451. The nearly entire left ramus of the mandible of a large (*Fig.*) individual, showing the incisor, pm. 4, the hinder half of $\overline{\text{m. 2}}$, and $\overline{\text{m. 3}}$; from Queensland. Figured by Owen in the 'Extinct Mammals of Australia,' pl. lxxxv. fig. 6. The slightly worn condition of the cheek-teeth makes the length of the series relatively longer than in the preceding specimen. *Presented by Dr. George Bennett*, 1886.

35967. (*Fig.*) Hinder part of the right ramus of the mandible, containing the last three true molars ($\overline{\text{m. 2}}$ imperfect); from Gowrie. Figured by Owen in the 'Phil. Trans.' 1874, pl. xxv. figs. 11–13, and also in the 'Extinct Mammals of Australia,' pl. lxxxv. figs. 11–13. The talon to $\overline{\text{m. 4}}$ is very slight. *Presented by Sir D. Cooper, Bart.*, 1866.

M. 48. Fragment of the left ramus of the mandible, containing the last three true molars; from Queensland. The talon to $\overline{\text{m. 4}}$ is scarcely perceptible.
Presented by G. F. Bennett, Esq., 1880.

40003. Fragment of the right ramus of the mandible, showing $\overline{\text{m. 3}}$ and $\overline{\text{m. 4}}$; from Queensland. The talon of $\overline{\text{m. 4}}$ is almost imperceptible. *Presented by Sir D. Cooper, Bart.*, 1866.

M. 42. The anterior part of the left ramus of the mandible, showing $\overline{\text{pm. 4}}$ and the alveolus of the incisor; from Queensland.
Presented by G. F. Bennett, Esq., 1880.

M. 41. Fragment of the left ramus of the mandible of an immature individual, showing $\overline{\text{pm. 3}}$, $\overline{\text{mm. 4}}$, $\overline{\text{m. 1}}$, and $\overline{\text{m. 2}}$, and $\overline{\text{pm. 4}}$ *in alveolo*; from Queensland. *Same history.*

50063. Part of the right ramus of the mandible of a young individual, showing $\overline{\text{pm. 3}}$ (imperfect), $\overline{\text{mm. 4}}$, $\overline{\text{m. 1}}$, and $\overline{\text{m. 2}}$ in use, and $\overline{\text{pm. 4}}$ *in alveolo*; from Queensland.
Presented by G. F. Bennett, Esq., 1879.

Macropus minor (Owen [1]).

Syn. *Sthenurus minor*, Owen [2].

Although the name *M. minor* is preoccupied by a synonym of *Potorous tridactylus* (*Hypsiprymnus murinus*) it seems inadvisable to apply a new one to the present form.

This species is of smaller size than the last, and is distinguished by the proportionately shorter pm. 4, and the relatively narrower upper true molars.

Hab. New South Wales and Queensland.

[1] Proc. Zool. Soc. 1877, p. 353.—*Sthenurus*. [2] *Loc. cit.*

48409. The imperfect palatal region of the cranium of an immature
(*Fig.*) individual; from the Pleistocene of County Phillip, New South Wales. This specimen is the type, and is figured by Owen in the 'Proc. Zool. Soc.' 1877, pl. xxxvii., and pl. xxxviii. figs. 1–3. The four true molars are shown on both sides, but while on the left side pm. 4 is in position, on the right it is still *in alveolo*, pm. 3 and mm. 4 being in use. *Presented by Rev. W. B. Clarke*, 1877.

M. 47. Fragment of the left maxilla, showing the first three cheek-teeth; from the Pleistocene of Queensland.
Presented by G. F. Bennett, Esq., 1880.

M. 47 a. Fragment of the right maxilla of an immature individual, showing mm. 4, m. 1, and m. 2 in use, and pm. 4 *in alveolo*; from Queensland. *Same history.*

40380. The imperfect left ramus of a mandible, with the four true molars and the base of pm. 4, probably belonging to this species; from the Condamine River.
Presented by Dr. F. Campbell, 1867.

50063 a. Part of the left ramus of the mandible of a young individual, probably belonging to this species; from Queensland. The first three true molars and pm. 4 are shown.
Presented by G. F. Bennett, Esq., 1879.

Macropus ualabatus (Lesson and Garnot[1]).

Syn. *Kangurus ualabatus*, Lesson and Garnot[2].
Halmaturus lessoni, Gray[3].
Halmaturus nemoralis, Wagner[4].
Halmaturus ualabatus, auct.

This species is smaller than any of the preceding ones, but appears to be their nearest living ally, having the fourth upper premolar usually equal in length to one and a half times that of the first true molar, and the upper true molar of a similar structure; the third upper incisor is vertically grooved, and somewhat larger than the first; there are large palatal vacuities. The lower true molars are of a short and simple type, like those of the preceding extinct forms. The length of the upper series of cheek-teeth averages 0,039, and that of the lower jaw 0,0395.

Hab. New South Wales.

[1] Voyage de la Coquille, Zool. vol. i. p. 161 (1826).—*Kangurus.*
[2] *Loc. cit.*
[3] Charlesworth's Mag. Nat. Hist. vol. i. p. 583 (1837).
[4] In Schreber's 'Säugethiere,' Suppl. vol. iii. p. 114 (1843).

43854. The imperfect right ramus of the mandible, containing the five cheek-teeth; from a cave in the Wellington Valley, New South Wales.

Presented by the Trustees of the Australian Museum, 1870.

42591. Part of the right ramus of the mandible of a young individual, showing $\overline{\text{pm. 3}}$, $\overline{\text{mm. 4}}$, and $\overline{\text{m. 1}}$ in use, and $\overline{\text{pm. 4}}$ and $\overline{\text{m. 2}}$ *in alveolo*, together with the basal half of the incisor; from the same locality. *Same history.*

42668. The right humerus; from the same locality. This specimen shows the ill-defined distal trochleæ so characteristic of the species, whereby it differs widely from the corresponding bone of the *M. bennetti* type (No. 42068, page 240). *Same history.*

42643 a. The left calcaneum; from the same locality. *Same history.*

Macropus parryi, Bennett [1].

Syn. *Halmaturus parryi*, Gould [2].
? *Osphranter gouldi*, Owen [3].
? *Macropus gouldi*, Flower [4].

Of somewhat larger size than *M. ualabatus*, with the fourth premolar decidedly shorter than the first true molar; and the true molars (especially those of the lower jaw) more elongated and rather more complex than in that species.

Hab. New South Wales (recent) and Queensland (Pleistocene).

43345. The greater part of the right maxilla of a nearly adult individual, showing the four true molars; from the Pleistocene of Darling Downs, Queensland. This specimen presents no characters by which it can be distinguished from the corresponding part of the cranium of a recent example.

Presented by Dr. George Bennett, 1872.

43345 a. Part of the right ramus of the mandible, containing the four true molars; from the same locality. This specimen cannot be distinguished from the type of the so-called *Osphranter gouldi* figured by Owen in the 'Phil. Trans.' 1874, pl. xxiii. figs. 15, 16. The latter is indeed stated

[1] Proc. Zool. Soc. 1834, p. 151.
[2] Monograph of Macropodidæ, pl. xix. (1842).
[3] Phil. Trans. 1874, p. 261.
[4] Cat. Vert. Anim. Mus. R. Coll. Surg. pt. 2, p. 719 (1884).

to belong to an immature animal, but there is no evidence that such is the case, and it might well have belonged to an individual slightly older than the present specimen.

Presented by Dr. George Bennett, 1872.

Macropus penicillatus, Gray[1].

Syn. *Petrogale penicillata*, Gray[2].
Heteropus penicillatus, auct.

Of rather smaller size than the preceding, with true molars of the same structure, but with the fourth premolar longer than the first true molar. Since no molar characters generically distinguish this, the type species of *Petrogale*, from the preceding and other forms, it is advisable that in a palæontological classification it should be included in the same genus.

Hab. New South Wales.

42658. The imperfect palate, showing all the cheek-teeth of the left side; from a cave in the Wellington Valley, New South Wales.

Presented by the Trustees of the Australian Museum, 1870.

42661 a. Fragment of the right maxilla, containing the first and second true molars; from the same locality.

Same history.

42591. The greater portion of the right ramus of the mandible, containing the five cheek-teeth; from the same locality.

Same history.

32316. Part of the left ramus of the mandible, showing a portion of the incisor and the first four cheek-teeth; from the same locality. *Purchased*, 1857.

32316 a. Two imperfect mandibular rami of young individuals, showing $\overline{\text{pm. 3}}$, $\overline{\text{mm. 4}}$, $\overline{\text{m. 1}}$, and $\overline{\text{m. 2}}$; from the same locality. The outer wall of one specimen has been cut away in order to exhibit $\overline{\text{pm. 4}}$ in its alveolus.

Same history.

42660 b. Six imperfect mandibular rami of young individuals, showing the same teeth as the preceding specimens; from the same locality.

Presented by the Trustees of the Australian Museum, 1870.

42663. Anterior portion of the left ramus of the mandible of a

[1] In Griffith's 'Animal Kingdom,' vol. v. p. 204 (1827).
[2] Charlesworth's Mag. Nat. Hist. vol. i. p. 583 (1837).

young individual, showing the incisor, $\overline{\text{pm. 3}}$, $\overline{\text{mm. 4}}$, and $\overline{\text{m. 1}}$ in use, and $\overline{\text{pm. 4}}$ *in alveolo*; from the same locality.
Presented by the Trustees of the Australian Museum, 1876.

42664. Fragment of the left ramus of the mandible of a young individual, showing the incisor and $\overline{\text{pm. 3}}$; from the same locality. *Same history.*

42660 a. Fragment of the left ramus of the mandible, containing the first three cheek-teeth; from the same locality. *Same history.*

Some or all of the following specimens probably belong to this species; all are from the Wellington Valley.

42668. The imperfect right humerus. *Same history.*

42669. The proximal half of the left ulna. *Same history.*

42595. The imperfect right innominate. *Same history.*

32321. Two specimens of the femur. *Purchased*, 1857.

43939. Eight specimens of the calcaneum.
Presented by the Trustees of the Australian Museum, 1870.

43927. Three specimens of the fourth metatarsal. *Same history.*

Macropus robustus, Gould[1].

Syn. *Petrogale robusta*, Gould[2].
Osphranter robustus, Gray[3].

In this species, which is one of the largest existing Kangaroos, the upper true molars (as in all the preceding species) have no longitudinal bridge between the anterior ridge and talon; pm. 4 is shorter than m. 1; $\overline{\text{pm. 4}}$, which has a simple secant crown, is frequently shed in old individuals; the alveolar border of the mandibular diastema descends rapidly in advance of the last-named tooth; and $\underline{\text{i. 3}}$ is of moderate length. *M. erubescens*, Sclater, cannot be palæontologically separated from this species. The molars in this and the following species are more elongated and somewhat more complex than in all the preceding forms.

Hab. New South Wales and Queensland.

50065. The palatal region of a nearly adult individual; from the Pleistocene of Gowrie, Queensland. All the cheek-teeth are

[1] Proc. Zool. Soc. 1840, p. 92.
[2] Monograph of Macropodidæ, pl. v. (1841).
[3] List Spec. Mamm. Brit. Mus. p. 91 (1843).

shown, except the right m. 4; and the specimen presents no characters by which it can be distinguished from a recent male cranium; it has a somewhat more recent appearance than most of the other Gowrie fossils.

Presented by G. F. Bennett, Esq., 1879.

35972. The anterior portion of the right ramus of the mandible of a nearly adult individual, probably belonging to the present species, and containing m. 2 and m. 3; from Gowrie.

Presented by Sir D. Cooper, Bart., 1861.

Macropus altus (Owen[1]).

Syn. *Phascolagus altus*, Owen[2].

This species is a larger form, closely allied to the preceding, to which it apparently presents the same relation as is borne by *M. titan* to *M. giganteus*; it is, however, distinguished by pm. 4 being longer than m. 1.

Hab. New South Wales and Queensland.

32296. The imperfect palate of a nearly adult individual, showing all the cheek-teeth except the right m. 4; from a cave in the Wellington Valley, New South Wales. Described by Owen in the 'Phil. Trans.' 1874, p. 264.

Purchased, 1857.

32317. Fragment of the right maxilla, showing the first three true molars; from the same locality. Described by Owen, *op. cit.* p. 263. *Purchased*, 1857.

38610. Part of the left maxilla, containing the four true molars and the alveolus of pm. 4; from the Pleistocene of Eton Vale, Queensland. *Presented by Sir D. Cooper, Bart.*, 1864.

32297. Part of the right maxilla, showing the last three true molars; from the Wellington Valley. *Purchased*, 1857.

50064 d. Part of the left maxilla, containing the last three true molars; from Gowrie.

Presented by G. F. Bennett, Esq., 1879.

43224. The extremity of the premaxillæ, showing the first and second incisors, of a Kangaroo not improbably belonging to the present species; from Queensland.

Presented by Sir R. Daintree, 1871

[1] Phil. Trans. 1874, p. 261.—*Phascolagus*. [2] *Loc. cit.*

46918. The nearly entire right ramus of the mandible of a nearly adult individual; from the Pleistocene of Kerban, near Mendoran, New South Wales. All the teeth are shown with the exception of $\overline{\text{pm. 4}}$, which has been shed, but $\overline{\text{m. 4}}$ is not fully protruded. The specimen agrees very closely in structure with the mandible of *M. robustus*, but accords in relative size with the maxillæ of the present form.
Presented by W. L. R. Gipps, Esq., 1875.

32302. Part of the right ramus of the mandible, showing the last three true molars; from a cave in the Wellington Valley.
Purchased, 1857.

32299. Part of the left ramus of the mandible, containing the first three true molars; from the same locality.
Same history.

40010. The anterior portion of the left ramus of the mandible, containing the incisor, $\overline{\text{pm. 4}}$, $\overline{\text{m. 1}}$, and $\overline{\text{m. 2}}$; from Queensland. This specimen agrees precisely with No. 46918; the premolar resembles in structure the corresponding tooth of *M. robustus*, but is longer than $\overline{\text{m. 1}}$.
Presented by Sir D. Cooper, Bart., 1866.

32887. The greater part of the right ramus of a mandible probably belonging either to a small individual of the present form, or to a large one of *M. robustus*; from the Pleistocene of the Condamine River, Queensland. *Purchased*, 1857.

Macropus cooperi (Owen[1]).

Syn. *Osphranter cooperi*, Owen[2].

This imperfectly known form is apparently allied to *M. robustus*, but is distinguished both from that species and *M. altus* by the much slighter downward inclination of the alveolar border of the mandibular diastema in advance of the premolars.

Hab. Queensland.

32886. (*Fig.*) The anterior portion of the left ramus of the mandible of an old individual, containing the first three cheek-teeth; from the Pleistocene of the Condamine River, Queensland. This specimen is the type, and is figured by Owen in the 'Phil. Trans.' 1874, pl. xxiv. figs. 17, 18, and also in the 'Extinct Mammals of Australia,' pl. lxxxiv. figs. 17, 18.
Purchased, 1857

[1] Phil. Trans. 1874, p. 261.—*Osphranter*. [2] *Loc. cit.*

Macropus giganteus (Zimmermann [1]).

Syn. *Yerboa gigantea*, Zimmermann [2].
Macropus major, Shaw [3].

This is one of the largest existing species. The fourth premolar is shorter than the first true molar; the upper true molars have a longitudinal bridge connecting the anterior talon with the first ridge; the third upper incisor is much larger than either of the others; the mandibular diastema is long, the ramus being very shallow at this part. The fourth premolar is always shed before the last true molar comes into use, and soon after the first true molar is also frequently lost. The crown of the fourth lower premolar is divided by a vertical groove. The length of the upper true molar series in an individual in which $\underline{\text{m. }4}$ is just coming into use is 0,048.

Hab. Australia and Tasmania.

42642. Fragment of the right maxilla of an immature individual, showing the first four cheek-teeth ($\underline{\text{pm. }4}$ broken) in use, and $\underline{\text{m. }4}$ partially protruded from its alveolus; from a cave in the Wellington Valley, New South Wales. Since this specimen agrees precisely in dimensions with existing examples, it is referred to the present species.
Presented by the Trustees of the Australian Museum, 1870.

The following specimens not improbably belong to this species.

M. 1515 a. Hinder part of the right ramus of the mandible, with the last three true molars; from King's Creek, Clifton, Queensland. *Presented by C. H. Hartmann, Esq.*, 1884.

42591. Two fragments of right mandibular rami of immature individuals, one with the last two, and the other with the last three true molars; from a cave in the Wellington Valley, New South Wales.
Presented by the Trustees of the Australian Museum, 1870.

Macropus titan, Owen [4].

This species is closely allied to *M. giganteus*, but is distinguished by its superior size, and the very general occurrence of one or more vertical grooves on the hinder surfaces of the lower true molars; it

[1] Spec. Zool.-Geogr. p. 526 (1777).—*Yerboa*. [2] *Loc. cit.*
[3] Gen. Zoology, vol. i. pt. 2, p. 505 (1800).
[4] In Mitchell's 'Three Expeditions into Australia,' 2nd ed. vol. ii. p. 360 (1838).

is, however, probable that the two forms pass imperceptibly into one another. The inner tubercle of pm. 4 is comparatively large.

Hab. Australia.

44118. (*Fig.*) The imperfect cranium, showing the four true molars (the left m. 1 absent) in a middle stage of wear; from the Pleistocene of King's Creek, Clifton, Queensland. Figured by Owen in the 'Extinct Mammals of Australia,' pls. lxxvi.–lxxix.

Presented by Dr. George Bennett, 1873.

35959–60. (*Fig.*) The imperfect palatal region, with the four true molars (m. 4 being just touched by wear) on the right side, and m. 3 and m. 4 on the left; from the Pleistocene of Gowrie, Queensland. The length of the space occupied by the four teeth is 0,061. The right half is figured by Owen in the 'Phil. Trans.' 1874, pl. xxi. figs. 15–17, and also in the 'Extinct Mammals of Australia,' pl. lxxxi. figs. 15–17. *Presented by Sir D. Cooper, Bart.*, 1861.

48421. Part of the right maxilla, showing the four true molars in a slightly earlier stage of wear; from Queensland. The length of the molar series is 0,056.

Presented by Dr. George Bennett, 1877.

50064 c. Part of the right maxilla, showing the last three true molars; from Gowrie.

Presented by G. F. Bennett, Esq., 1879.

M. 44. Part of the left maxilla, showing the four true molars; from Queensland.

Presented by G. F. Bennett, Esq., 1880.

M. 46. The right half of the posterior portion of the palate, showing the four true molars and the alveolus of pm. 4; from Queensland. *Same history.*

38609. Fragment of the left maxilla, containing the last three true molars; from Gowrie.

Presented by Sir D. Cooper, Bart., 1864.

M. 2266. Part of the right maxilla of a small individual, showing m. 3 and m. 4; from Queensland.

Presented by Dr. George Bennett, 1884.

39997. Part of the right maxilla, showing the last three true molars; from Queensland.

Presented by Sir D. Cooper, Bart., 1866.

39998. Fragment of the left maxilla, with m. 3 and m. 4; from Queensland. *Same history.*

35961. Fragment of the right maxilla, containing m. 3 and m. 4; from Gowrie. *Presented by Sir D. Cooper, Bart.*, 1861.

32294. (*Fig.*) The left half of the posterior portion of the palate of a nearly adult individual, containing the four true molars; from a cave in the Wellington Valley, New South Wales. Figured by Owen in the 'Phil. Trans.' 1874, pl. xxi. fig. 10, and also in the 'Extinct Mammals of Australia,' pl. lxxxi. fig. 10. *Purchased*, 1857.

M. 3669. Fragment of the left maxilla of an immature individual, containing pm. 4 and the first three true molars; from the Pleistocene of South Australia. *No history.*

32295. The middle portion of the cranium of a young individual, showing pm. 3, mm. 4, and m. 1 in use, and pm. 4 *in alveolo*; from a cave in the Wellington Valley. *Purchased*, 1857.

45935. (*Fig.*) Part of the left maxilla of a young individual, showing pm. 3, mm. 4, and the first three true molars in use, and pm. 4 *in alveolo*; from Queensland. Figured by Owen in the 'Phil. Trans.' 1874, pl. xxi. figs. 6–9, and also in the 'Extinct Mammals of Australia,' pl. lxxxi. figs. 6–9. *Presented by Sir R. Owen, K.C.B.*, 1874.

50064 f. Part of the right maxilla of a young individual, showing mm. 4, m. 1, and m. 2 in use and pm. 4 *in alveolo*; from Queensland. *Presented by G. F. Bennett, Esq.*, 1879.

38752. Part of the right maxilla of a young individual, having pm. 3, mm. 4, m. 1, and m. 2 in use, and pm. 4 *in alveolo*; from Queensland. Described by Owen in the 'Phil. Trans.' 1874, p. 263, and referred to *M. altus.* *Presented by Dr. George Bennett*, 1861.

50064 g. Part of the right maxilla of a young individual, showing the first three true molars protruded, and pm. 4 in its alveolus; from Queensland. *Presented by G. F. Bennett, Esq.*, 1879.

M. 1894. (*Fig.*) The imperfect mandible, showing the four true molars of either side; from Queensland. Figured by Owen in the 'Phil. Trans.' 1874, pl. xxi. figs. 12–14, and pl. xxvi. fig. 9. *Presented by Sir R. Owen, K.C.B.*, 1884.

32298. The greater part of the left ramus of the mandible of an immature individual, showing the first three true molars in use, and m. 4 *in alveolo*; from a cave in the Wellington Valley, New South Wales. *Purchased*, 1857.

38611. (*Fig.*) The greater portion of the right and the anterior extremity of the left ramus of the mandible; from Darling Downs. Figured by Owen in the 'Phil. Trans.' 1874, pl. xxvi. figs. 11, 12, and also in the 'Extinct Mammals of Australia,' pl. lxxxvi. figs. 11, 12. All the teeth except $\overline{\text{pm. 4}}$ are shown on the right side.
Presented by Sir D. Cooper, Bart., 1864.

48421 a. The greater part of the right ramus of the mandible, showing the last three true molars; from Queensland. Associated with the maxilla No. 48421.
Presented by Dr. George Bennett, 1874.

M. 3670. Cast of the greater portion of the right ramus of the mandible, showing the incisor and the four true molars. The original, of which the locality is unknown, is preserved in the Museum at Oxford, and is figured by Owen in the 'Phil. Trans.' 1874, pl. xxii. figs. 13–16.
Presented by Prof. J. Phillips, 1872.

50055. The greater part of the right ramus of the mandible of a fully adult individual, showing the last three true molars; from Gowrie. *Presented by G. F. Bennett, Esq.*, 1879.

50056. The greater part of the left ramus of the mandible, showing the last three true molars; from Gowrie. *Same history.*

32884. (*Fig.*) Middle portion of the left ramus of the mandible, showing the four true molars; from the Condamine River. Figured by Owen in the 'Phil. Trans.' 1874, pl. xxvi. figs. 13–15, and also in the 'Extinct Mammals of Australia,' pl. lxxxvi. figs. 13–15. *Purchased*, 1857.

43581 a. (*Fig.*) Middle portion of the left ramus of the mandible of an old individual, showing the last three true molars; from the Condamine River. Figured by Owen in the 'Phil. Trans.' 1874, pl. lxxvi. figs. 3, 5, 6, and also in the 'Extinct Mammals of Australia,' pl. lxxxix. figs. 3, 5, 6.
Presented by Dr. George Bennett, 1872.

43344. (*Fig.*) Part of the right ramus of the mandible of an old individual, with $\overline{\text{m. 4}}$; from Darling Downs. Figured by Owen in the 'Phil. Trans.' 1874, pl. lxxvi. figs. 2, 4, and also in the 'Extinct Mammals of Australia,' pl. lxxxix. figs. 2, 4.
Presented by Dr. George Bennett, 1872.

40000. Part of the right ramus of the mandible, containing the four true molars; from Queensland.
Presented by Sir D. Cooper, Bart., 1866.

35966. Part of the left ramus of the mandible, with $\overline{\text{m. 3}}$; from Gowrie. *Presented by Sir D. Cooper, Bart.*, 1861.

43958. Part of the left ramus of the mandible, showing the four true molars; from Darling Downs.
Presented by Dr. George Bennett, 1874.

50057. Part of the left ramus of the mandible, showing the four true molars; from Gowrie.
Presented by G. F. Bennett, Esq., 1879.

50053. The greater part of the right ramus of the mandible of a small individual, showing the four true molars; from Gowrie. *Same history.*

46071. Fragment of the left ramus of the mandible, with $\overline{\text{m. 3}}$ and $\overline{\text{m. 4}}$; from Queensland.
Presented by Dr. George Bennett, 1874.

M. 2267. Fragment of the right ramus of the mandible, with the last three true molars; from Queensland.
Presented by Dr. George Bennett.

32300. Fragment of the right ramus of the mandible of a small individual, with the last three true molars; from a cave in the Wellington Valley. The last molar is slightly larger than in the mandible (No. M. 1515) provisionally referred to *M. giganteus*. *Purchased*, 1857.

M. 3671. Part of the left ramus of the mandible of a large individual, showing $\overline{\text{m. 3}}$ and $\overline{\text{m. 4}}$; from Australia. *No history.*

M. 3672. Part of the left ramus of the mandible of a small individual, containing the last three true molars; from Queensland. *No history.*

M. 36. Part of the right ramus of the mandible, with $\overline{\text{m. 1}}$, $\overline{\text{m. 2}}$, and $\overline{\text{m. 3}}$; from Darling Downs.
Presented by G. F. Bennett, Esq., 1880.

38754. Part of the right ramus of the mandible of an immature individual, showing $\overline{\text{m. 2}}$ and $\overline{\text{m. 3}}$, and the broken bases of the preceding teeth; from Darling Downs.
Presented by F. N. Isaac, Esq., 1861

50062. The greater part of the right ramus of the mandible of a young individual, showing $\overline{\mathrm{pm.\,3}}$, $\overline{\mathrm{mm.\,4}}$, $\overline{\mathrm{m.\,1}}$, and $\overline{\mathrm{m.\,2}}$ in use, and $\overline{\mathrm{pm.\,4}}$ *in alveolo*; from Gowrie.
Presented by G. F. Bennett, Esq., 1879.

38755. Part of the left ramus of the mandible of a young individual, showing $\overline{\mathrm{m.\,1}}$ and $\overline{\mathrm{m.\,2}}$ in use, and $\overline{\mathrm{m.\,3}}$ emerging from its alveolus; from Darling Downs.
Presented by F. N. Isaac, Esq., 1861.

M. 3454. The greater part of the mandible of a young individual, showing $\overline{\mathrm{pm.\,3}}$, $\overline{\mathrm{mm.\,4}}$, and $\overline{\mathrm{m.\,1}}$ in use, and $\overline{\mathrm{m.\,2}}$ in its alveolus; from Queensland. This specimen was named *Macropus elatus*, Owen, MS.
Presented by Sir R. Owen, K.C.B., 1883.

32889. Fragment of the right ramus of the mandible of a young individual, showing $\overline{\mathrm{mm.\,4}}$; from Queensland.
Purchased, 1857.

32901. A lower incisor, probably belonging to this species; from Queensland. *Purchased*, 1857.

47854. Two similar specimens; from Queensland.
Presented by Dr. George Bennett, 1875.

Macropus ferragus, Owen [1].

Syn. *Pachysiagon ferragus*, Owen [2].
Leptosiagon gracilis, Owen [3].

This species, which was originally founded upon a portion of the mandible, appears allied to *M. titan*, but is of rather larger size, and has the upper true molars with their ridges placed somewhat more obliquely, the hind talon of $\underline{\mathrm{m.\,4}}$ of different form, and the crowns of both upper and true molars of slightly more complex structure.

Hab. Queensland and New South Wales.

32895. The hinder portion of the palate, showing the last three true molars of either side: from the Pleistocene of the Condamine River, Queensland. The resemblance between the teeth of this specimen and those of No. 32903 is such as to leave little, if any, doubt as to their specific identity. The length of the space occupied by the three teeth is 0,052, against 0,049 in *M. titan* No. 35959.
Purchased, 1857.

[1] Phil. Trans. 1874, p. 784.
[2] Extinct Mammals of Australia, p. 449 (1877).
[3] Phil. Trans. 1874, p. 785.

32896. Fragment of the left maxilla, containing the last three true molars; from the same locality. *Purchased*, 1857.

32903. (*Fig.*) Hinder part of the right ramus of the mandible, containing the last two true molars; from the same locality. This specimen is the type, and is figured by Owen in the 'Phil. Trans.' 1874, pl. lxxxi. fig. 4, and pl. lxxxii. figs. 3, 4, and also in the 'Extinct Mammals of Australia,' pl. xcvi. fig. 4, and pl. xcvii. figs. 3, 4. *Same history.*

40005. Fragment of the right ramus of the mandible, containing $\overline{\text{m. 2}}$ and $\overline{\text{m. 3}}$; from the Pleistocene of Queensland. This specimen is the type of *Leptosiagon gracilis* of Owen, and is figured by him in the 'Phil. Trans.' 1874, pl. lxxvi. figs. 11–15, and also in the 'Extinct Mammals of Australia,' pl. lxxxix. figs. 11–15. Allowing for the difference in their wear, the third molar of this specimen presents no characters by which it can be distinguished from the corresponding tooth of No. 32903.
Presented by Sir D. Cooper, Bart., 1866.

46917. Fragment of the right ramus of the mandible, showing the imperfect $\overline{\text{m. 2}}$ and $\overline{\text{m. 3}}$ in use and $\overline{\text{m. 4}}$ *in alveolo*; from the Pleistocene of Kirban, Mendoran, New South Wales. The teeth of this specimen agree precisely with those of the last. *Presented by W. L. R. Gibbs, Esq.*, 1875.

46916. Fragment of a left maxilla, from the same locality as the preceding specimen, containing the last three true molars in an imperfect condition, and not improbably belonging to the same species. *Same history.*

Genus **STHENURUS**, Owen [1].

Syn. *Protemnodon*, Owen [2] (*in parte*).

The fourth upper premolar develops a complete inner and the lower one a corresponding outer lobe, so that the worn crowns of these teeth present oval, flat surfaces, and have no secant edge. The true molars have no vertical enamel-folds, and are short and

[1] Phil. Trans. 1874, p. 264.

[2] *Ibid.* p. 274. In describing the immature maxillæ (Phil. Trans. 1874, pl. xxiii. figs. 4–9) on which *Protemnodon* was founded, Owen mistook the inner lobe of $\underline{\text{pm. 4}}$ (which was the only part he saw) for the inner surface of the outer lobe, and consequently concluded that there was no inner ridge or tubercle to this tooth. He therefore regarded the tooth as being of a more secant type than $\underline{\text{pm. 4}}$ of *Macropus*, and consequently associated it with the secant lower

wide; the longitudinal bridge connecting the ridges being very imperfect, and the anterior talon of the upper molars unconnected by such a bridge with the first ridge. The mandibular symphysis is not anchylosed, and the lower incisors are of the macropine type.

The genus connects *Macropus* with *Procoptodon*.

Sthenurus atlas, Owen [1].

Syn. *Macropus atlas*, Owen [2].
Protemnodon anak, Owen [3] (*in par e*).

This is the type and only species [4], and is nearly of the size of *Macropus anak*. The type mandible is figured in the 'Phil. Trans.' 1874, pl. xxii. figs. 3, 4, and in the 'Extinct Mammals of Australia,' pl. lxxxii. figs. 3, 4.

Hab. Queensland and New South Wales.

42592. (*Fig.*) Part of the left maxilla of a young individual, showing pm. 3, mm. 4, and m. 1 in use, and the germ of m. 2 *in alveolo*; from a cave in the Wellington Valley, New

Fig. 37.

Sthenurus atlas.—Part of left maxilla of a young individual; from a cave in the Wellington Valley. ½.

South Wales. This specimen (fig. 37) agrees with the examples figured in the 'Phil. Trans.' 1874, pl. xxiii. figs. 4–9 ('Extinct Mammals of Australia,' pl. lxxxiii. figs. 4–9), under the name of *Protemnodon anak*. The third premolar, although relatively shorter, agrees in

tooth of *M. anak*, instead of with the present form, to which he referred the upper jaw of *M. anak*. In his earlier 'Catalogue of Fossil Mammalia in the Museum of the College of Surgeons,' the above-mentioned maxillæ were correctly referred to *M. atlas*. Owing to this confusion the upper jaws of several species of *Macropus* were referred to *Sthenurus*, and their lower jaws to the so-called *Protemnodon*, which was really founded upon the upper jaw of *Sthenurus atlas*.

[1] In Mitchell's 'Three Expeditions into Australia,' 2nd ed. vol. ii. p. 359 (1838).—*Macropus*.
[2] *Loc. cit.*
[3] Phil. Trans. 1874, p. 275.
[4] The other species referred by Owen to this genus belong to *Macropus*.

character with $\overline{\text{pm. 4}}$ of the following mandibles, with the corresponding teeth of which the true molars also agree in size and structure.

Presented by the Trustees of the Australian Museum, 1870.

42663. Fragment of the left maxilla, containing the four true molars (the first and fourth imperfect); from the same locality. *Same history.*

40001. (*Fig.*) Part of the left ramus of the mandible, containing all the cheek-teeth except $\overline{\text{m. 4}}$; from the Pleistocene of Gowrie, Queensland. Figured by Owen in the 'Phil. Trans.' 1874, pl. xxii. fig. 9, and pl. xxiv. figs. 7, 8, and also in the 'Extinct Mammals of Australia,' pl. lxxxii. fig. 9, and pl. lxxxiv. figs. 7, 8.

Presented by Sir D. Cooper, Bart., 1866.

M. 3673. Cast of the slightly imperfect left ramus of the mandible of a nearly adult individual, showing all the teeth; $\overline{\text{pm. 4}}$ being only partially protruded. The original, which is preserved in the Museum at Oxford, was obtained from the Pleistocene of Queensland, and is figured by Owen in the 'Phil. Trans.' 1874, pl. xxii. figs. 5–8, and also in the 'Extinct Mammals of Australia,' pl. lxxxii. figs. 5–8.

Presented by Prof. J. Phillips, 1872.

42663. Fragment of the left ramus of the mandible of an immature individual, showing $\overline{\text{m. 3}}$ fully protruded and $\overline{\text{m. 4}}$ *in alveolo*; from a cave in the Wellington Valley.

Presented by the Trustees of the Australian Museum, 1870.

Genus **PROCOPTODON,** Owen [1].

The mandibular symphysis is anchylosed in the adult, and the ramus of the mandible short and deep, the diastema being also short. The premolars resemble those of *Sthenurus* in structure; but the true molars are elongated, and usually have their enamel thrown into a series of vertical folds. There are large palatal vacuities; and the lower incisors are subcylindrical.

The metatarsals referred by Owen to this genus (*infrà*, p. 248) are of a shorter type than those of the existing species of *Macropus*, and, if rightly associated, indicate that the disproportion between the anterior and posterior limbs was less marked than in the latter.

[1] Phil. Trans. 1874, p. 788.

Procoptodon goliah, Owen [1].

Syn. *Măcropus goliah*, Owen [2].

This is the type and largest species. The vertical folds in the enamel of the true molars are strongly marked.

Hab. Queensland and New South Wales.

46311. (*Fig.*) Part of the left maxilla, showing the five cheek-teeth in an imperfect condition; from the Pleistocene of Queensland. Figured by Owen in the 'Phil. Trans.' 1874, pl. lxxix. figs. 2–4, and also in the 'Extinct Mammals of Australia,' pl. xciv. figs. 2–4. The length of the space occupied by the five cheek-teeth is 0,100, and that by the four true molars 0,082. *Presented by Dr. George Bennett.*

M. 1896. (*Fig.*) Part of the right maxilla, containing the last three true molars; from Queensland. This specimen is the type, and is figured by Owen in the 'Phil. Trans.' 1874, pl. lxxix. fig. 1, and also in the 'Extinct Mammals of Australia,' pl. xciv. fig. 1.

Presented by Sir R. Owen, K.C.B., 1884 [3].

46312. (*Fig.*) Part of the left maxilla, containing the imperfect $\underline{\text{pm. }4}$ and the first three true molars; from Queensland. Figured by Owen in the 'Phil. Trans.' 1874, pl. lxxix. figs. 5, 6, 7, 9, and also in the 'Extinct Mammals of Australia,' pl. xciv. figs. 5, 6, 7, 9. *Presented by Dr. George Bennett.*

M. 3674. Cast of the hinder part of the left ramus of the mandible, showing $\overline{\text{m. }3}$ and $\overline{\text{m. }4}$. The original, the precise locality of which is unknown, is preserved in the Australian Museum, Sydney, and is figured by Owen in the 'Phil. Trans.' 1874, pl. lxxix. fig. 8, and pl. lxxx. figs. 3, 4.

Presented by Sir R. Owen, K.C.B., 1884.

M. 1897. (*Fig.*) Part of the right ramus of the mandible of a nearly adult individual, showing the four true molars in an early stage of wear. Figured by Owen in the 'Phil. Trans.' 1874, pl. lxxx. figs. 1, 2, and also in the 'Extinct Mammals of Australia,' pl. xcv. figs. 1, 2.

Presented by Sir R. Owen, K.C.B., 1884.

M. 3675. Part of the right ramus of the mandible, containing portions of $\overline{\text{m. }1}$ and $\overline{\text{m. }2}$, and the entire $\overline{\text{m. }3}$ and $\overline{\text{m. }4}$; from Queensland. The premolar has been shed and its alveolus obliterated. *Presented by Sir R. Owen, K.C.B.*, 1884.

[1] In Waterhouse's Mammalia, vol. i. p. 59 (1846).—*Macropus.*

[2] *Loc. cit.*

[3] Said to have been received from Sir T. Mitchell in 1844.

32902. Part of the left ramus of the mandible, showing the hinder molars in a broken condition; from Queensland.
Purchased, 1857.

32902 a. The second left lower true molar; from Queensland.
Purchased, 1857.

Procoptodon rapha, Owen [1].

Syn. *Macropus rapha*, Flower [2].
Procoptodon pusio, Owen [3] (*in parte*).

This form is mainly, if not entirely, distinguished from the preceding species by its inferior dimensions, and since there is a considerable amount of variation in the size of the following specimens, it is not improbable that it will eventually prove merely a small race of that species.

Hab. Queensland and New South Wales.

39996. (*Fig.*) The imperfect palate of a small immature individual apparently belonging to this species; from the Pleistocene of Queensland. Shows the first three true molars in use, with pm. 4 *in alveolo*. Figured by Owen in the 'Phil. Trans.' 1874, pl. lxxvii. figs. 2–6, and 'Extinct Mammals of Australia,' pl. xc. figs. 2–6, as *P. pusio*, of which it is the type; it cannot, however, be specifically distinguished from No. M. 1519. *Presented by Sir D. Cooper, Bart.*, 1866.

46836. (*Fig.*) The nearly entire mandible, showing the left incisor, and the entire series of cheek-teeth on both sides; from the Pleistocene of Queensland. Figured by Owen in the 'Extinct Mammals of Australia,' pl. cxxviii. The length of the series of cheek-teeth is 0,092, and that of the four true molars 0,078.
Presented by Dr. George Bennett, 1875.

48420. The symphysis and nearly entire left ramus of the mandible, showing the five cheek-teeth, and the alveolus of the incisor; from Queensland.
Presented by Dr. George Bennett, 1877.

M. 3676. Cast of the nearly entire right ramus of the mandible, showing the five cheek-teeth (pm. 4 imperfect). The original, of which the exact locality is unknown, is preserved in the Australian Museum, Sydney, and is figured by Owen in the 'Phil. Trans.' 1874, pl. lxxviii. figs. 1–3.
Presented by Sir R. Owen, K.C.B.

[1] Phil. Trans. 1874, p. 788.
[2] Cat. Vert. Anim. Mus. R. Coll. Surg. pt. ii. p. 721 (1884).
[3] Phil. Trans. 1874, p. 788.

M. 38. The greater part of the right ramus of the mandible, containing the five cheek-teeth; from Queensland. This specimen shows the premolar in an early stage of wear.
Presented by G. F. Bennett, Esq., 1880.

M. 1519. Cast of part of the left ramus of the mandible of a small individual, showing the four true molars. The original was obtained from a cave in the Wellington Valley, New South Wales, and is preserved in the Australian Museum.
Presented by the Trustees of the Australian Museum, 1883.

47012. Two portions of the mandibular rami of a single individual; from Queensland. The left ramus shows $\overline{\text{m. 3}}$ and $\overline{\text{m. 4}}$, while the right ramus contains all the cheek-teeth except $\overline{\text{m. 3}}$. *Presented by Dr. George Bennett*, 1876.

32885. The anterior extremity of the left ramus of the mandible (*Fig.*) of an immature individual, showing a portion of the incisor, and $\overline{\text{pm. 4}}$ in its alveolus; from the Pleistocene of the Condamine River, Queensland. Figured by Owen in the 'Phil. Trans.' 1874, pl. lxxvii. figs. 8–12, and also in the 'Extinct Mammals of Australia,' pl. xc. figs. 8–12.
Purchased, 1857.

50052. Part of the left ramus of a mandible, apparently belonging to a small individual of this species; from Gowrie, Queensland. This specimen (which was referred by Owen to *P. pusio*) is so slightly smaller than No. M. 1519, with which it agrees in dental characters, that there appears no good reason for referring it to a different species.
Presented by G. F. Bennett, Esq., 1879.

Procoptodon otuel (Owen [1]).

Syn. *Pachysiagon otuel*, Owen [2].
Procoptodon pusio, Owen [3] (*in parte*).

The type specimen of this species was identified by its describer with *P. pusio*, but the structure of the teeth differs widely from the type specimen of the latter, which is here referred to *P. rapha*.

The species is of generally smaller size than *P. rapha*, from which it is distinguished by the almost entire absence of enamel-folds on the cheek-teeth.

Hab. Queensland.

[1] Phil. Trans. 1874, p. 784.—*Pachysiagon*. [2] *Loc. cit.*
[3] Extinct Mammals of Australia, p. 455 (1877).

46310. Hinder part of the right ramus of the mandible, containing
(*Fig.*) the last three true molars and a fragment of $\overline{\text{m. 1}}$; from the Pleistocene of King's Creek, Clifton, Queensland. This specimen is the type, and is figured by Owen in the 'Phil. Trans.' 1874, pl. lxxvi. figs. 7–10, under the name of *Pachysiagon otuel*, and also in the 'Extinct Mammals of Australia,' pl. lxxxix., as *Procoptodon pusio*.

Presented by G. B. King, Esq., 1874.

46310 a. The greater part of the symphysis and right ramus of the
(*Fig.*) mandible, showing portions of $\overline{\text{m. 2}}$ and $\overline{\text{m. 3}}$ and the entire $\overline{\text{m. 4}}$; from Queensland. Figured by Owen in the 'Extinct Mammals of Australia,' pl. xci. figs. 2, 3, as *P. pusio*.

Same history.

38759. Part of the right ramus of the mandible, containing the
(*Fig.*) five cheek-teeth; from Queensland. Figured by Owen, *op. cit.* pl. xci. fig. 5, as *P. pusio*.

Presented by F. N. Isaac, Esq., 1861.

38760. Fragment of the left ramus of the mandible, showing $\overline{\text{m. 3}}$ and $\overline{\text{m. 4}}$, in which the enamel-folds are more strongly marked than in the preceding specimens; from Darling Downs, Queensland.

Presented by F. N. Isaac, Esq., 1861.

Genus **PALORCHESTES**, Owen[1].

The mandibular symphysis is anchylosed, and both this and the ramus are elongated; the diastema is long. The premolars are like those of *Sthenurus*; the true molars are without corrugations of the enamel, and have no distinct anterior talon; in the lower molars the median longitudinal bridge is tall, and forms the continuation of the curved outer extremity of the hinder ridge. There are no palatal vacuities; and the lower incisors are spatulate.

In the absence of an anterior talon the upper true molars resemble those of the existing *Macropus magnus*, Owen; while the lower ones approximate to those of some specimens of *Nototherium*.

Palorchestes azael, Owen[2].

Syn. *Palorchestes crassus,* Owen[3].

This is the type and only species, and is the largest known member

[1] Phil. Trans. 1874, p. 797. [2] *Ibid.* p. 798.
[3] Trans. Zool. Soc. vol. xi. p. 7 (1880).

of the family, the length of the entire cranium being estimated at sixteen inches.

Hab. Victoria, Queensland, and New South Wales.

46316. (*Fig.*) The anterior portion of the cranium, showing the alveoli of the incisors and the five cheek-teeth in an imperfect condition; from the Pleistocene of Victoria. This specimen is the type, and is figured by Owen in the 'Phil. Trans.' 1874, pls. lxxxi., lxxxii., lxxxiii., and 1876, pl. xx.; as well as in the 'Extinct Mammals of Australia,' pl. xcvii. fig. 1. It was discovered in 1851 by Dr. L. Becker, by whom it was sent to Dr. Kaup.

Presented by Sir R. Owen, K.C.B.

M. 2573. Cast of the palatal region of the cranium, showing the five cheek-teeth of either side. The original was obtained from a cave in the Wellington Valley, New South Wales, and is preserved in the Australian Museum, Sydney. The specimen, which is labelled *Macropus ajax* in Sir R. Owen's writing, presents no characters by which it can be distinguished from the type.

Presented by Prof. Ramsay, 1885.

40034. (*Fig.*) Two portions of the mandibular rami of one individual; from the Pleistocene of Darling Downs, Queensland. The right ramus, which shows the entire first three true molars and the anterior lobe of the fourth, is figured by Owen in the 'Phil. Trans.' 1876, pl. xix. figs. 1–4, and the left ramus, in which $\overline{\text{m. 2}}$ remains, in figs. 5–8 of the same plate; all the figures being reproduced in pl. cvi. of the 'Extinct Mammals of Australia.'

Presented by Sir D. Cooper, Bart., 1866.

M. 34. (*Fig.*) The greater portion of the mandible of a large individual; from Queensland. On the left side the four true molars, and on the right side the last three teeth of that series are shown. This specimen is the type of *P. crassus*, the right ramus being figured by Owen in the 'Trans. Zool. Soc.' vol. xi. pl. x. under that name. In the left ramus the four teeth are in their normal position; but in the right it appears that $\overline{\text{m. 3}}$ and $\overline{\text{m. 4}}$ have changed places, and it was from considering the teeth occupying the places of $\overline{\text{m. 3}}$ and $\overline{\text{m. 4}}$ as really being these teeth, that Owen was mainly led to specifically distinguish the specimen from *P. azael*. *Presented by Dr. George Bennett*, 1879.

Generically undetermined Specimens.

A considerable number of the following specimens of limb-bones and vertebræ belong in all probability to *Macropus*, but others doubtless to the three preceding genera. With the exception of some of the bones of the hind foot there appears in the majority of cases no evidence for referring individual specimens to one genus or species rather than to another, although the very large size of the skull of *Palorchestes* indicates that the largest limb-bones probably belong to that genus.

The following series comprises merely a selection of the less imperfect specimens, from the very large collection preserved in the Museum.

44198. The glenoidal extremity of the right scapula of a large species; from King's Creek, Clifton, Queensland.
Presented by Dr. George Bennett, 1873.

44198 a. The corresponding portion of a similar left scapula; from the same locality. *Same history.*

43378. The corresponding portion of a slightly smaller right scapula; from the same locality. *Same history.*

43384. The corresponding portion of a rather smaller left scapula; from the Condamine River. *Same history.*

43964. The corresponding portion of a still smaller right scapula; from King's Creek. *Same history.*

38775. The proximal half of the right humerus of a very large form; from Queensland. This specimen agrees in relative size with the femur No. 38619*, which is referred by Owen to *Palorchestes*.
Presented by F. N. Isaac, Esq., 1861.

43370. (*Fig.*) The entire right humerus of a slightly smaller animal; from Clifton, Queensland. Figured by Owen in the 'Extinct Mammals of Australia,' pl. cxii., and referred to *Macropus titan*; it agrees, however, much more nearly in relative size with the femora referred by the same writer to *Palorchestes*, and could not possibly have belonged to the same species as the femur No. 47830, which Owen refers to the former. *Presented by Dr. George Bennett*, 1872.

44199. (*Fig.*) The proximal extremity of a similar right humerus; from Queensland. Figured by Owen, *op. cit.* pl. cxiii. figs. 1–3.
Presented by Dr. George Bennett, 1873.

M. 3677. The distal extremity of a left humerus, of larger size than (*Fig.*) No. 43370, and therefore probably agreeing with No. 38775; from Queensland. Figured by Owen, *op. cit.* pl. cxii. figs. 4–6, as *M. titan.*

Presented by Dr. George Bennett.

35932. A rather smaller left humerus, wanting the proximal extremity; from Queensland.

Presented by Sir D. Cooper, Bart., 1861.

M. 3678. A left humerus (wanting the proximal extremity), agreeing in size with the last specimen, but belonging to a different species; from Queensland. The dorsal aspect of the distal articular surface differs markedly from that of No. 35932. *Presented by Dr. George Bennett.*

43585. The distal portion of a left humerus of the same type as the preceding specimen; from the Condamine River.

Presented by G. F. Bennett, Esq., 1872.

46301. A left humerus, imperfect distally, of rather smaller size than No. 43370; from Queensland. This specimen would apparently accord in size with *Macropus brehus.*

No history.

40020. A left humerus (wanting the distal extremity) agreeing in size with the preceding, but apparently specifically distinct; from Queensland.

Presented by Sir D. Cooper, Bart., 1866.

35939. The entire left humerus of a considerably smaller form; (*Fig.*) from King's Creek, Queensland. Figured by Owen in the 'Extinct Mammals of Australia,' pl. cxi., and referred to *Macropus altus*; it agrees in proportionate size with the femur (No. 47830) referred by Owen to *M. titan.*

Presented by F. N. Isaac, Esq., 1861.

35940. The distal extremity of a left humerus agreeing in size with the preceding, but belonging to a different species; from Queensland. *Presented by Sir D. Cooper, Bart.*, 1861.

32320. The proximal half of the right humerus of a form rather smaller than *M. bennetti*; from a cave in the Wellington Valley. *Purchased*, 1857.

42068. The distal two thirds of a rather smaller left humerus, agreeing in structure with the corresponding bone of *M. bennetti*; from the same locality.

Presented by the Trustees of the Australian Museum, 1870.

43940. A left humerus, agreeing in size and general structure with that of *M. ualabatus* (*suprà*, p. 220, No. 42668), but belonging to a different species; from the same locality. *Same history.*

M. 3687. The proximal portion of the left ulna of a very large species; from Queensland. *No history.*

20684. The proximal half of the right ulna of an equally large species; from the Pleistocene of Torres Straits. *Presented by Prof. J. B. Jukes*, 1847.

45893. The olecranal process of the left ulna of a large species; from Queensland. *Presented by Dr. G. Bennett*, 1874.

43587. The proximal portion of a right ulna, agreeing in size with the preceding; from Queensland. *Presented by G. F. Bennett, Esq.*, 1872.

32873. The proximal portion of the right ulna of a smaller species; from Queensland. Agrees in relative size with the humerus No. 35989. *Purchased*, 1857.

43908. The imperfect proximal extremity of a similar left ulna; from a cave in the Wellington Valley. *Presented by the Trustees of the Australian Museum*, 1870.

42665. The proximal portions of the right and left ulnæ of a smaller form; from the same locality. These specimens are considerably larger than the ulna of *M. bennetti*. *Same history.*

42669. Two imperfect right ulnæ, nearly of the size of the corresponding bone of *M. bennetti*; from the same locality. *Purchased*, 1853.

40022. The left radius, imperfect distally, of a very large species; from Queensland. *Presented by Sir D. Cooper, Bart.*, 1866.

40023. The proximal half of a similar radius; from Queensland. *Same history.*

35930. (*Fig.*) The imperfect right innominate of a very large species; from Queensland. Figured by Owen in the 'Extinct Mammals of Australia,' pl. cxxx., and referred to *Palorchestes*. *Presented by Sir D. Cooper, Bart.*, 1861.

44205. The acetabular portion of a similar right innominate; from
(*Fig.*) Queensland. Figured by Owen, *op. cit.*, under the above name. *Presented by Dr. George Bennett*, 1873.

M. 2574. Cast of the left innominate of a rather smaller form. The original is from a cave in the Wellington Valley, and is preserved in the Museum at Sydney.
Presented by Prof. Ramsay, 1885.

38770. An imperfect right innominate, agreeing very closely with the last specimen; from Darling Downs.
Presented by F. N. Isaac, Esq., 1861.

M. 3688. The acetabular portion of a right innominate, agreeing in size with the preceding, but differing somewhat in structure; from Queensland. *Presented by Dr. George Bennett.*

M. 3689. The acetabular portion of an apparently similar right innominate; from Queensland. *Same history.*

44206. A slightly imperfect smaller left innominate, agreeing in relative size with the femur No. 47830; from Darling Downs. *Presented by Dr. George Bennett*, 1873.

36078. The acetabular portion of a right innominate agreeing in size with the last specimen; from the Condamine River.
Presented by J. H. Hood, Esq., 1861.

38619*. The proximal portion of the right femur of a very large
(*Fig.*) species; from Queensland. Figured by Owen in the 'Phil. Trans.' 1876, pl. xxiii. fig. 1, and also in the 'Extinct Mammals of Australia,' pl. cxv. fig. 1, under the name of *Palorchestes*.
Presented by Sir D. Cooper, Bart., 1861.

M. 3690. The imperfect proximal extremity of a similar right femur; from Queensland. *Presented by Dr. George Bennett.*

40013. The distal extremity of a left femur, agreeing in size with
(*Fig.*) the preceding; from Queensland. Figured by Owen in the 'Phil. Trans.' 1876, pl. xxiii. fig. 2, and in the 'Extinct Mammals of Australia,' pl. cxv. fig. 2, and referred to *Palorchestes*. *Presented by Sir D. Cooper, Bart.*, 1866.

M. 3691. The distal portion of a similar left femur; from Queensland. *Presented by Dr. George Bennett.*

36078. The imperfect distal extremity of a similar left femur; from Queensland. *Presented by J. H. Hood, Esq.*, 1861.

M. 2260. The slightly imperfect distal extremity of a left femur,

agreeing in size with the preceding; from Queensland. Provisionally referred by Owen, in the 'Extinct Mammals of Australia,' p. 494, to *Macropus brehus*.
Presented by Dr. George Bennett.

43216. The distal portion of a rather smaller left femur; from Maryvale Creek, Queensland. Referred by Owen, *op. cit.* p. 494, to *Procoptodon goliah*.
Presented by Sir R. Daintree, 1871.

43589. The distal portion of a similar right femur; from the Condamine River. *Presented by G. F. Bennett, Esq.*, 1872.

35931. The imperfect distal extremity of a similar right femur; from Queensland.
Presented by Sir D. Cooper, Bart., 1861.

40015. The distal extremity of a rather smaller right femur; from
(*Fig.*) Queensland. Figured by Owen in the 'Phil. Trans.' 1876, pl. xxiii. fig. 3, and also in the 'Extinct Mammals of Australia,' pl. cxv. fig. 3, under the name of *Procoptodon goliah*. *Presented by Sir D. Cooper, Bart.*, 1866.

45887. The left femur, wanting the distal extremity, of an animal agreeing approximately in size with the preceding; from Queensland. Referred by Owen, in the 'Extinct Mammals of Australia,' p. 495, to *Macropus brehus* [*Protemnodon mimas*]. *Presented by Dr. George Bennett*, 1874.

43588. Portions of the proximal extremities of two femora, agreeing nearly in size with the preceding.
Presented by G. F. Bennett, Esq., 1872.

M. 3692. The imperfect distal extremity of a right femur, agreeing nearly in size with No. 40015, but with the inner ridge of the anterior trochlear surface much more prominent; from Queensland. *Presented by Dr. George Bennett.*

45388. The proximal portion of a left femur, rather smaller than No. 43588; from Queensland.
Presented by Dr. George Bennett, 1874.

43383. The shaft of a very similar right femur; from King's Creek, Clifton. *Presented by Dr. George Bennett*, 1872.

43959. Part of the shaft of a similar right femur; from Queensland.
Presented by Dr. George Bennett, 1872.

47830. Cast of the entire right femur of a somewhat smaller animal. The original was obtained from Queensland, and is figured by Owen in the 'Extinct Mammals of Australia,' pl. cxiv., where it is referred to *Macropus titan*.
Presented by Dr. George Bennett, 1876.

40016. The distal portion of a slightly larger right femur; from Queensland. *Presented by Sir D. Cooper, Bart.*, 1866.

36275. The imperfect distal portion of a very similar femur; from the Pleistocene of Melbourne.
Presented by F. M. Raynall, Esq., 1862.

38780. The distal portion of a right femur, somewhat smaller than No. 47830; from Queensland.
Presented by F. N. Isaac, Esq., 1861.

42665*. The distal portion of the right femur of an animal equal in size to a small individual of *Macropus giganteus*; from a cave in the Wellington Valley.
Presented by the Trustees of the Australian Museum, 1871.

32322. The distal portion of a smaller left femur; from the same locality. *Purchased*, 1857.

43220. The distal portion of a still smaller right femur, agreeing in size with the corresponding bone of *M. bennetti*; from Maryvale Creek. *Presented by Sir R. Daintree*, 1871.

44128. (*Fig.*) The proximal portion of the left tibia of a very large species; from Queensland. Figured by Owen in the 'Phil. Trans.' 1876, pl. xxiv., and also in the 'Extinct Mammals of Australia,' pl. cxxxi., where it is referred to *Palorchestes*.
Presented by Dr. George Bennett, 1873.

M. 3693. Part of the middle of the shaft of a left tibia agreeing approximately in size with the last specimen; from Queensland. *Presented by Dr. George Bennett.*

M. 3694. The distal portion of the left tibia of a large species; from Queensland. *Same history.*

38619. The distal portion of a similar right tibia; from Eton Vale.
Presented by Sir D. Cooper, Bart., 1861.

38783. The proximal epiphysis of a left tibia, of rather smaller size than No. 44128; from Darling Downs.
Presented by F. N. Isaac, Esq., 1861.

43597. The middle portion of the shaft of a right tibia, agreeing approximately in size with the preceding; from Queensland. *Presented by Dr. George Bennett*, 1872.

45934. The greater portion of the shaft of the left tibia of an animal of rather larger size than full-grown males of *M. giganteus*. *Presented by Sir D. Cooper, Bart.*, 1866.

M. 3695. The middle portion of the shaft of a left tibia, agreeing in diameter with the last specimen, but evidently of a much shorter type; from Queensland. This specimen may have belonged to *Procoptodon*. *Presented by Dr. George Bennett.*

43590. The left calcaneum of a very large species; from Queensland.
(*Fig.*) Figured by Owen in the 'Phil. Trans.' 1876, pl. xxiii. fig. 5, and also in the 'Extinct Mammals of Australia,' pl. cxv. fig. 5, where it is referred to *Palorchestes*. This bone is of the elongated type characteristic of the large existing species of *Macropus*.
Presented by Dr. George Bennett, 1872.

50078. The right calcaneum of an equally large but distinct species; from Queensland. This bone is of a shorter type, and approximates to the calcaneum (No. 47829) provisionally referred by Owen to *Procoptodon*.
Presented by G. F. Bennett, Esq., 1879.

50077. The imperfect right calcaneum of a very similar type; from Queensland. *Same history.*

47829. The right calcaneum of a rather smaller form and of a very
(*Fig.*) short type; from Queensland. Figured by Owen in the 'Phil. Trans.' 1876, pl. xxiii. fig. 4, and also in the 'Extinct Mammals of Australia,' pl. cxv. fig. 4, where it is referred to *Procoptodon goliah*.
Presented by Dr. George Bennett, 1876.

47851. The imperfect left calcaneum of a very similar type; from Darling Downs. *Same history.*

38789. A left calcaneum of similar type, but belonging to a smaller species; from Darling Downs.
Presented by F. N. Isaac, Esq., 1861.

M. 1515. The imperfect right calcaneum of the same type, but belonging to a smaller and perhaps immature animal; from Queensland. *Presented by C. H. Hartmann, Esq.*, 1884.

32879. The right calcaneum, approximately equal in size with the preceding, but of the elongated type of the corresponding bone of *Macropus giganteus* and *M. robustus*; from the Condamine River. This bone agrees approximately in size with the tibia No. 45934. *Purchased*, 1857.

42665. A very similar imperfect left calcaneum; from a cave in the Wellington Valley. *Presented by the Trustees of the Australian Museum*, 1871.

42665 a. The right calcaneum, wanting the epiphysis, of a species about the size of *M. antilopinus*, and of a shorter type than the preceding specimen; from the same locality. *Same history.*

50085. A left calcaneum (without the epiphysis) of a very similar type; from Gowrie. *Presented by G. F. Bennett, Esq.*, 1879.

32331. Three calcanea of rather smaller animals, agreeing approximately in size with the corresponding bone of *M. bennetti*; from the caves of the Wellington Valley. *Purchased*, 1857.

32331 a. Two calcanea, of rather larger size than the corresponding bone of *M. ualabatus*; from the same locality. *Same history.*

38785. The right astragalus of a very large species, agreeing approximately in size with the calcaneum No. 50078; from Darling Downs. *Presented by F. N. Isaac, Esq.*, 1861.

38786. The right astragalus of a considerably smaller species; from Darling Downs. *Presented by F. N. Isaac, Esq.*, 1861.

M. 1515 a. An imperfect right astragalus, agreeing in size with the preceding; from Queensland. *Presented by C. H. Hartmann, Esq.*, 1884.

M. 3696. A smaller right astragalus, characterized by the great prominence of the ridges on the tibial surface; from Queensland. *Presented by Dr. George Bennett.*

43920. A smaller left astragalus, agreeing approximately in size with the astragalus of *M. antilopinus*; from a cave in the Wellington Valley. *Presented by the Trustees of the Australian Museum*, 1870.

43920 a. A right astragalus, rather larger than that of *M. ualabatus*; from the same locality. *Same history.*

43920 b. A considerably smaller right astragalus; from the same locality. *Same history.*

The following metatarsals are of the elongated type of those of existing species of Macropus.

35942. The left fourth metatarsal of a very large species; from Queensland. The extreme length is 0,220.
Presented by Sir D. Cooper, Bart., 1861.

35943. A slightly smaller right fourth metatarsal; from Gowrie.
(*Fig.*) Figured by Owen in the 'Extinct Mammals of Australia,' pl. cxviii. figs. 1–5, and referred to *Macropus titan*.
Presented by Sir D. Cooper, Bart., 1861.

35944. A rather smaller right fourth metatarsal; from Gowrie.
(*Fig.*) Figured by Owen, *op. cit.* pl. cxviii. figs. 1–7, and referred to *M. altus*. *Presented by Sir D. Cooper, Bart.*, 1861.

35942 a. The left metatarsus of a species equal in size to that to
(*Fig.*) which No. 35942 belonged, but with the fourth metatarsal of a rather shorter type; from Queensland. Figured by Owen, *op. cit.* pl. cxvi. figs. 1–3, and referred to *Palorchestes*. *Presented by Sir D. Cooper, Bart.*

45889. A right fourth metatarsal, of rather smaller size than
(*Fig.*) No. 35944; from Queensland. Figured by Owen, *op. cit.* pl. cxx. figs. 1–6, under the name of *Protemnodon anak*.
Presented by Dr. George Bennett, 1874.

43964. A right fourth metatarsal, imperfect distally, but apparently agreeing precisely with the preceding; from Queensland.
Presented by Dr. George Bennett, 1872.

35945. The proximal portion of a similar right fourth metatarsal;
(*Fig.*) from Queensland. Figured by Owen, *op. cit.* pl. cxvii. figs. 7–9, under the name of *M. affinis*[1].
Presented by Sir D. Cooper, Bart., 1861.

40030. The proximal two thirds of a similar left fourth metatarsal; from Queensland.
Presented by Sir D. Cooper, Bart., 1866.

43592. The proximal half of a slightly larger left fourth metatarsal; from Queensland.
Presented by Dr. George Bennett, 1872.

[1] Owen, Cat. Foss. Mamm. and Aves, Mus. R. Coll. Surg. p. 328 (1845)

40028. A fourth left metatarsal of a large form, but relatively shorter than any of the preceding, and intermediate in this respect between the preceding and the metatarsals referred by Owen to *Procoptodon*; from Darling Downs.
Presented by Sir D. Cooper, Bart., 1866.

40026. A left fifth metatarsal, agreeing approximately in size with No. 35942, and wanting the proximal extremity; from Queensland. *Presented by Sir D. Cooper, Bart.*, 1866.

40027. A rather smaller left fifth metatarsal, agreeing in size with (*Fig.*) No. 35943; from Queensland. Figured by Owen, *op. cit.* pl. cxviii. figs. 6, 7, and referred to *M. titan.*
Presented by Sir D. Cooper, Bart., 1866.

The following metatarsals are of a shorter type, and are referred by Owen to Procoptodon.

35946. The fourth right metatarsal of a large species; from (*Fig.*) Gowrie. Figured by Owen in the 'Extinct Mammals of Australia,' pl. cxxi. figs. 6, 7, 8, 11, as *P. goliah.*
Presented by Sir D. Cooper, Bart., 1861.

35947. A slightly smaller fourth left metatarsal; from Gowrie.
Same history.

43371. A similar fourth left metatarsal; from Clifton, Queensland.
Presented by Dr. George Bennett, 1872.

38792. A similar fourth right metatarsal; from Darling Downs.
Presented by F. N. Isaac, Esq., 1861.

40029. A similar specimen, wanting the distal extremity; from Queensland. *Presented by Sir D. Cooper, Bart.*, 1866.

35942 b. The associated fourth and fifth left metatarsals of a (*Fig.*) somewhat smaller animal; from Queensland. Figured by Owen in the 'Phil. Trans.' 1876, pl. xxxi. figs. 1–3, and also in the 'Extinct Mammals of Australia,' pl. cxix. figs. 1–3, as *P. rapha.*
Presented by Sir D. Cooper, Bart., 1861.

35942 c. The fourth left metatarsal of a slightly smaller animal; (*Fig.*) from Queensland. Figured by Owen in the 'Extinct Mammals of Australia,' pl. cxix. figs. 10–12, as *P. pusio.*
Same history.

43355. A similar fourth right metatarsal; from Queensland.
Presented by Dr. George Bennett, 1872.

36078. The fifth left metatarsal of a large species; from the Condamine River, Queensland.
Presented by J. H. Hood, Esq., 1861.

40025. The fifth right metatarsal of a rather smaller form; from
(*Fig.*) Queensland. Figured by Owen, *op. cit.* pl. cxviii. figs. 8–11, as *P. goliah.*
Presented by Sir D. Cooper, Bart., 1866.

35948. A nearly similar fifth left metatarsal; from Queensland.
(*Fig.*) Figured by Owen, *op. cit.* pl. cxix. fig. 7.
Presented by Sir D. Cooper, Bart., 1861.

M. 2499. A very similar fourth right metatarsal; from Queensland.
Presented by Sir R. Owen, K.C.B., 1884.

43374. A fifth right metatarsal, agreeing with the corresponding bones of No. M. 35942 *b*.
Presented by Dr. George Bennett, 1872.

32324. A similar fifth right metatarsal; from a cave in the Wellington Valley. *Purchased*, 1853.

The following phalangeals of the hind foot are of the elongated type characteristic of the existing species of Macropus.

50084. The first phalangeal of the fourth digit of a very large species, characterized by the excessive lateral expansion of the two extremities; from Queensland.
Presented by G. F. Bennett, Esq., 1879.

M. 3697. The homologous bone of an equally large species, with the extremities less expanded; from Queensland. The extreme length is 0,071. *Same history.*

M. 3698. A rather smaller bone of similar type; from Queensland.
Same history.

47852. A first phalangeal of the fourth digit, agreeing in size with
(*Fig.*) the preceding, but with less lateral expansion of the proximal articular surface; from Queensland. Figured by Owen in the 'Extinct Mammals of Australia,' pl. cxvii. fig. 10. This specimen agrees approximately in size with the fourth metatarsal No. 35943.
Presented by Dr. George Bennett, 1876.

32880. A very similar bone; from Queensland. *Purchased*, 1857.

M. 1515 b. A rather smaller homologous bone; from Queensland.
Presented by C. H. Hartmann, Esq., 1884.

50084 a. An homologous bone of equal length, but of much more slender proportions; from Queensland.
Presented by G. F. Bennett, Esq., 1879.

38793. A slightly shorter and wider homologous bone; from Queensland. *Presented by F. N. Isaac, Esq.*, 1861.

46073. A rather shorter first phalangeal of the fourth digit; from Queensland. This specimen agrees approximately in size with the fourth metatarsal No. 45889; its length being 0,056. *Presented by Dr. George Bennett*, 1874.

46073 a. A similar bone; from Queensland. *Same history.*

50084 b. A very similar homologous bone; from Queensland.
Presented by G. F. Bennett, Esq., 1879.

50084 c. A similar specimen; from Queensland. *Same history.*

42597. A very similar homologous bone; from a cave in the Wellington Valley. *Presented by the Trustees of the Australian Museum*, 1870.

46073 b. A rather smaller and more slender homologous bone, agreeing in shape with No. 50084 *a*; from Queensland.
Presented by Dr. George Bennett, 1874.

46073 c. A considerably smaller first phalangeal of the fourth digit; from Queensland. Length 0,049. *Same history.*

50084 d. A similar bone; from Queensland.
Presented by G. F. Bennett, Esq., 1879.

32293. A very similar bone; from a cave in the Wellington Valley.
Purchased, 1857.

42554. An imperfect first phalangeal of the fourth digit, agreeing approximately in length with the preceding, but with a much wider distal extremity; from the same locality.
Presented by the Trustees of the Australian Museum, 1870.

42597 a. A smaller homologous bone; from the same locality. Length 0,038. *Same history.*

46839. An homologous bone of a slightly shorter and thicker type; from Queensland. *Presented by Dr. George Bennett*, 1875.

46073 d. A right first phalangeal of the fifth digit of a large species; from Queensland. This bone agrees approximately in size with the first phalangeal of the fourth digit No. 47852.
Presented by Dr. George Bennett, 1874.

46073 h. The homologous bone of the opposite side belonging to a considerably smaller species; from Queensland.
Presented by Dr. George Bennett, 1874.

43335. A slightly smaller homologous bone of the right side; from Queensland. *Same history.*

43596. The second phalangeal of the fourth digit of a very large form; from the Condamine River. This bone agrees in relative size with the first phalangeal No. M. 3697.
Presented by Dr. George Bennett, 1872.

46073 e. Two similar bones; from Queensland.
Presented by Dr. George Bennett, 1874.

46839 a. A rather smaller homologous bone agreeing approximately with the first phalangeal No. 47852.
Presented by Dr. George Bennett, 1875.

46073 f. An homologous bone of a rather shorter and wider type; from Queensland.
Presented by Dr. George Bennett, 1874.

M. 3701. A rather smaller homologous bone; from the same region.
Same history.

M. 3702. A still smaller homologous bone; from the same region.
Same history.

43925 a. The homologous bone of a considerably smaller species; from the Wellington Valley.
Presented by the Trustees of the Australian Museum, 1870.

43924. A slightly smaller homologous bone; from the same locality.
Same history.

40040. The imperfect terminal phalangeal of the fourth digit of a very large species; from Queensland.
Presented by Sir D. Cooper, Bart., 1866.

46839 b. The homologous bone of a smaller species; from Queensland. *Presented by Dr. George Bennett*, 1875.

40024. A smaller homologous bone; from Queensland.
Presented by Sir D. Cooper, Bart., 1866.

43346. A slightly larger but imperfect homologous bone; from Queensland. *Presented by Dr. George Bennett*, 1872.

46073 g. A smaller homologous bone; from Queensland.
Presented by Dr. George Bennett, 1874.

50084. A still smaller homologous bone; from Queensland.
Presented by G. F. Bennett, Esq., 1879.

The following phalangeals of the pes are of a shorter type, and correspond in this respect with the metatarsals referred by Owen to Procoptodon.

M. 3699. The first phalangeal of the fourth digit of a very large species ; from Queensland. Length 0,050, width proximally 0,036. *No history.*

42597 b. A very similar homologous bone ; from a cave in the Wellington Valley.
Presented by the Trustees of the Australian Museum, 1870.

35949. The homologous bone of a slightly smaller form ; from Gowrie. *Presented by Sir D. Cooper, Bart.,* 1861.

42597 c. The homologous bone of a still smaller form ; from the Wellington Valley.
Presented by the Trustees of the Australian Museum, 1870.

46839 c. A rather smaller homologous bone ; from Queensland.
Presented by Dr. George Bennett, 1875.

32881. The second phalangeal of the fourth digit of a very large species; from the Condamine River. *Purchased,* 1857.

43331. A similar specimen ; from Queensland.
Presented by Dr. George Bennett, 1872.

46839 d. The terminal phalangeal of the fourth digit of a large species, from Queensland.
Presented by Dr. George Bennett, 1875.

35951. A similar bone; from Queensland.
Presented by Sir D. Cooper, Bart., 1861.

M. 3700. Three imperfect specimens of the homologous bone, which are of a still shorter type; from Queensland.
Presented by Dr. George Bennett.

The following specimens are from the vertebral column.

50079. The imperfect atlas vertebra of a very large species; from Queensland. The transverse diameter across the occipital articular cups is 0,095.
Presented by G. F. Bennett, Esq., 1879.

35952. A dorsal vertebra of a large species ; from Queensland.
Presented by Sir D. Cooper, Bart., 1861.

M. 3703. A very similar dorsal vertebra; from Queensland.
Presented by Dr. George Bennett.

35955. A dorsal vertebra of a species of the size of *Macropus bennetti*; from Queensland.
Presented by Sir D. Cooper, Bart., 1861.

M. 18. A lumbar vertebra of a large species; from Queensland.
Presented by Sir G. Verdon, 1877.

43382. The sacrum and first caudal vertebra of a very large species;
(*Fig.*) from King's Creek, Clifton, Queensland. Figured by Owen in the 'Extinct Mammals of Australia,' pl. cxxix., and referred to *Palorchestes*.
Presented by Dr. George Bennett, 1872.

43582. The sacrum of an equal-sized species; from the Condamine River. *Presented by Dr. George Bennett*, 1872.

44204. The sacrum of a smaller species; from Darling Downs.
Presented by Dr. George Bennett, 1873.

38617. The sacrum and first caudal vertebra of a slightly smaller form; from Eton Vale, Queensland.
Presented by Sir D. Cooper, Bart., 1861.

M. 20. A first caudal vertebra, agreeing in size with the preceding; from Queensland. *Presented by Sir G. Verdon*, 1877.

38616. Six associated early caudal vertebræ of a very large species; from Queensland.
Presented by Sir D. Cooper, Bart., 1861.

38766. An early caudal vertebra of a very large species; from Darling Downs. *Presented by F. N. Isaac, Esq.*, 1861.

38767. An imperfect early caudal vertebra of an equally large species; from the same locality. *Same history.*

50076. An early caudal vertebra of a medium-sized species; from Queensland. *Presented by G. F. Bennett, Esq.*, 1879.

45892. An early caudal vertebra of a very large species; from Queensland. Length 0,100.
Presented by Dr. George Bennett, 1874.

45892 a. A slightly smaller late caudal vertebra; from Queensland.
Same history.

35957. A late caudal vertebra of a slightly smaller species; from Gowrie. *Presented by Sir D. Cooper, Bart.*, 1861.

38768. A shorter late caudal vertebra of an equally large species; from Darling Downs. *Presented by F. N. Isaac, Esq.*, 1861.

M. 3704. An imperfect late caudal vertebra of a large form; from Queensland. *Presented by Dr. George Bennett.*

40170. The centrum of a rather smaller late caudal vertebra; from Darling Downs. *Presented by Sir D. Cooper, Bart.*, 1866.

36079 *. An imperfect late caudal vertebra of a smaller form; from the Condamine River. *Presented by J. H. Hood, Esq.*, 1861.

32330. A late caudal vertebra, with a shorter centrum than any of the preceding; from a cave in the Wellington Valley. *Purchased*, 1857.

42667. A late caudal vertebra of similar proportions; from the same locality. *Presented by the Trustees of the Australian Museum*, 1870.

42667 a. A late caudal vertebra of a smaller form; from the same locality. *Same history.*

Suborder *MARSUPIALIA POLYPROTODONTIA.*

Incisors (in existing genera) $\frac{(4-5)}{(3-4)}$, small, and subequal; canines larger; molars usually with sharp cusps. In existing genera the premolars are normally three, and correspond to the first, third, and fourth of the typical Eutherian series; but in the Mesozoic *Triconodon* and *Amblotherium* there are four. Five upper and usually six lower true molars are present in *Myrmecobius*, at least six in *Amphitherium*, and seven in *Amblotherium*. The fourth premolar is smaller than the first true molar, and is never simply secant.

Family PERAMELIDÆ.

Dentition:—I. $\frac{(4-5)}{3}$, C. $\frac{1}{1}$, Pm. $\frac{3}{3}$, M. $\frac{4}{4}$. Upper incisors small, with short broad crowns; lower moderate, narrow and proclivous. Premolars compressed and pointed; the fourth preceded by a milk-tooth. First three upper true molars with squared crowns, usually consisting of a pair of V-shaped outer columns and a long conjoint inner column; first three lower true molars composed of two lobes, of which the first carries three cusps arranged in a triangle; last molar in both jaws simpler. Hind feet syndactylous, as in the *Macropodidæ*; terminal phalangeals cleft; an entepicondylar foramen to the humerus; canines sometimes with grooved or double roots.

Genus **PERAMELES,** Geoffroy[1].

Limbs subequal in length; crowns of molars short and nearly straight; auditory bulla single.

Perameles nasuta, Geoffroy[2].

Syn. *Perameles tenuirostris*, Owen[3].

A large species with elongated muzzle.

Hab. New South Wales.

The following specimens are from the caves of the Wellington Valley, New South Wales, and were presented by the Trustees of the Australian Museum, 1870.

42639. Part of the middle of the cranium, showing the last six cheek-teeth of the left side, in a damaged condition.

42671. Fragment of the left maxilla, containing the unworn m. 3 and m 4.

42670. Part of the right ramus of the mandible, showing the canine and cheek-teeth. This specimen, which cannot be distinguished from the mandible of the existing form, agrees precisely with the fossil ramus from the Wellington Caves, figured by Owen in the 'Extinct Mammals of Australia,' pl. v. figs. 10, 11, under the name of *P. tenuirostris*.

42670 a. Four fragments of mandibular rami with teeth.

43884. Three fragments of mandibular rami without teeth.

42675. Five specimens of the humerus. The length of this bone is 0,042.

42672. The right tibia, wanting the distal epiphysis.

42672 a. Three imperfect specimens of the tibia.

[1] Bull. Soc. Philom. Paris, vol. iii. No. 80, p. 249—*errorim* 149—(1803).

[2] Ann. d. Muséum, vol. iv. p. 62 (1804).

[3] Extinct Mammals of Australia, p. 107, pl. v. figs. 10, 11 (1877).

Perameles obesula (Shaw[1]).

Syn. *Didelphis obesula*, Shaw[2].

Rather smaller than the preceding, and with a shorter muzzle.
Hab. Australia and Tasmania.

42675 a. Five specimens of the humerus, agreeing in size with those of the present species; from the caves of the Wellington Valley, New South Wales. Their length is 0,0365.
Presented by the Trustees of the Australian Museum, 1870.

Perameles bougainvillei, Quoy & Gaimard[3].

Including *Perameles fasciata*, Gray[4].
Considerably smaller than the preceding.
Hab. South and West Australia and New South Wales.

42675 b. A left humerus, agreeing in size with the corresponding bone of a skeleton of the West-Australian variety in the Museum; from a cave in the Wellington Valley, New South Wales. Length 0,028.
Presented by the Trustees of the Australian Museum, 1870.

42668. A slightly larger left humerus; from the same locality.
Same history.

Perameles, sp.

The lower molars equal in size to those of *P. nasuta*, but with an external cingulum.
Hab. New South Wales.

42607 b. Fragment of the right ramus of the mandible, containing two true molars; from a cave in the Wellington Valley, New South Wales.
Presented by the Trustees of the Australian Museum, 1870.

Genus **PERAGALE**, Gray[5].

Syn. *Macrotis*, Reid[6].

Anterior limbs considerably shorter than the posterior; crowns of molars short, or elongated and curved; auditory bulla double.

[1] Naturalist's Miscellany, vol. viii. pl. 298 (*circa* 1793).—*Didelphis.*
[2] *Loc. cit.*
[3] Voyage d'Uranie, Zool. p. 56 (1824).
[4] Appendix to Grey's Journal of Two Expeditions in Australia, vol. ii. p. 407 (1841).
[5] *Ibid.* p. 401.—Amended.
[6] Proc. Zool. Soc. 1836, p. 131.—Preoccupied by a genus of Coleoptera.

Peragale lagotis (Reid[1]).

Syn. *Perameles* (*Macrotis*) *lagotis*, Reid[2].

Of large size, with the crowns of the molars elongated and curved.

Hab. Western Australia (recent), and New South Wales (Pleistocene).

42663. Part of the right maxilla, containing the last three true molars; from a cave in the Wellington Valley, New South Wales.

Presented by the Trustees of the Australian Museum, 1870.

43884. Hinder part of the right ramus of the mandible; showing the last two molars; from the same locality.

Same history.

Family TRICONODONTIDÆ[3].

Dentition:—I. $\frac{?}{3}$, C. $\frac{?}{1}$, Pm. $\frac{?}{4}$, M. $\frac{?}{(3\text{–}4)}$. The lower true molars are without an inner cingulum or cusp, and consist of three subequal secant cusps or cones, of which the first two correspond to the blade of the homologous teeth of the *Dasyurinæ*, and the third to the talon of the same; the upper true molars resemble the lower, and have no inner tubercle; the premolars are much simpler than the true molars, $\overline{\text{pm. 4}}$ being taller than $\overline{\text{m. 1}}$; and the lower incisors are in contact. The last premolar was preceded by a milk-tooth; the cheek-teeth are of large relative size, and the mandibular ramus is deep. The inflection of the angle of the mandible is strongly marked; and the canines have grooved or double roots.

The one English genus of the present family appears to be a highly specialized primitive form, of which the lower true molars come nearest to those of *Thylacinus*, and bear the same relation to the lower molars of *Dasyurus* as is presented by the inferior carnassial tooth of *Icticyon* to that of *Canis*. The specialization of the dentition of this early form is parallelled by the case of *Plagiaulax* among the Diprotodontia.

[1] Proc. Zool. Soc. 1836, p. 129.—*Perameles.*

[2] *Loc. cit.*

[3] The division into families of the Mesozoic members of the Suborder, as well as the serial arrangement of such families, must be regarded as provisional.

Genus **TRICONODON**, Owen[1].

Including *Triacanthodon*, Owen[2].

The characters of the genus are those of the family. The discovery of specimens of *Triconodon mordax* with four true molars has shown that the so-called *Triacanthodon* is not separable.

In the American Upper Jurassic a species of this genus is recorded by Marsh[3], the genus *Priacodon* being made for an allied form.

Triconodon mordax, Owen[4].

Syn. (?) *Triacanthodon serrula*, Owen[5].

This is the type species; the length of the space occupied by $\overline{m.\ 1} + \overline{m.\ 2}$ in the type specimen is 0,095, and that by $\overline{m.\ 1} + \overline{m.\ 2} + \overline{m.\ 3}$ 0,008. At least in many adult specimens there were four lower true molars, but it is not improbable that in other examples $\overline{m.\ 4}$ was never developed.

Hab. Europe (England).

The following specimens (or their originals) were obtained from the Middle Purbeck group of Durdlestone Bay, Swanage, Dorsetshire.

47766. (*Fig.*) The nearly entire left ramus of the mandible of a subadult individual, containing the anterior teeth and the first three true molars. This specimen (woodcut, fig. 38) is the type,

Fig. 38.

Triconodon mordax.— The left ramus of the mandible (reversed): from the Purbeck of Swanage. $\frac{1}{1}$.

and is figured by Owen in the 'Encyclopædia Britannica,' 8th ed. vol. xvii. p. 161, fig. 86, and in his 'Mesozoic Mammalia,' pl. iii. figs. 7, 7*a*.

Beckles Collection. Purchased, 1876.

[1] Encyclopædia Britannica, 8th ed. vol. xvii. p. 161 (1859).
[2] Mesozoic Mammalia (Mon. Pal. Soc.), p. 72 (1871).
[3] Amer. Journ. ser. 3, vol. xxxiii. p. 343 (1887).
[4] Encyclopædia Britannica, *loc. cit.*
[5] Mesozoic Mammalia, *loc. cit.*

47764. Portion of the left ramus of the mandible, showing $\overline{\text{pm. 2}}$ and (*Fig.*) the first three true molars. Figured by Owen in the 'Mesozoic Mammalia,' pl. iii. figs. 9, 9 *a*; there is no trace of $\overline{\text{m. 4}}$. *Same history.*

47767. Part of the left ramus of the mandible, containing $\overline{\text{pm. 4}}$ (*Fig.*) and $\overline{\text{m. 1}}$. Figured by Owen, *op. cit.* pl. iii. figs. 10, 10 *a*. *Same history.*

M. 19. Cast of the right ramus of the mandible, showing $\overline{\text{pm. 3}}$, $\overline{\text{pm. 4}}$, and the four true molars. The original (woodcut, fig. 39) is preserved in the Museum of Practical Geology, Jermyn Street, and is described and figured by Willett in the 'Quart. Journ. Geol. Soc.' vol. xxxvii. p. 378 (woodcut):

Fig. 39.

Triconodon mordax.—The right ramus of the mandible; from the Purbeck of Swanage. $\frac{1}{1}$.

it was suggested in the description that the first of the four trilobed teeth might be $\overline{\text{mm. 4}}$; this is, however, disproved by the large size of the two preceding teeth, which must be $\overline{\text{pm. 3}}$ and $\overline{\text{pm. 4}}$, since $\overline{\text{pm. 2}}$ is shown by other specimens to be a much smaller tooth; the length of $\overline{\text{m. 1}}$ + $\overline{\text{m. 2}}$ is 0,006, and that of $\overline{\text{m. 1}}$ + $\overline{\text{m. 2}}$ + $\overline{\text{m. 3}}$ 0,009. *Presented by H. Willett, Esq.*, 1881.

M. 3750. Cast of the right ramus of the mandible, showing the homologous teeth to those of the preceding specimen. The original is in the collection of Mr. Willett. *Made in the Museum.*

The following specimens are referred by Owen to Triacanthodon serrula *(47763 being the type), but apparently present no characters by which they can be distinguished from the present species.*

M. 47763. A slab of Purbeck limestone, containing the left ramus (*Fig.*) of the mandible and some other imperfect bones. The mandible has been figured by Owen in the 'Mesozoic Mammalia,' pl. iv. figs. 7, 8; it shows the third incisor, the canine, the first three premolars, $\overline{\text{mm. 4}}$, $\overline{\text{m. 1}}$ and $\overline{\text{m. 2}}$, the impress of $\overline{\text{m. 3}}$ on the limestone, and $\overline{\text{m. 4}}$ on the inner side of the coronoid process. Since the specimen was figured, the outer lamina of the ramus has been cut away to exhibit the germ of $\overline{\text{pm. 4}}$ *in alveolo*. The length of $\overline{\text{m. 1}}$ + $\overline{\text{m. 2}}$ is 0,005, and that of $\overline{\text{m. 1}}$ + $\overline{\text{m. 2}}$ + $\overline{\text{m. 3}}$ 0,008. The position of $\overline{\text{m. 4}}$ on the inner side of the coronoid indicates that this tooth may be present in No. 47766.

Beckles Collection. Purchased, 1876.

48395. The imperfect mandible of an immature individual, apparently showing $\overline{\text{pm. 2}}$, $\overline{\text{pm. 3}}$, $\overline{\text{mm. 4}}$, $\overline{\text{m. 1}}$, and $\overline{\text{m. 2}}$.

Same history.

The following specimens not improbably belong to this species.

47771. Part of a right mandibular ramus, containing three true (*Fig.*) molars. Figured by Owen, *op. cit.* pl. iv. fig. 6 (as *T.* (?) *minor*). *Same history.*

47784. Fragment of the right maxilla, containing the last true (*Fig.*) molar. Figured by Owen, *op. cit.* pl. iv. fig. 5.

Same history.

47768. Fragment of the right mandibular ramus of an immature (*Fig.*) individual, showing $\overline{\text{pm. 4}}$ half protruded from its alveolus and the imperfect $\overline{\text{m. 1}}$ and $\overline{\text{m. 2}}$. Figured by Owen, *op. cit.* pl. iii. fig. 15. *Same history.*

Triconodon ferox, Owen[1].

Syn. (?) *Triconodon occisor*, Owen[2].

Distinguished from the preceding species by its superior size and the contour of the angle of the mandible. None of the specimens show any trace of $\overline{\text{m. 4}}$. The length of $\overline{\text{m. 1}}$ + $\overline{\text{m. 2}}$ in the type specimen is 0,008, and that of $\overline{\text{m. 1}}$ + $\overline{\text{m. 2}}$ + $\overline{\text{m. 3}}$ 0,013.

Hab. Europe (England).

The following specimens were obtained from the Middle Purbeck group of Durdlestone Bay, Swanage, Dorsetshire.

47775. The nearly entire left ramus of the mandible, containing (*Fig.*) the canine, the broken base of $\overline{\text{pm. 1}}$, the remaining three

[1] Mesozoic Mammalia (Mon. Pal. Soc.), p. 64 (1871).
[2] *Ibid.* p. 69.

premolars, and the three anterior true molars of the typical series. This specimen is the type, and is figured by Owen in his 'Mesozoic Mammalia' (Mon. Pal. Soc.), pl. iii. figs. 11, 11*a*. There is no trace of $\overline{\text{m. 4}}$ on the exposed inner surface of the coronoid.

Beckles Collection. Purchased, 1876.

47776. Part of the left ramus of the mandible, showing the three
(*Fig.*) true molars ($\overline{\text{m. 3}}$ imperfect). Figured by Owen, *op. cit.* pl. iii. figs. 12, 12 *a*. *Same history.*

47777. Hinder part of the left ramus of the mandible, containing
(*Fig.*) $\overline{\text{pm. 4}}$, part of $\overline{\text{m. 1}}$, and the entire $\overline{\text{m. 2}}$ and $\overline{\text{m. 3}}$. Figured by Owen, *op. cit.* pl. iii. figs. 13, 13 *a*. *Same history.*

47780. Fragment of the right ramus of the mandible, containing the
(*Fig.*) third true molar. Figured by Owen, *op. cit.* pl. iii. figs. 19, 19 *a*. *Same history.*

47778. Fragment of the left maxilla, showing $\underline{\text{pm. 3}}$, $\underline{\text{pm. 4}}$, $\underline{\text{m. 1}}$, and
(*Fig.*) $\underline{\text{m. 2}}$. Figured by Owen, *op. cit.* pl. iii. figs. 18, 18 *a*.
Same history.

47779. Part of the right maxilla, containing $\underline{\text{pm. 3}}$, $\underline{\text{pm. 4}}$, $\underline{\text{m. 1}}$, and
(*Fig.*) part of $\underline{\text{m. 2}}$. Figured by Owen, *op. cit.* pl. iii. figs. 17, 17 *a*.
Same history.

47781. Portions of the cranium, the mandibular rami, and other
(*Fig.*) bones. Figured by Owen, *op. cit.* pl. iv. fig. 1.
Same history.

47765. Part of the left mandibular ramus of a subadult individual,
(*Fig.*) showing the three true molars. Figured by Owen, *op. cit.* pl. iii. fig. 8, under the name of *T. mordax*; the teeth agree, however, in size with those of the present species, the length of $\overline{\text{m. 1}}$ + $\overline{\text{m. 2}}$ being 0,008. *Same history.*

47791. The hinder part of the right ramus of the mandible, containing $\overline{\text{m. 2}}$ and $\overline{\text{m. 3}}$. The anterior lobe of $\overline{\text{m. 2}}$ is imperfectly developed. *Same history.*

The following specimen is the type of T. occisor, *Owen, but apparently presents no characters by which it can be distinguished specifically from the preceding specimens.*

47782. (*Fig.*) The imperfect mandible of a subadult individual, showing nearly all the teeth. Figured by Owen, *op. cit.* pl. iv. figs. 2, 2 *a*; $\overline{\mathrm{pm.\ 4}}$ is not fully protruded from its alveolus, and the anterior lobe of the left $\overline{\mathrm{m.\ 1}}$ is aborted.
Same history.

The following specimens not improbably belong to this species.

47772. (*Fig.*) Fragment of the right ramus of the mandible, showing $\overline{\mathrm{pm.\ 4}}$ and $\overline{\mathrm{m.\ 1}}$. Figured by Owen, *op. cit.* pl. iii. fig. 21.
Same history.

47773. (*Fig.*) The greater part of the right ramus of the mandible with one imperfect cheek-tooth. Figured by Owen, *op. cit.* pl. iv. fig. 4. *Same history.*

47774. (*Fig.*) Part of a mandibular ramus of an old individual, containing an entire premolar and a worn true molar. Figured by Owen, *op. cit.* pl. iii. figs. 14, 14 *a*. *Same history.*

47774 a. The greater part of the right ramus of the mandible, with the teeth in a crushed condition. *Same history.*

47788. Fragment of a maxilla, showing the canine, two premolars, and what is apparently part of $\overline{\mathrm{m.\ 1}}$. *Same history.*

48404 a. A fourth premolar. *Same history.*

Triconodon major, Owen[1].

This species is estimated to have been fully one-third larger than *T. ferox*, and about equal in size to the existing *Dasyurus viverrinus*.
Hab. Europe (England).

40722. (*Fig.*) Part of the right ramus of the mandible, containing the slightly imperfect $\overline{\mathrm{m.\ 2}}$ and portions of the anterior cheek-teeth; from the Middle Purbeck group of Durdlestone Bay, Swanage, Dorsetshire. This specimen, which is the type and only known example, is figured by Owen in his 'Mesozoic Mammalia,' pl. iv. fig. 3.
Beckles Collection. Purchased, 1876.

[1] Mesozoic Mammalia (Mon. Pal. Soc.), p. 70 (1871).

Specifically undetermined Specimens.

The following specimens from the Middle Purbeck group of Durdlestone Bay, Swanage, are referred by Owen to the present genus.

47770. The two terminal extremities and the impression of the (*Fig.*) middle portion of a mandibular ramus of a small form. Figured by Owen, *op. cit.* pl. iii. fig. 20.

Beckles Collection. Purchased, 1876.

47797. Fragments of a mandible, showing the bases of some of the teeth. *Same history.*

48396. Small fragment of a mandibular ramus, showing one true molar. *Same history.*

47769. The greater portion of a mandibular ramus of a very small (*Fig.*) form, showing some broken cheek-teeth. Figured by Owen, *op. cit.* pl. iii. figs. 16, 16 *a*. *Same history.*

Family DASYURIDÆ.

Dentition:—I. $\frac{4}{3}$, C. $\frac{1}{1}$, Pm. + M. various. There are never more than three premolars in each jaw; the lower true molars are without an inner cingulum, and are either multicuspidate or differentiated into a blade and talon, the latter being never simply secant in all the teeth of the series and always smaller than the hinder lobe of the blade; the upper true molars are cusped and either subquadrangular, or triangular with an inner tubercle. The hind feet have the hallux without a claw, and generally either rudimentary or absent, while the four outer digits are well developed.

Subfamily Dasyurinæ.

Dentition:—Pm. + M. not exceeding $\frac{7}{7}$, and of relatively large size; the premolars are differentiated from the true molars; the lower incisors are in contact; the lower true molars are differentiated into a bilobed blade, which frequently has an inner cusp, and a talon which is usually tubercular, but may be secant in one tooth of the series, in which case it is always smaller than the hinder outer cusp of the blade; the upper true molars are triangular with a well-marked inner tubercle; and the fourth premolar when well-developed is preceded by a milk-tooth.

The upper true molars are simpler than in the *Peramelidæ*, the elongated inner column of the latter forming a rounded tubercle, and the two outer V-shaped columns either in contact (*Dasyurus*) or fused together (*Sarcophilus*); the hinder lobe of the lower true molars of the *Peramelidæ* is represented by the talon[1].

Genus **THYLACINUS**, Temminck[2].

Dentition:—I. $\frac{4}{3}$, C. $\frac{1}{1}$, Pm. $\frac{3}{3}$, M. $\frac{4}{4}$. The cheek-teeth resemble those of *Sarcophilus*, but with total absence of the external cusp in the upper, and of the internal cusp in the blade of the lower true molars, and with the talons of the latter well developed. Axis vertebra long; and an entepicondylar foramen to the humerus.

Thylacinus spelæus, Owen[3].

Syn. *Thylacinus major*, Owen[4] (*in parte*).

This species apparently presents the same relation to the existing Tasmanian *T. cynocephalus* as is borne by *Sarcophilus laniarius* to *S. ursinus*, but in this case the excess in size of the fossil form seems to have been still more marked. The smaller specimens included under the present head cannot be distinguished from large male individuals of *T. cynocephalus*, but it is probable that they belong to the female of *T. spelæus*, since the existing species exhibits a great difference in the size of the two sexes; these smaller specimens are, however, referred by Owen[5] to the existing species.

Hab. New South Wales and Queensland.

The following specimens may be referred to male individuals.

32306. Fragment of the right ramus of the mandible of a subadult individual, containing the last true molar; from a cave in the Wellington Valley, New South Wales. The length of the tooth is 0,019, against 0,016 in the corresponding tooth of a large male of the existing species; it agrees in relative size with the tooth of the present species

[1] The type of lower true molar obtaining in *Dasyurus* agrees precisely with those of the *Proviverridæ* (*vide* Supplement), and also with the lower carnassial of the *Viverridæ*.

[2] Monographies de Mammalogie, vol. i. p. 60 (1827).

[3] Cat. Foss. Mamm. et Aves Mus. R. Coll. Surg. p. 335 (1845).

[4] 'Extinct Mammals of Australia,' p. 106 (1877). This name is apparently given by inadvertence for *T. spelæus*; the mandible represented in pl. v. fig. 8 is apparently drawn from the last three true molars of *Sarcophilus laniarius* added to the hinder part of a mandible of *Thylacinus cynocephalus*.

[5] *Ibid.* p. 106.

figured in vol. ii. pl. xxxii. fig. 7 of Mitchell's 'Three Expeditions into Australia' (see Owen, *op. cit.* p. 335).
Purchased, 1854.

32874. The proximal portion of the right ulna; from the Pleistocene of Queensland. *Purchased*, 1853.

43938. Two specimens of the right calcaneum; from the caves of the Wellington Valley. The length of the larger specimen is 0,048.
Presented by the Trustees of the Australian Museum, 1870.

The following smaller specimens are provisionally regarded as belonging to the female.

43887. A left upper true molar; from the Wellington Valley.
Presented by the Trustees of the Australian Museum, 1870.

42596. The distal portion of the left humerus; from the same locality. *Same history.*

M. 3706. The axis vertebra; from Queensland. *No history.*

Genus **SARCOPHILUS**, F. Cuvier[1].

The number and homology of the teeth the same as in *Dasyurus*, but the outer columns of the upper true molars fused together, not forming distinct V's, and with only a rudiment of the antero-external cusp of the posterior column; inner cusp of blade of lower true molars rudimentary, and talon very small. Axis vertebra short; no entepicondylar foramen to the humerus.

Sarcophilus laniarius, Owen[2].

Syn. *Dasyurus laniarius*, Owen[3].

This species presents a considerable excess in size over the existing Tasmanian *S. ursinus*, but may probably be regarded merely as the ancestral form of the latter, which, from inhabiting a continental area, attained superior dimensions. The interdental pits in the maxilla are slightly deeper in the fossil form, and Owen notes a small variation in the contour of the palatal vacuities and the mandible. The smaller specimens identified by Owen[4] with *S. ursinus* probably belong to the female of the present form.

Hab. New South Wales and Queensland.

[1] Hist. Nat. des Mammifères, pl. 70 (1838).

[2] In Mitchell's 'Three Expeditions into Australia,' 2nd ed. vol. ii. p. 363 (1838).—*Dasyurus.* [3] *Loc. cit.*

[4] Extinct Mammals of Australia, p. 105 (1877).

Unless otherwise stated, the following specimens are from the caves of the Wellington Valley, New South Wales, and were presented by the Trustees of the Australian Museum, 1870.

42555. The anterior portion of the cranium, showing the greater number of the teeth in a damaged condition. The length of the series of cheek-teeth is 0,056, against 0,049 in a large male of *S. ursinus*.

42556. Part of the brain-case.

42543. The middle portion of the cranium, showing the alveoli of the last three true molars.

42559. Part of the right maxilla, showing the first three true
(Fig.) molars. Figured by Owen in the 'Extinct Mammals of Australia,' pl. v. fig. 3. The length of the space occupied by the three teeth is 0,0395, the corresponding dimension in a large male of the existing species being 0,036.

42560. Part of the left maxilla, containing the four true molars.

42584. A similar specimen.

42557. Fragment of the left maxilla, with the last three true molars.

42583. A similar specimen.

42561. Fragment of the right maxilla, with $\underline{\text{m. 3}}$ and $\underline{\text{m. 4}}$.

43303. Fragment of the left maxilla of a very large individual, containing $\underline{\text{m. 2}}$ and $\underline{\text{m. 3}}$. The length of the space occupied by the two teeth is 0,029, against 0,0246 in the above-mentioned specimen of *S. ursinus*.

43303 a. Fragment of the left maxilla, containing $\underline{\text{m. 2}}$ and $\underline{\text{m. 3}}$.

42585. Fragment of the left maxilla, containing $\underline{\text{m. 1}}$ and $\underline{\text{m. 2}}$.

42586. Fragment of the left maxilla of a small (? female) individual, containing $\underline{\text{m. 1}}$.

43951. Part of the right maxilla, showing the two premolars and the basal half of the canine.

50068. Part of the left maxilla, containing the cheek-teeth in a broken condition; from the Pleistocene of Gowrie, Queensland. *Presented by G. F. Bennett, Esq.*, 1879.

42549. Three canines.

42582. The symphysis of the mandible, showing the cutting-teeth
(*Fig.*) and the right premolars. Figured by Owen, *op. cit.* pl. v. figs. 5, 6.

42564. The symphysis and part of the right ramus of the mandible, showing $\overline{\text{pm. 4}}$ and the first three true molars.

42562. Part of the right ramus of the mandible, containing the last three true molars.

42565. The corresponding portion of the left ramus of the mandible.

42566. Part of the left ramus of the mandible, containing $\overline{\text{m. 2}}$ and $\overline{\text{m. 3}}$.

42568. Fragment of the right ramus of the mandible, showing $\overline{\text{pm. 4}}$, $\overline{\text{m. 1}}$, and $\overline{\text{m. 2}}$.

42569. Fragment of the right ramus of the mandible, with $\overline{\text{m. 3}}$ and $\overline{\text{m. 4}}$.

42571. Part of the right ramus of the mandible, containing the four true molars.

42572. Part of the left ramus of the mandible, containing the last three true molars.

42574. Fragment of the left ramus of the mandible, showing the two premolars and $\overline{\text{m. 1}}$.

42575. Fragment of the left ramus of the mandible, containing $\overline{\text{m. 2}}$ and $\overline{\text{m. 3}}$.

42576. Part of the left ramus of the mandible of a small individual, showing the six cheek-teeth.

42581. Part of the right ramus of the mandible, containing $\overline{\text{pm. 4}}$, $\overline{\text{m. 1}}$, and $\overline{\text{m. 2}}$.

43387. Fragment of the right ramus of the mandible, with $\overline{\text{m. 2}}$ and $\overline{\text{m, 3}}$.

48425. Part of the left ramus of the mandible, containing the last three true molars; from the Pleistocene of Queensland. The length of the space occupied by the three teeth is 0,036, against 0,0325 in the above-mentioned specimen of *S. ursinus.* The specimen is labelled by Owen *Thylacinus major* [1]. *Presented by Dr. George Bennett,* 1877.

[1] See page 264, note 4.

42596. The distal portion of the left humerus.

42596 a. Two specimens of the right calcaneum.

Genus **DASYURUS,** E. Geoffroy[1].

Dentition:—I. $\frac{4}{3}$, C. $\frac{1}{1}$, Pm. $\frac{2}{2}$, M. $\frac{4}{4}$. The two premolars have been shown by Thomas to correspond to the first and third of the typical series. The upper true molars have the two outer V-shaped columns distinct, and with an antero-external cusp on the hinder one; the blade (anterior lobe) of the lower true molars with three cusps arranged in a triangle, and the talon of the same well marked and cusped. The axis vertebra is elongated; and there is an entepicondylar foramen to the humerus.

Dasyurus viverrinus (Shaw[2]).

Syn. *Didelphis viverrina*, Shaw[3].

Hab. Australia and Tasmania.

The following specimens, with the exception of No. M. 1906, *are from the caves of the Wellington Valley, New South Wales, and were presented by the Trustees of the Australian Museum*, 1870.

43880. Fragment of the right maxilla, containing the last three true molars.

43880 a. Fragment of the left maxilla, with $\underline{\text{m. 2}}$ and $\underline{\text{m. 3}}$.

43880 b. The symphysis and part of the left ramus of the mandible.

43880 c. Several fragments of mandibular rami with teeth.

M. 1906. The nearly entire right ramus of the mandible, showing many of the teeth; from the Pleistocene of Gowrie, Queensland. *Presented by G. F. Bennett, Esq.*

42641. The right humerus of a large individual.

42668. The left humerus of a much smaller (? female) individual.

42669. The left ulna of an individual agreeing in size with the preceding specimen.

[1] Bull. Soc. Philom. Paris, sér. 1, vol. i. p. 106 (1796).
[2] General Zoology, vol. i. pt. 2, p. 491 (1800).—*Didelphis.*
[3] *Loc. cit.*

Subfamily MYRMECOBIINÆ.

(Although unrepresented in a fossil state, it is important that this group should be mentioned here.)

Dentition:—Pm.+M. $\frac{8}{9}$, and of small size; the three premolars are not well differentiated from the true molars; the lower incisors are spaced; the true molars are subquadrangular and multicuspidate, the lower ones not being differentiated into a blade and talon; the mandibular ramus is shallow, with a mylohyoid groove. There is no evidence of the existence of a milk-molar preceding the last premolar.

Family AMPHITHERIIDÆ[1].

The genera provisionally included in this family comprise a number of small primitive mammals, apparently connected on the one hand with the Myrmecobiine branch of the *Dasyuridæ*, and on the other with the *Didelphyidæ*. It may be eventually necessary to divide the family; and it is difficult with the present imperfect evidence to give a distinctive diagnosis which will embrace all the forms. They all, however, have a mylohyoid groove in the mandible; their lower true molars, which (with the possible exception of *Phascolotherium*) exceed four, are either multicuspidate, tricuspidate, or differentiated into a blade and talon; the premolars may be reduced to three or perhaps two, but are frequently four, and have been thought in some instances to exceed that number; the lower incisors are either three or four; and the canines have grooved or double roots. It is not known whether the last premolar had a milk predecessor.

The difference in the characters of the lower molars in the various genera is not greater than obtaining in the *Dasyuridæ*. If the number of premolars in some members of the first section exceed four, it is a unique feature among heterodont mammals.

SECTION A.—The lower true molars multicuspidate, or tricuspidate and cingulated; the lower incisors spaced, and either three or four in number; and the premolars in some instances reduced below the latter number.

The American Triassic *Dromatherium*[2] may not improbably be included in this section; the dental formula being I. $\frac{?}{3}$, C. $\frac{?}{1}$, Pm. $\frac{?}{3}$, M. $\frac{?}{7}$.

[1] Owen, British Fossil Mammals and Birds, p. 29 (1846).

[2] See Osborn, Proc. Amer. Phil. Soc. vol. xxiv. p. 109 (1887).

Genus **PHASCOLOTHERIUM,** Owen[1].

Dentition:—I. $\frac{?}{4}$, C. $\frac{?}{1}$, Pm.+M. $\frac{?}{7}$. Molars and premolars very similar in structure; a diastema between lower canine and premolars; true molars tricuspidate, with inner cingulum, which has no cusp on the inner side of the crown, but develops a distinct accessory cusp at each extremity of the crown; mandibular ramus deep.

Owen considered that there were probably only three lower incisors, but the writer is convinced, from an examination of the type specimen, that there were four of these teeth.

Phascolotherium bucklandi (Broderip[2]).

Syn. *Didelphis bucklandi*, Broderip[3].
Thylacotherium bucklandi, Valenciennes[4].

This is the type and only named species; and was about equal in size to *Phascogale penicillata*; the length of the lower series of cheek-teeth being 0,012.

Hab. Europe (England).

112. (*Fig.*) The nearly entire right ramus of the mandible, containing all the teeth except i. 4; from the Lower Jurassic Slate of

Fig. 40.

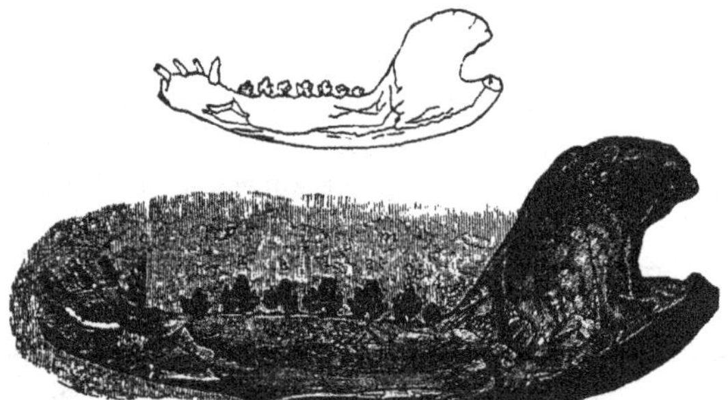

Phascolotherium bucklandi.—Inner view of the right ramus of the mandible; from the Stonesfield Slate. $\frac{3}{1}$. The outline figure is nat. size. (From Owen's 'Palæontology.')

Stonesfield, Oxfordshire. This specimen (woodcut, fig. 40) is the type, and is figured by Broderip in the 'Zool. Journ.' vol. iii. pl. xl.; by Buckland in his 'Bridgewater Treatise,'

[1] Trans. Geol. Soc. ser. 2, vol. vi. pt. 1, p. 58 (1841).
[2] Zool. Journ. vol. iii. p. 411 (1828).—*Didelphis*. [3] *Loc. cit.*
[4] Comptes Rendus, vol. vii. p. 580 (1838).

pl. ii. (under the name of *Didelphis*); and also by Owen in the 'Trans. Geol. Soc.' ser. 2, vol. vi. pl. vi. fig. 2, in his 'Odontography,' pl. xcix. fig. 4, and also in his 'Mesozoic Mammalia' (Mon. Pal. Soc.), pl. i. figs. 26, 26*a*.

Presented by W. J. Broderip, Esq.

Genus **AMPHILESTES**, Owen[1].

Distinguished from *Phascolotherium* by the larger number of lower cheek-teeth, and by at least some of the premolars being simpler than the true molars; in the type species (fig. 41) the cingulum of the tricuspidate true molars encircles the crown, and has minute fore-and-aft cusps; the mandibular condyle is apparently sessile and placed on the level of the molars. In the type specimen[2] (fig. 41) there are nine teeth remaining (the crown

Fig. 41.

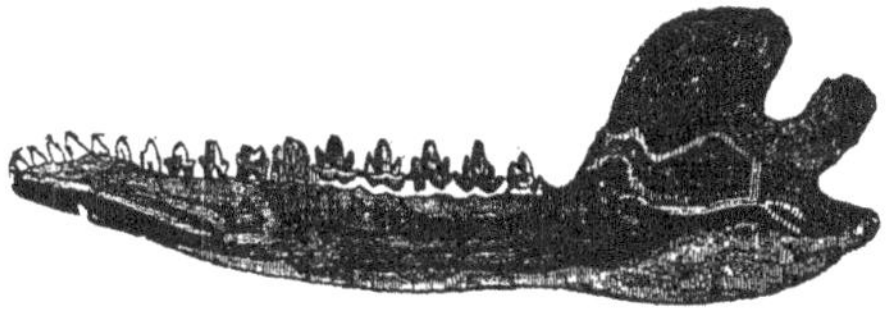

Amphilestes broderipi, Owen.—Reversed inner view of the left ramus of the mandible; from the Stonesfield Slate. $\frac{2}{1}$. The restoration of the anterior teeth is conjectural; and the condyle is placed too high. (From Owen's 'Palæontology'[3].)

of the sixth from the posterior end broken off), of which the first two are clearly premolars; the third tooth may possibly be a milk-molar, and it is not certain that there may not be an additional molar in its alveolus behind the ninth tooth.

[1] Encyclopædia Britannica, 8th ed. vol. xvii. p. 157 (1859). In his 'Mesozoic Mammalia' (Mon. Pal. Soc.) Owen provisionally included the type species in *Amphitherium*, but the observations of Osborn (kindly communicated in MS. to the writer) apparently indicate its right to distinction.

[2] Compare Owen's 'Odontography,' pl. xcix. fig. 3. Owen regarded the lower dental formula as I. $\frac{}{3}$, C. $\frac{}{1}$, Pm. $\frac{}{6}$, M. $\frac{}{6}$; but it is quite probable that there may have been four incisors, which would reduce the premolars to five; while, if the specimen be adult, the seventh tooth from the end might be a true molar, in which case there would be only four premolars; and there may have been a diastema behind the canine.

[3] Wrongly given in that work (fig. 112) and in the 'Encyclopædia Britannica,' *op. cit.*, as *Amphitherium prevosti*, and so reproduced in several later works; it is really reduced from pl. vi. of the 'Trans. Geol. Soc.' ser. 2, vol. vi. pl. vi. fig. 1, and pl. xcix. fig. 3 of Owen's 'Odontography.'

Amphilestes (?) sp.

This form is of larger size than *A. broderipi*; the length of the space occupied by the last four lower true molars being 0,0085, against 0,006 in the latter.

Hab. Europe (England).

M. 2300. Cast of a left mandibular ramus, showing the last four true molars. The original was obtained from the Lower Jurassic Slate of Stonesfield, Oxfordshire, and is preserved in the Museum at Oxford. It is figured by Phillips in his 'Geology of Oxford and the Valley of the Thames,' p. 235, diagram 81, under the name of *Phascolotherium bucklandi*; but the true molars differ from those of that species in the small development of the accessory fore-and-aft cusps, and come nearer in this respect to those of *Amphilestes*, although the contour of the ramus is more like that of *Phascolotherium*.

Presented by Prof. Prestwich, 1885.

GENUS *non. det.*

M. 2299. Cast of a mandibular ramus showing all the teeth, with the exception of the fourth, in a worn or damaged condition. The original was obtained from the Lower Jurassic Slate of Stonesfield, Oxfordshire, and is preserved in the Museum at Oxford. It is figured (under the name of *Amphitherium prevosti*) by Owen in his 'Odontography,' pl. xcix. fig. 1, in his 'British Fossil Mammals and Birds,' p. 29, fig. 15, and in his 'British Mesozoic Mammals' (Mon. Pal. Soc.), pl. i. figs. 23, 23 *a*. It is, however, regarded by Osborn [1] as decidedly distinct from that form, with molars approximating to those of *Amphilestes*, but with the cingulum confined to the inner side, and with the condyle pedunculate and placed above the molar level. Owen gives the lower dental formula as I. $\frac{\ }{3}$, C. $\frac{\ }{1}$, Pm. $\frac{\ }{6}$, M. $\frac{\ }{6}$, but there appears equal reason for rendering it as I. $\frac{\ }{4}$, C. $\frac{\ }{1}$, Pm. $\frac{\ }{4}$, M. $\frac{\ }{7}$, or the same as in *Amblotherium* [2].

Presented by Prof. Prestwich, 1885.

[1] See note 1, preceding page.

[2] The superiority in the size of the fourth over the fifth remaining tooth would seem to indicate the probability of the former being a canine rather than a premolar; and that the canine was inserted by two roots is probable from the observations of Marsh, Amer. Journ. ser. 3, vol. xxxiii. p. 335 (1887).

M. 2297. Cast of a mandibular ramus with seven cheek-teeth of a species belonging to the present section. The original, which is from the Stonesfield Slate, is preserved in the Museum at Oxford, and is figured by Phillips (with an added molar) in his 'Geology of Oxford and the Valley of the Thames,' p. 234, diagram No. 79, under the name of *Amphitherium prevosti*. There are three premolars and a diastema behind the canine; the fourth cheek-tooth may be $\overline{\text{mm. 4}}$. *Same history.*

SECTION B.—The true molars usually differentiated into a bilobed blade (which may have a small inner cusp) and a hind talon; four premolars present in all the English genera of which the entire lower dentition is known; and generally four lower incisors.

Those forms which have an inner cusp to the blade of the lower true molars approximate to the *Didelphyidæ*. It is probable that several of the American genera included by Marsh [1] in his *Dryolestidæ* belong to the present section; *Dryolestes* [2] itself being apparently very close to *Amblotherium*. In *Asthenodon* [3], which Marsh regards as connecting *Dryolestes* with *Stylodon*, the lower true molars have no distinct hind talon.

Genus **AMPHITHERIUM,** Blainville [4].

Syn. *Amphigonus*, Agassiz [5].
Thylacotherium, Valenciennes [6] (*in parte*).

It is uncertain, owing to the imperfect nature of the type specimen, whether this genus (as restricted by Osborn [7]) should be referred to the present or the preceding section; if, however, No. 36822 be rightly referred to it, it undoubtedly belongs to the present one.

Amphitherium prevosti, Blainville [8] (*ex* Cuv. MS.).

Syn. *Didelphis prevosti*, Blainville [9].
Thylacotherium prevosti, Valenciennes [10].

This is now taken as the type species of the genus, and is of minute size [11].

Hab. Europe (England).

[1] Amer. Journ. ser. 3, vol. xxxiii. p. 334 (1887).
[2] *Ibid.* pl. ix. fig. 2. [3] *Ibid.* pl. ix. fig. 7.
[4] Comptes Rendus, vol. vii. p. 417 (1838).
[5] Neues Jahrb. 1835, p. 185.
[6] Comptes Rendus, vol. vii. p. 580 (1838). [7] See note 1, p. 271.
[8] Comptes Rendus, vol. vii. p. 403 (1838).—*Didelphys*. [9] *Loc. cit.*
[10] Comptes Rendus, vol. vii. p. 580 (1838).
[11] Originally *Phascolotherium* was included.

M. 2298. Cast of the hinder portion of a mandibular ramus. The original is from the Lower Jurassic Slate of Stonesfield, Oxfordshire, and is preserved in the Museum at Oxford. It is the type specimen, and is figured by Prevost in the 'Ann. Sci. Nat.' vol. iv. p. 389, pl. xviii. fig. 2; by Buckland in his 'Bridgewater Treatise,' pl. ii. (as *Didelphys*); by Owen in the 'Trans. Geol. Soc.' ser. 2, vol. vi. pl. v. fig. 3, in his 'British Fossil Mammals and Birds,' p. 44, fig. 16, in his 'Odontography,' pl. xcix. fig. 2, and in his 'Mesozoic Mammalia' (Mon. Pal. Soc.), pl. i. fig. 21; and by Phillips in his 'Geology of Oxford and the Valley of the Thames,' p. 233, diagram no. 77. The jaw contains four premolars and six true molars.

Presented by Prof. Prestwich, 1884.

36822. Part of a left mandibular ramus, containing the last premolar and five true molars, provisionally referred to this spécies. The true molars are of the type of those of *Amblotherium*; the length of the space occupied by the five teeth of this series being 0,005. *Morris Collection. Purchased*, 1862.

Genus **AMBLOTHERIUM**, Owen [1].

Including:—*Peraspalax*, Owen [2].
Peralestes (*Phascolestes*), Owen [3] (*in parte*).

Dentition:—I. $\frac{?}{4}$, C. $\frac{?}{1}$, Pm. $\frac{?}{4}$, M. $\frac{?}{(7-8)}$. The lower incisors are spaced, the lower true molars may have a small inner cusp to the blade [4], the anterior lobe of the blade of the latter is of moderate height, and $\overline{\text{pm. 3}}$ and $\overline{\text{pm. 4}}$ are slender.

Amblotherium soricinum, Owen [5].

This is the type species; the hinder lower true molars have a minute inner cusp to the blade, and that series of teeth did not apparently exceed seven in number; the length of the space occupied by the first six of these teeth is 0,005.

Hab. Europe (England).

[1] Mesozoic Mammalia (Mon. Pal. Soc.), p. 29 (1871).
[2] *Ibid.* p. 40. [3] *Ibid.* p. 35.
[4] Owen regarded the presence of this cusp as distinguishing *Peraspalax* from *Amblotherium*; it is, however, distinctly present in the hinder molars of *A. soricinum*.
[5] *Op cit.* p. 29.

47752. The nearly entire right ramus of the mandible; containing
(*Fig.*) all the anterior teeth and the first six true molars; from the Middle Purbeck group of Durdlestone Bay, Swanage, Dorsetshire. This specimen is the type, and is figured by Owen in his 'Mesozoic Mammalia' (Mon. Pal. Soc.), pl. ii. fig. 1. Owen considers that there should be a seventh tooth behind the sixth true molar.

Beckles Collection. Purchased, 1876.

Amblotherium talpoides (Owen [1]).

Syn. *Peraspalax talpoides*, Owen [2].

Apparently agreeing in the structure of the teeth with the preceding species, but of larger size.

Hab. Europe (England).

47738. Part of the right ramus of the mandible, containg $\overline{\text{pm. 4}}$ and
(*Fig.*) six true molars; from the Middle Purbeck group of Durdlestone Bay, Swanage, Dorsetshire. This specimen is the type, and is figured by Owen in his 'Mesozoic Mammalia' (Mon. Pal. Soc.), pl. ii. fig. 9, under the generic name of *Peraspalax*, of which it is the type.

Beckles Collection. Purchased, 1876.

Amblotherium mustelula, Owen [3].

Syn. *Peralestes (Phascolestes) longirostris*, Owen [4] (*in parte*).

Of considerably larger size than *A. soricinum*, and furnished with eight lower true molars, which apparently had no inner cusp. There appears no characters by which the mandible figured by Owen under the name of *Peralestes (Phascolestes) longirostris* can be distinguished from the type specimen of the present species.

Hab. Europe (England).

47753. The nearly entire right ramus of the mandible, showing
(*Fig.*) most of the cheek-teeth in a mutilated condition; from the Middle Purbeck group of Durdlestone Bay, Swanage, Dorsetshire. This specimen is the type, and is figured by Owen in his 'Mesozoic Mammalia' (Mon. Pal. Soc.), pl. ii. fig. 2. *Beckles Collection. Purchased*, 1876.

47808. The greater part of a slightly deeper left mandibular ramus, which appears specifically identical with the preceding;

[1] Mesozoic Mammalia (Mon. Pal. Soc.), p. 40 (1871).—*Peraspalax.*

[2] *Loc. cit.*

[3] *Ibid.* p. 31.

[4] *Ibid.* p. 35.

from the same locality. Most of the anterior teeth and the eight true molars (except the 6th) are shown; the true molars are less worn than in the preceding specimen, and thereby show a taller anterior lobe of the blade, which approaches that of *Achyrodon*; the length of the space occupied by the eight true molars is 0,008.

Beckles Collection. Purchased, 1876.

47741. The greater part of the left ramus of a very similar mandible,
(*Fig.*) containing the incisors, canine, the four premolars, and the first five true molars; from the same locality. Figured by Owen, *op. cit.* pl. ii. fig. 4, under the name of *Peralestes (Phascolestes) longirostris*, but apparently presenting no characters by which it can be distinguished from the type specimen of the present form. The three impressions in the matrix behind the sixth molar mentioned by Owen as being made by teeth appear to be of an entirely different origin. *Same history.*

Genus **ACHYRODON**, Owen [1].

This genus appears closely allied to the preceding, but is distinguished by the greater height and more oblique direction of the anterior lobe of the blade of the lower true molars, and the stouter $\overline{\text{pm. }3}$ and $\overline{\text{pm. }4}$.

Achyrodon nanus, Owen [2].

This (the type) species is rather smaller than *Amblotherium soricinum*.

Hab. Europe (England).

47745. The middle portion of the right ramus of the mandible,
(*Fig.*) containing $\overline{\text{pm. }3}$, $\overline{\text{pm. }4}$, and the eight true molars; from the Middle Purbeck group of Durdlestone Bay, Swanage, Dorsetshire. This specimen, which is the type, is figured by Owen in his 'Mesozoic Mammalia,' pl. ii. fig. 5.

Beckles Collection. Purchased, 1876.

47746. Part of the left ramus of the mandible, containing $\overline{\text{pm. }1}$,
(*Fig.*) $\overline{\text{pm. }4}$, and the first five true molars; from the same locality. Figured by Owen, *op. cit.* pl. ii. fig. 6.

Same history.

47783. The greater part of the left ramus of the mandible, con-
(*Fig.*) taining four true molars; from the same locality. Figured by Owen, *op. cit.* pl. ii. fig. 7. *Same history.*

[1] Mesozoic Mammalia (Mon. Pal. Soc.), p. 37 (1871).
[2] *Loc. cit.*

47785. Hinder part of the right ramus of the mandible, showing five true molars; from the same locality.

Beckles Collection. Purchased, 1876.

Achyrodon pusillus, Owen[1].

Distinguished from the preceding species by the structure of the mandible and teeth.

Hab. Europe (England).

47747. (*Fig.*) Fragment of the left ramus of the mandible, containing one tooth; from the Middle Purbeck group of Durdlestone Bay, Swanage, Dorsetshire. This specimen is the type, and is figured by Owen in his 'Mesozoic Mammalia' (Mon. Pal. Soc.), pl. ii. fig. 8; at that time there were four cheek-teeth preserved.

Beckles Collection. Purchased, 1876.

Specifically undetermined specimen.

48401. Fragment of a mandibular ramus with four true molars, probably belonging to a small species of the present genus; from the Middle Purbeck group of Durdlestone Bay, Swanage, Dorsetshire.

Beckles Collection. Purchased, 1876.

Genus **PERAMUS,** Owen[2].

The lower true molars are relatively stouter, and their blade proportionally wider than in either of the preceding genera; the complete dental formula is unknown.

Peramus tenuirostris, Owen[3].

This is the type species.

Hab. Europe (England).

47742. (*Fig.*) The left ramus of the mandible, containing the last four cheek-teeth; from the Middle Purbeck group of Durdlestone Bay, Swanage, Dorsetshire. This specimen is the type, and is figured by Owen in his 'Mesozoic Mammalia' (Mon. Pal. Soc.), pl. ii. fig. 10.

Beckles Collection. Purchased, 1876.

[1] Mesozoic Mammalia (Mon. Pal. Soc.), p. 39 (1871).

[2] *Ibid.* p. 41 (1871).

[3] *Loc. cit.*

47744. Part of a left mandibular ramus, containing four cheek-
(*Fig.*) teeth, provisionally referred to this species; from the same locality. Figured by Owen, *op. cit.* pl. ii. fig. 12.
Beckles Collection. Purchased, 1876.

47754. A left mandibular ramus with one cheek-tooth, which is
(*Fig.*) provisionally referred to this species; from the same locality. Figured by Owen, *op. cit.* pl. ii. fig. 13. If eventually proved distinct, the name *P. minor* is proposed by Owen for this specimen. *Same history.*

The following specimen was provisionally referred by Owen to this species, but is probably generically distinct.

47743. Part of the symphysis and left ramus of a mandible con-
(*Fig.*) taining the anterior teeth. Figured by Owen, *op. cit.* pl. ii. fig. 11. *Same history.*

Family DIDELPHYIDÆ.

Dentition:—I. $\frac{5}{4}$, C. $\frac{1}{1}$, Pm. $\frac{3}{3}$, M. $\frac{4}{4}$. Incisors very small and sharp; canines large; premolars compressed, the fourth being preceded by a milk-tooth which persists till a late period; true molars formed on the general type of those of *Dasyurus*, the lower ones being differentiated into blade and talon. Five complete digits to each foot; an entepicondylar foramen to the humerus.

Genus **DIDELPHYS**, Linn.[1]

Including:—*Peratherium*, Aymard[2].
Amphiperatherium, Filhol[3].

The characters of this genus usually available in the case of fossils are identical with those of the family[4]. In existing forms the lower true molars may either decrease[5] or increase[6] in size from the first to the last, and the fourth lower premolar may be either lower than or equal to each of the preceding ones. In the Tertiary

[1] Syst. Nat. ed. 12, vol. i. p. 71 (1766).—Amended from *Didelphis*.
[2] Ann. Soc. Agric. Sci. Le Puy, vol. xii. p. 248 (1848), *teste* Filhol.
[3] Ann. Sci. Nat. vol. x. art. 3, p. 201 (1879).
[4] *Chironectes*, except in the case of the existing species, could not be distinguished in a fossil state from *Didelphys*.
[5] *D. lanigera*, *D. philander*.
[6] *D. pusilla*, *D. crassicaudata*, *D. turneri*, *Chironectes*.

forms[1] similar variations occur in the size of the true molars, and pm. 4 may be either equal to or taller than either of the two preceding teeth. The talon of the last lower true molar may be either bi- or tricuspid. The resemblance between lower jaws of existing species from the Brazilian caves and those of certain Tertiary forms is so close as apparently to leave little or no doubt as to their generic identity. From the White-River group of North America certain forms (fig. 42) have been referred to *Peratherium*, but are said to

Fig. 42.

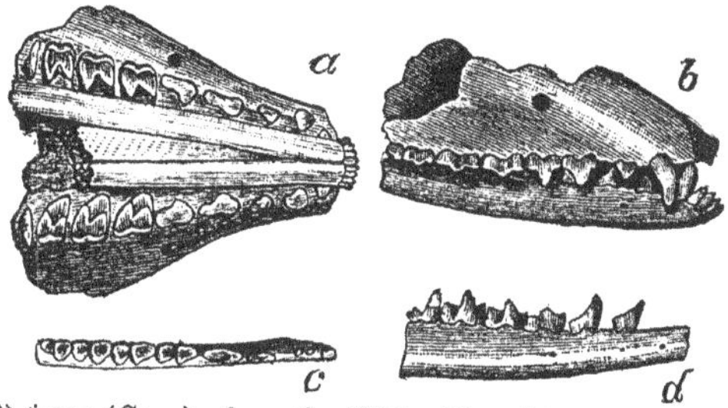

Didelphys (?) *fugax* (Cope); from the White-River Miocene of Colorado. *a, b*, inferior and lateral views of skull; *c, d*, superior and lateral views of right mandibular ramus. $\frac{3}{1}$. (From the 'Amer. Nat.')

differ from *Didelphys* by the non-inflection of the angle of the mandible[2], although this feature is well marked in the European Tertiary species. The figured American specimens do not, however, show the angle; but if the alleged absence of inflection be eventually proved correct, it will indicate a very remarkable deviation from the ordinary type. It should be observed that the number of incisors in the European Tertiary forms is not known, but it was doubtless the same as in existing species[3].

[1] *Peratherium* was proposed on the grounds that the lower true molars increased in size from the first to the last, which has a biscuspid talon, and that pm. 4 was larger than either of the others; while *Amphiperatherium* (see Filhol, 'Ann. Sci. Géol.' vol. xii. art. 3, pp. 67–68) was made to include those forms in which the lower true molars decreased in size, and pm. 4 was not the tallest of the series. The latter forms are therefore identical in these respects with species like *D. lanigera*. Another group is proposed by Filhol (see p. 54 of memoir cited) for species intermediate in these characters between *Peratherium* and *Amphiperatherium*. Gaudry ('Les Enchaînements, etc.—Mamm. Tert.' p. 11) regards *Peratherium* as undistinguishable from *Didelphys*.

[2] See Cope, Amer. Nat. vol. xviii. p. 687 (1884); it is there stated that the inflection is also absent in the European forms.

[3] It is almost certain that the number of named Tertiary species will have to be reduced, but the specimens in the Museum are not sufficiently numerous or perfect to admit of this work being undertaken. In several instances a provisional element enters into the following specific determinations.

A. South-American Pleistocene Species.

The following specimens are from the Caverns of Minas Geraes, Brazil; and belong to the Claussen Collection. Purchased, 1845.

As was pointed out by Waterhouse[1], they apparently all belong to existing species; but the large number of the latter inhabiting Brazil renders it extremely difficult in many instances to make precise specific determinations on the evidence of fossil lower jaws. The following references must therefore be regarded in many instances as provisional, and can only be taken as indicating that the specimens cannot be distinguished from the species under which they are placed; and it is highly probable that the fossils really belong to a larger number of species than those under which they are provisionally arranged. Lund in all cases distinguished the fossil forms by the prefix *affinis*, and in cases where there has been doubt to which species a particular fossil should be referred, the species which Lund records in a fossil state has been generally adopted.

Didelphys azaræ, Temminck[2].

Syn. *Didelphys aurita*, Maximilian[3].
Didelphys affinis-auritæ, Lund[4].
Didelphys albiventris, Lund[5].
Didelphys affinis-albiventri, Lund[6].

It is probable that this form cannot be specifically distinguished from the widely spread *D. marsupialis*, Linn. (*D. cancrivora* and *D. virginiana*). In a recent specimen in the Museum the length of the lower series of cheek-teeth is 0,035, and that of the true molars 0,021.

Hab. Brazil, Paraguay, Bolivia, &c.

18889. The cranium.

18890. Three portions of mandibular rami.

Didelphys crassicaudata, Desmarest[7].

Length of lower series of cheek-teeth in a recent specimen 0,024, and of the true molars 0,014; $\overline{\text{pm. 4}}$ is much lower than $\overline{\text{pm. 3}}$, and $\overline{\text{m. 4}}$ longer than $\overline{\text{m. 3}}$; talon of $\overline{\text{m. 4}}$ bicuspid.

Hab. Brazil and Paraguay.

[1] Mammalia, vol. i. p. 529.
[2] Monogr. de Mammalogie, vol. i. p. 30 (1827).
[3] Beiträge z. Naturgesch. v. Brasil, vol. ii. p. 395 (1832).
[4] K. Danske Vid. Selsk. Skr. vol. viii. p. 293 (1841).
[5] *Ibid.* p. 236 (1841).
[6] *Ibid.* p. 293 (1841).
[7] Nouv. Dict. d'Hist. Nat. vol. ix. p. 425 (1817).

18890 a. Fragment of the left ramus of a mandible, apparently belonging to this species. All the cheek-teeth except $\overline{\text{pm. 1}}$ are preserved; the length of the space occupied by the true molars being 0,0135.

Didelphys nudicaudata, Geoffroy [1].

Syn. *Didelphys myosura*, Temminck [2].
Didelphys affinis-myosuræ, Lund [3].

Length of lower series of cheek-teeth in two recent examples 0,023–0,022, and of the true molars 0,013; $\overline{\text{pm. 4}}$ and $\overline{\text{pm. 3}}$ are subequal in height, and $\overline{\text{m. 4}}$ is shorter than $\overline{\text{m. 3}}$.

It would be very difficult to distinguish this form in the fossil state from *D. quica* and *D. opossum*, which inhabit the same area; the undermentioned fossil is referred provisionally to *D. nudicaudata* on account of Lund's reference of other specimens to that type (*D. affinis-myosuræ*).

Hab. Brazil and Guiana.

18890 b. The nearly entire left ramus of the mandible, showing all the teeth except the incisors. The length of the cheek-series is 0,0215, and that of the true molars 0,013.

Didelphys cinerea, Temminck [4].

Length of series of lower cheek-teeth in a recent specimen 0,017, and of the true molars 0,010; $\overline{\text{pm. 3}}$ is taller than $\overline{\text{pm. 4}}$.

Hab. Brazil.

18890 c. The greater part of the right ramus of the mandible, showing all the teeth except the incisors. The length of the series of cheek-teeth is 0,017, and that of the true molars 0,0105.

Didelphys grisea, Desmarest [5].

Syn. *Didelphys incana*, Lund [6].
Didelphys affinis-incanæ, Lund [7].

Somewhat smaller than *D. cinerea*, the length of the lower

[1] Nouv. Dict. d'Hist. Nat. vol. ix. p. 424 (1817).
[2] Monogr. d. Mammalogie, vol. i. p. 38 (1827).
[3] K. Danske Vid. Selsk. Skr. vol. viii. p. 293 (1841).
[4] Monogr. de Mammalogie, vol. i. p. 46 (1827).
[5] Dict. Sci. Nat. vol. xlvii. p. 393 (1827).
[6] K. Danske Vid. Selsk. Skr. vol. viii. p. 237 (1841).
[7] *Ibid.* p. 293 (1841).

series of cheek-teeth being 0,0145, and that of the true molars 0,008. It would apparently be difficult to distinguish *D. hunteri* in a fossil state.

Hab. Brazil.

18890 d. Nine mandibular rami, probably belonging to this species. These specimens appear identical with those to which Lund applied the name *D. affinis-incanæ*, the living form called *D. incana* being described from Minas Geraes.

18890 e. A slightly larger left mandibular ramus.

Didelphys murina, Linn.[1]

Syn. *Didelphys affinis-murinæ*, Lund [2].

In two recent examples of this species (with which *D. dorsigera* is apparently identical) the lengths of the lower series of cheek-teeth are respectively 0,014 and 0,0125, while that of the true molars is 0,008 in both.

Hab. Brazil, Guiana, Peru, and Mexico.

18890 f. Three mandibular rami, probably belonging to this species. In one specimen the length of the four true molars is 0,0082.

Didelphys elegans, Waterhouse [3].

Syn. *Didelphys affinis-eleganti*, Lund [4].

Length of the lower series of cheek-teeth in a recent specimen 0,011, and of the true molars 0,007. It may be doubtful whether some of the following specimens do not belong to the short-tailed *D. brasiliensis* (*tristriata*) of Brazil, which is said to agree in size with *D. elegans*.

Hab. Chili, Bolivia (Recent), and (?) Brazil (Pleistocene).

18890 g. Four mandibular rami, agreeing in size with the present species. The length of the four true molars in one specimen is 0,007.

Didelphys pusilla, Desmarest [5].

Syn. *Didelphys affinis-pusillæ*, Lund [6].

The length of the lower cheek-series in a recent example is 0,011,

[1] Syst. Nat. ed. 12, vol. i. p. 72 (1766).
[2] K. Danske Vid. Selsk. Skr. vol. viii. p. 298 (1841).
[3] Zoology of the Voyage of the 'Beagle,' pt. 2, p. 95 (1840).
[4] K. Danske Vid. Selsk. Skr. vol. ix. p. 133 (1842).
[5] Nouv. Dict. d'Hist. Nat. vol. ix. p. 430 (1817).
[6] K. Danske Vid. Selsk. Skr. vol. viii. p. 293 (1841).

and that of the true molars 0,007; the mandibular ramus is more slender, and the crowns of the molars are lower than in *D. elegans*.

Hab. Paraguay and South Brazil.

18890 h. Three mandibular rami, probably belonging to this species. The length of the cheek-series in the perfect specimen is 0,0105, and that of the true molars 0,0069.

Didelphys, sp. *a.*

Of still smaller size than the preceding, the length of the four lower true molars being 0,006.

18890 i. Six mandibular rami.

Of uncertain Species.

18890 k. Five fragments of maxillæ with teeth.

B. European Tertiary Species.

Didelphys affinis, Gervais [1].

Syn. *Peratherium affine*, Gervais [2].

The length of the space occupied by the seven lower cheek-teeth is 0,014, that by the last five cheek-teeth 0,010, and that by the four true molars 0,007–75. The third and fourth premolars are subequal; the true molars increase in size from the first to the last, and have bicuspid talons. Gervais doubts if this form is distinct from *D. cuvieri*, Fischer [3], of the Paris gypsum, but has given the length of the lower cheek-series of the latter as 0,016, which apparently comes nearer to *D. antiqua*.

Hab. Europe (France).

26792. Part of the left ramus of the mandible, containing the last five cheek-teeth; from the Upper Eocene (Lower Oligocene) of Débruge, near Apt (Vaucluse), France. This specimen apparently agrees precisely with the type mandible figured by Gervais in the 'Zool. et Pal. Françaises,' pl. xlv. fig. 4. *Pomel Collection. Purchased*, 1851.

Didelphys antiqua, Gervais [4].

Syn. *Peratherium antiquum*, Gervais [5].

Closely allied to the preceding, but of rather larger size, the length of the space occupied by the last five lower cheek-teeth being 0,012.

Hab. Europe (France).

[1] Zool. et Pal. Françaises, 1st ed. vol. i. Exp. no. 45 (1848–52).
[2] *Ibid.* 2nd ed. p. 265 (1859).
[3] Synopsis Mammalium, p. 268 (1829).
[4] Zool. et Pal. Françaises, 1st ed. vol. i. Exp. no. 45 (1848–52).
[5] *Ibid.* 2nd ed. p. 266 (1859).

28134. Fragment of the left ramus of the mandible, showing two true molars; from the Upper Eocene (Lower Oligocene) of Débruge near Apt (Vaucluse), France.
Bravard Collection. Purchased, 1852.

28131. Anterior portion of the left ramus of the mandible, showing the outermost incisor, the canine, and the three premolars, $\overline{\text{pm. 4}}$ being not protruded from its alveolus; from the same locality. *Same history.*

Didelphys arvernensis, Croizet [1].

Syn. *Peratherium arvernense*, Gervais [2].

About the size of *D. affinis*, from which it is stated to differ by the talons in the lower true molars being tricuspid. Gervais gives the length of the lower series of cheek-teeth as 0,0135; the mandibular ramus is slender; $\overline{\text{pm. 1}}$ is small and in contact with $\overline{\text{pm. 3}}$, the latter being nearly as tall as $\overline{\text{pm. 4}}$.

Hab. Europe (France).

27700. The nearly entire right ramus of a mandible containing the canine and most of the cheek-teeth, apparently belonging to this species; from the Lower Miocene (Upper Oligocene) of Sauvetat (Puy-de-Dôme), France. The talon of $\overline{\text{m. 3}}$ is tricuspid, and the length of the series of cheek-teeth 0,0128. *Croizet Collection. Purchased*, 1848.

27810. (*Fig.*) Part of a left mandibular ramus with the canine, $\overline{\text{pm 3}}$, $\overline{\text{pm. 4}}$, and $\overline{\text{m. 1}}$, apparently belonging to the same species as the preceding; from the Lower Miocene of Cournon (Puy-de-Dôme). Figured by Gervais in the 'Zool. et Pal. Françaises,' 2nd ed. p. 264, fig. 33, and referred to *D. exilis*; it differs from the type specimen of the latter (No. 27806) by the much more slender ramus. *Same history.*

Didelphys blainvillei, Croizet [3].

Syn. *Peratherium blainvillei*, Gervais [4].

Of somewhat larger size than *D. arvernensis*, but with the molars apparently rather taller ($\overline{\text{pm. 4}}$ is not shown in the type) and $\overline{\text{pm. 1}}$ relatively larger. The length of the space occupied by the four true molars in the type is 0,009, and that by the last three 0,007.

Hab. Europe (France).

[1] In Gervais's Zool. et Pal. Françaises, 1st ed. vol. i. p. 134 (1848–52).
[2] *Ibid.* 2nd ed. p. 263 (1859).
[3] *Ibid.* 1st ed. vol. i. p. 134 (1848–52).
[4] *Ibid.* 2nd ed. p. 263 (1859).

27823. Hinder part of the right ramus of the mandible, containing the three true molars; from the Lower Miocene (Upper Oligocene) of Cournon (Puy-de-Dôme), France. The length of the space occupied by the three teeth is 0,007, and the specimen apparently agrees precisely with the type mandible figured by Gervais in the 'Zool. et Pal. Françaises,' pl. xlv. fig. 2.

Croizet Collection. Purchased, 1848.

Didelphys exilis, Gervais[1].

Syn. (?) *Didelphys lemanensis*, Pomel[2].
Peratherium exile, Gervais[3].

Considerable confusion has existed in regard to this species owing to the figure of the type specimen (No. 27806) given by Gervais being stated to be enlarged, whereas it is really almost of the natural size.

The species is nearly of the size of *D. affinis* and *D. arvernensis*; in the type mandible (No. 27806) the length of the series of cheek-teeth is 0,013 (not 0,009 as stated by Gervais), $\overline{\text{pm. 1}}$ is of considerable size and separated by an interval from $\overline{\text{pm. 3}}$, which is longer than $\overline{\text{pm. 4}}$. The lower canine is directed nearly vertically; the lower true molars increase in length from the first to the fourth, and have their anterior cusp equal in height to the hinder outer cusp; and the mandibular ramus is deep.

Hab. Europe (France).

27806. (*Fig.*) The nearly entire right ramus of the mandible, containing the canine and cheek-teeth; from the Lower Miocene (Upper Oligocene) of Cournon (Puy-de-Dôme), France. This specimen is the type, and is figured by Gervais in the 'Zool. et Pal. Françaises,' 2nd ed. p. 264, fig. 34, the figure being very slightly above the natural size.

Croizet Collection. Purchased, 1848.

Didelphys lemanensis (Filhol[4]).

Syn. *Amphiperatherium lemanense*, Filhol[5].

The length of the lower series of cheek-teeth is 0,012, and that of the four true molars 0,007; $\overline{\text{pm. 1}}$ is small, $\overline{\text{pm. 3}}$ and $\overline{\text{pm. 4}}$ are

[1] Zool. et Pal. Françaises, 2nd ed. p. 264 (1859).—*Peratherium*.

[2] Catalogue Méthodique, p. 118 (1853). [3] *Loc. cit.*

[4] Ann. Sci. Géol. vol. x. art. 3, p. 201 (1879).—*Amphiperatherium*. The specific name is preoccupied by *D. lemanensis*, Pomel; but as the latter appears to be a synonym of *D. exilis* it seems unadvisable to make a change.

[5] *Loc. cit.*

subequal and have concave posterior borders; the true molars have tall talons, so that in an inner profile view they appear trilobed; $\overline{\mathrm{m.\,4}}$ is shorter than $\overline{\mathrm{m.\,3}}$. The types are from St. Gérand-le-Puy (Allier).

Hab. Europe (France).

26706. The greater part of the right ramus of the mandible, containing $\overline{\mathrm{m.\,3}}$ and $\overline{\mathrm{m.\,4}}$; from the Lower Miocene (Upper Oligocene) of Allier, France. This specimen agrees very closely with the one figured by Filhol in the 'Ann. Sci. Géol.' vol. xi. pl. xix. figs. 1, 2.

Pomel Collection. Purchased, 1851.

Didelphys, sp. *b* (? nov.).

Apparently allied to the preceding, but of smaller size, and with $\overline{\mathrm{m.\,4}}$ longer than $\overline{\mathrm{m.\,3}}$; premolars of similar structure, but not in contact; length of lower cheek-series 0,010, and of the true molars 0,0062. These dimensions equally distinguish it from *D. ronzoni* and *D. ambigua* (Filhol [1]).

Hab. Europe (France).

27811. The greater part of the mandible, showing the canine and cheek-teeth of both sides; from the Lower Miocene (Upper Oligocene) of the Auvergne (Puy-de-Dôme), France. The specimen is broken in two, one fragment being shown on one face of the block and the other on the opposite.

Croizet Collection. Purchased, 1848.

Didelphys, sp. *c*.

Of rather larger size than the next species.

Hab. Europe (England).

30350. Fragment of the right ramus of the mandible, containing the last three true molars; from the Upper Eocene (Lower Oligocene) of Hordwell, Hampshire. The length of the space occupied by the three teeth is 0,006.

Hastings Collection. Purchased, 1855.

Didelphys lamandini, Filhol [2].

Syn. *Peratherium lamandini*, Filhol [3].

A very small species in which the length of the lower cheek-

[1] See Ann. Sci. Géol. vol. xii. art. 3, p. 68; where both species are referred to *Amphiperatherium*.

[2] Comptes Rendus, vol. lxxxii. p. 289 (1876).

[3] Ann. Sci. Géol. vol. viii. art. 1, p. 256 (1877). The reference to the figure is given as 387 in place of 385.

series is 0,011, and that of the true molars 0,006; the molars increase in size posteriorly, and their talons are tricuspid. It is not easy to see how this form[1] differs from *D. laurillardi* of the Paris gypsum, or *D. parva* of Vaucluse; in the latter of which the length of the last three true molars is 0,005.

Hab. Europe (France).

M. 2388 d. Fragment of the right ramus of the mandible; from the Phosphorites of Caylux (Tarn-et-Garonne), France. The last three true molars remain, and have a united length of 0,005. *Purchased*, 1885.

Didelphys aymardi (Filhol[2]).

Syn. *Peratherium aymardi*, Filhol[3].

About the size of *D. antiqua*. Lower molars increasing in size from first to fourth; $\overline{\text{pm. 1}}$ very minute, separated by intervals from canine and $\overline{\text{pm. 3}}$; $\overline{\text{pm. 4}}$ taller than $\overline{\text{pm. 3}}$; talon of $\overline{\text{m. 4}}$ tricuspid. In the specimens described by Filhol[4] the length of the series of lower cheek-teeth is 0,014, and that of the true molars 0,008; but in the one figured by Gaudry in 'Les Enchaînements, etc.—Mamm. Tert.' p. 11, fig. 1, these dimensions are rather larger.

Hab. Europe (France).

M. 2388. The nearly entire right ramus of the mandible, showing all the teeth except the incisors; from the Phosphorites of Caylux (Tarn-et-Garonne), France. This specimen agrees precisely with one figured by Gaudry, *loc. cit.*; the length of the space occupied by the seven cheek-teeth is 0,015, and that by the true molars 0,009. *Purchased*, 1885.

M. 2388 a. Hinder part of a similar right ramus, showing the last three true molars; from the same deposits. *Purchased*, 1885.

M. 2388 b. Two rather smaller mandibular rami, in which the length of the four molars is 0,008; from the same deposits. *Purchased*, 1885.

M. 1489–90. Two similar mandibular rami; from the same deposits. *Purchased*, 1884.

[1] Owing to Gervais's error in regard to the dimensions of *D. exilis*, Filhol compares the present species to that form.

[2] Bull. Soc. Philom. Paris, sér. 6, vol. x. p. 89 (1874).—*Peratherium.*

[3] *Loc. cit.*

[4] Ann. Sci. Géol. vol. viii. art. 1, p. 251 (1877). The reference to the figure is given as 388 instead of 387.

Didelphys cadurcensis (Filhol[1]).

Syn. *Peratherium cadurcense*, Filhol[2].

Rather larger than the preceding, and distinguished by the longer interval between $\overline{m.\ 4}$ and the ascending ramus; in the type specimen the length of the space occupied by the series of lower cheek-teeth is 0,018.

Hab. Europe (France).

M. 2388 c. Three mandibular rami, provisionally referred to this species; from the Phosphorites of Caylux (Tarn-et-Garonne), France. *Purchased*, 1885.

OF UNCERTAIN SPECIES.

a. *From the Lower Miocene (Upper Oligocene) of Sauvetat and Cournon, Auvergne; all belonging to the Croizet Collection. Purchased,* 1848.

27809. The imperfect right half of the palatal region of the cranium, showing the hinder true molars and the alveoli of the anterior teeth. Nearly all the species being founded upon the evidence of the mandible, it is at present impossible to make any specific determinations in the case of upper jaws.

27807. Part of the right maxilla, showing the four true molars.

27699. The anterior portion of the right ramus of the mandible, showing the canine, premolars, and $\overline{m.\ 1}$. The premolars agree in structure with those of species *a*, and of *D. lemanensis* of St. Gerand-le-Puy, but the specimen is larger than the former, and does not apparently agree with the latter.

27699 a. Fragment of a mandibular ramus with the teeth broken.

b. *From the Upper Eocene (Lower Oligocene) of Hordwell, Hampshire.*

36801. Part of a left mandibular ramus, containing the four true molars and $\overline{pm.\ 3}$. *Presented by S. Laing, Esq.*, 1862.

c. *From the Phosphorites of Caylux (Tarn-et-Garonne), France.*

M. 2388 d. Four fragments of maxillæ, with teeth. *Purchased*, 1885.

M. 2388 e. Thirteen imperfect mandibular rami. *Purchased*, 1885.

[1] Ann. Sci. Géol. vol. viii. art. 1, p. 258 (1878).—*Peratherium.* The reference to the figure is given as 391 instead of 390. [2] *Loc. cit.*

Genus **CHIRONECTES,** Illiger[1].

Mainly distinguished from *Didelphys* by the structure of the feet.

Chironectes minimus (Zimmermann[2]).

Syn. *Lutra minima*, Zimmermann[3].
Chirontes variegatus, Illiger[4].

In two recent examples the length of the lower series of cheek-teeth is 0,030 and 0,027, and that of the true molars 0,0185 and 0,016.

Hab. Central and South America.

18890 l. The left ramus of the mandible of a subadult individual, showing the canine, $\overline{\text{pm. 3}}$, $\overline{\text{m. 1}}$, and $\overline{\text{m. 2}}$ in use, and $\overline{\text{pm. 4}}$ and $\overline{\text{m. 4}}$ *in alveolo*, the crowns of the other teeth being broken off; from a cave in Minas Geraes, Brazil.
Claussen Collection. Purchased, 1845.

Genus *non det.*

41995. The hinder half of the skeleton of a small Mammal, perhaps belonging to the *Didelphyidæ*, embedded in a slab of the Montmartre gypsum. The limb-bones are rather larger than those of the type of *Didelphys cuvieri*.

The affinities of the two following families require further elucidation.

Family STYLODONTIDÆ.

The following characters are found in the English genera:—

Dentition (in the type genus):—I. $\frac{?}{4}$, C. $\frac{?}{1}$, Pm. $\frac{?}{4}$, M. $\frac{?}{7}$. The hinder lower true molars form single cingulated cones; the lower premolars are laterally compressed; and the lower incisors are in contact. In the upper dentition referred by Owen to the type genus the true molars have narrow, triangular crowns[5]. The American Jurassic form described by Marsh[6] under the name of *Stylacodon*, and referred, with *Stylodon*, to his family *Dryolestidæ*, appears to be closely allied to, if not identical, with the latter; *Asthenodon* (see p. 273) is probably related.

[1] Prod. Syst. Mamm. et Avium, p. 76 (1811).
[2] Geogr. Geschicht, vol. ii. p. 317 (1780).—*Lutra*. [3] *Loc. cit.*
[4] Abh. k. Ak. Wiss. Berlin, 1811, p. 107.
[5] There is considerable doubt as to the correctness of this reference and the specimens may belong to another family.
[6] See Amer. Journ. ser. 3, vol. xxxiii. p. 334 (1887).

Genus **STYLODON**, Owen[1].

All the lower true molars are simply conical, have tall crowns, and are separated from one another by short intervals; while $\overline{\text{pm. 4}}$ is taller than $\overline{\text{m. 1}}$.

Stylodon pusillus, Owen[2].

This is the type species.

Hab. Europe (England).

All the following specimens are from the Middle Purbeck group of Durdlestone Bay, Swanage, Dorsetshire.

47756. (*Fig.*) Fragment of a mandibular ramus, containing five true molars. Figured by Owen in his 'Mesozoic Mammalia' (Mon. Pal. Soc.), pl. ii. fig. 16.
Beckles Collection. Purchased, 1876.

47757. (*Fig.*) The greater part of the left ramus of the mandible, containing most of the teeth. Figured by Owen, *op. cit.* pl. ii. fig. 17. *Same history.*

47758. (*Fig.*) The nearly entire mandible, with portions of the crushed cranium. Figured by Owen, *op. cit.* pl. ii. fig. 18. *Same history.*

47759. (*Fig.*) Part of the left ramus of the mandible containing several cheek-teeth. Figured by Owen, *op. cit.* pl. ii. fig. 19. *Same history.*

47760. (*Fig.*) Hinder part of the left ramus of the mandible, containing the last eight cheek-teeth. Figured by Owen, *op. cit.* pl. iii. fig. 2. *Same history.*

47761. (*Fig.*) Part of the left ramus of the mandible, containing $\overline{\text{pm. 4}}$ and the first six true molars. Figured by Owen, *op. cit.* pl. iii. fig. 3. *Same history.*

Stylodon robustus, Owen[3].

Distinguished from the preceding species by the deeper mandibular ramus, and a difference in the relative height of the lower true molars.

Hab. Europe (England).

[1] Geol. Mag. dec. i. vol. i. p. 199 (1866).
[2] *Loc. cit.*
[3] Mesozoic Mammalia (Mon. Pal. Soc.), p. 52 (1871).

47762. Part of the left ramus of the mandible, showing all the
(*Fig.*) cheek-teeth; from the Middle Purbeck group of Durdlestone Bay, Swanage, Dorsetshire. Figured by Owen, in his 'Mesozoic Mammalia' (Mon. Pal. Soc.), pl. iii. fig. 1.

Beckles Collection. Purchased, 1876.

Genus *non det.* (*Incertæ sedis.*)

The following specimens were referred by Owen to Stylodon pusillus. *All are from the Middle Purbeck, Durdlestone Bay.*

47786. A small block of Purbeck limestone, containing a fragment of a mandibular ramus with teeth, together with part of a right maxilla, with the last nine cheek-teeth. The association of these specimens is very strong evidence as to their specific identity; the teeth of the maxilla apparently agree in structure with those of No. 47755, but seem of rather smaller size.

Beckles Collection. Purchased, 1876.

48403. A right maxilla, with the cheek-teeth agreeing in size with No. 47786. *Same history.*

47755. Part of a rather larger right maxilla, containing two
(*Fig.*) incisors, the canine, and the cheek-teeth. Figured by Owen, *op. cit.* pl. ii. fig. 14. *Same history.*

48402. Part of a similar right maxilla, containing $\underline{\text{pm. 4}}$ and five true molars. *Same history.*

47787. Part of a jaw with teeth, perhaps generically identical with the preceding.

Genus **LEPTOCLADUS,** Owen[1].

The hinder lower true molars are simply conical, but the most anterior one[2] has antero-posterior cusps; the cheek-teeth are separated from one another by long intervals; and $\overline{\text{pm. 4}}$ is not taller than $\overline{\text{m. 1}}$.

[1] Mesozoic Mammalia (Mon. Pal. Soc.), p. 53 (1871). The name was applied merely as a provisional one; but its right to distinction is confirmed by the (MS.) observations of Osborn.

[2] Reckoned by Owen as a fifth premolar.

Leptocladus dubius, Owen[1].

This is the type species.
Hab. Europe (England).

47739. (*Fig.*) Part of the left ramus of the mandible, containing eight cheek-teeth, of which the four first are probably premolars; from the Middle Purbeck group of Durdlestone Bay, Swanage, Dorsetshire. This specimen is the type, and is figured by Owen in his 'Mesozoic Mammalia' (Mon. Pal. Soc.), pl. iii. fig. 4. *Beckles Collection. Purchased*, 1876.

Family SPALACOTHERIIDÆ.

Dentition (in type genus):—I. $\frac{?}{3}$, C. $\frac{?}{1}$, Pm. + M. $\frac{?}{10}$. The lower true molars of the type genus closely resemble in structure those of *Chrysochloris*, consisting of a single column surmounted by three cusps arranged in a V. The upper true molars described under the name of *Peralestes* likewise agree with those of the same genus. The lower true molars decrease in size posteriorly. Marsh[2] refers the American Jurassic genus *Menacodon* to this family.

Genus **SPALACOTHERIUM,** Owen[3].

The characters of this genus are the same as those of the family. The so-called *Peralestes* (in which there are six upper true molars) is probably founded upon the upper jaw of the present genus.

Spalacotherium tricuspidens, Owen[4].

This is the type species.
Hab. Europe (England).

46019. (*Fig.*) Fragment of the hinder part of the left ramus of the mandible, showing the last four true molars, together with the impress in the matrix of the anterior portion of the ramus; from the Middle Purbeck group of Durdlestone Bay, Swanage, Dorsetshire. This specimen (woodcut fig. 43) is the type, and is figured by Owen in the 'Quart. Journ. Geol. Soc.' vol. x. p. 426, fig. 1, and also in his 'Mesozoic Mammalia' (Mon. Pal. Soc.), pl. i. fig. 32. *Purchased*, 1874.

[1] Mesozoic Mammalia (Mon. Pal. Soc.), p. 53 (1871).
[2] Amer. Journ. ser. 3, vol. xxxiii. p. 340 (1887).
[3] Quart. Journ. Geol. Soc. vol. x. p. 426 (1854).
[4] *Loc. cit.*

47748. The anterior portion of the left ramus of the mandible,
(Fig.) showing the alveoli of three incisors and the canine, together with three imperfect premolars, the impression of the hinder part of the ramus being shown on the matrix; from the same locality. Figured by Owen in the 'Mesozoic Mammalia,' pl. i. fig. 35.

Beckles Collection. Purchased, 1876.

Fig. 43.

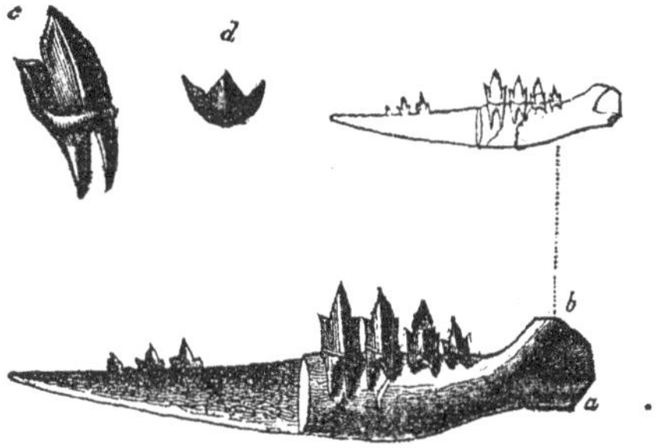

Spalacotherium tricuspidens.—*a*, *b*, the imperfect left ramus of the mandible, $\frac{2}{1}$ (the outline nat. size); *c*, *d*, lateral and upper views of a molar tooth, $\frac{3}{1}$; from the Middle Purbeck of Swanage. (From the 'Quart. Journ. Geol. Soc.')

47749 b. Fragment of the right mandibular ramus containing three
(Fig.) teeth, together with the impression of adjacent portions of the mandible; from the same locality. Figured by Owen, *op. cit.* pl. i. fig. 36. *Same history.*

47749 a. Fragment of the right ramus of the mandible containing
(Fig.) two true molars, and the impression of the anterior part of the ramus; from the same locality. Figured by Owen, *op. cit.* pl. i. fig. 37. *Same history.*

47750. Hinder part of the left ramus of the mandible, containing
(Fig.) the last five true molars; from the same locality. Figured by Owen, *op. cit.* pl. i. fig. 38. *Same history.*

Spalacotherium minus, Owen[1].

Of smaller size than the preceding; from which it is also distinguished by the form of the mandibular ramus.

Hab. Europe (England).

[1] Mesozoic Mammalia (Mon. Pal. Soc.), p. 28 (1871).

47751. The greater part of the left ramus of the mandible, containing
(*Fig.*) three true molars; from the Middle Purbeck group of Durdlestone Bay, Swanage, Dorsetshire. This specimen is the type, and is figured by Owen in his 'Mesozoic Mammalia' (Mon. Pal. Soc.), pl. i. fig. 39.

Beckles Collection. Purchased, 1876.

47799. Fragment of a maxillary ramus with the last two true molars, perhaps belonging to this species; from the same locality.

Same history.

Genus **PERALESTES**, Owen[1].

This genus was founded upon a maxilla, of which the true molars (6 in number) agree so closely in structure with those of *Chrysochloris* that there is every probability that the specimen belongs to *Spalacotherium tricuspidens*. The fourth upper premolar is of a Dasyurine type.

Owen regarded *Peralestes* as allied to *Amblotherium*, and associated with it a mandible which, as mentioned above (p. 275), cannot apparently be distinguished from *A. mustelula*.

Peralestes longirostris, Owen[2].

Hab. Europe (England).

47740. Part of the right maxilla, showing pm. 4 and the six true
(*Fig.*) molars; from the Middle Purbeck group of Durdlestone Bay, Swanage, Dorsetshire. This specimen is the type, and is figured by Owen in his 'Mesozoic Mammalia' (Mon. Pal. Soc.), pl. ii. fig. 3.

Beckles Collection. Purchased, 1876.

Specimens of which the Families are undetermined.

A. *From the Middle Purbeck group of Durdlestone Bay, Swanage, Dorsetshire.*

All the following specimens belong to the Beckles Collection. Purchased, 1876.

47737. Fragment of a left mandibular ramus containing two (? pre-
(*Fig.*) molar) teeth. Figured by Owen in his 'Mesozoic Mammalia' (Mon. Pal. Soc.), pl. i. fig. 41, and provisionally named *Phascolestes* (?) *dubius*.

[1] Mesozoic Mammalia (Mon. Pal. Soc.), p. 33 (1871). [2] *Loc. cit.*

47795. The left half of a mandibular symphysis, with the anterio teeth.

48404. Fragment of a mandibular ramus, with five cheek-teeth.

48406. Part of a mandibular ramus, showing three or four teeth.

48248. Part of a mandibular ramus, with one cheek-tooth.

48389. Part of a left mandibular ramus, with worn teeth (? *Amblotherium* or *Achyrodon*).

47796. Part of a mandibular ramus, with the cheek-teeth in a broken condition. The teeth have comparatively broad crowns, and the specimen may possibly belong to *Bolodon* (*suprà* p. 203).

48361. An imperfect humerus.

48249. A smaller humerus.

48250. Two femora, agreeing in relative size with the preceding.

B. *From the Lower Jurassic Slate of Stonesfield, Oxfordshire.*

32752. The femur and humerus of a small mammal. Figured by (*Fig.*) Seeley in the 'Quart. Journ. Geol. Soc.' vol. xxxv. p. 456, figs. 1, 2. *Presented by S. P. Pratt, Esq.*, 1857.

Subclass PROTOTHERIA.

Order MONOTREMATA.

Family ECHIDNIDÆ.

Genus **ECHIDNA,** Cuvier[1].

Syn. *Tachyglossus*, Illiger[2].

Proechidna (*Acanthoglossus*), Gervais, is included for palæontological purposes in the type genus.

Echidna oweni, Krefft[3].

Syn. *Echidna ramsayi*, Owen[4].

Considerably exceeding in size the existing *E. bruijnii* of New Guinea.

Hab. New South Wales.

1 Tabl. élément. d'Hist. Nat. p. 143 (1798).
2 Prod. Syst. Mamm. et Avium, p. 114 (1811).
3 Ann. & Mag. Nat. Hist. ser. 4, vol. i. p. 113 (1868).
4 Phil. Trans. 1884, p. 273.

The originals of the following casts were obtained from the Pleistocene breccia of the caves of the Wellington Valley, New South Wales; they are preserved in the Museum at Sydney. The casts were presented by the Trustees of the Australian Museum, 1884.

M. 1908. Cast of process of the humerus.

M. 1909. Cast of the left humerus. Figured by Owen in the 'Phil. Trans,' 1884, pl. xiv. figs. 1–3.

M. 1910. Cast of a scapula.

M. 1911. Cast of an imperfect femur.

M. 1531: Cast of an imperfect femur.

SUPPLEMENT.

[In the case of families and genera already mentioned in this work the references are given to the pages where they occur.]

Order PRIMATES.

Suborder *ANTHROPOIDEA.*

Family SIMIIDÆ (pt. i. p. 1).

Genus **TROGLODYTES,** E. Geoffroy [1].

Syn. *Palæopithecus*, Lydekker [2].

Troglodytes sivalensis, Lydekker [3].

Syn. *Palæopithecus sivalensis*, Lydekker [4].

Distinguished from *T. niger* by the anterior convergence of the two series of the upper cheek-teeth, and by the antero-posterior diameter of the premolars being less than is frequently the case in that species.

Hab. India.

M. 2014. Cast of the incomplete palate of a male. The original, which is the type and only known specimen, was obtained from the Pliocene Siwaliks of the Punjab, and is preserved in the Indian Museum, Calcutta (No. D. 1). It is described and figured by the present writer in the 'Rec. Geol. Surv. Ind.' vol. xii. p. 33, pl. facing p. 52, figs. 1, 5 (under the name of *Palæopithecus*), and also in the 'Palæontologia Indica' (Mem. Geol. Surv. Ind.), ser. 10, vol. iv. p. 2, pl. i. figs. 1, 1 *a*. *Presented by the Director of the Geological Survey of India*, 1886.

[1] Ann. du Muséum, vol. xix. p. 87 (1812).
[2] Rec. Geol. Surv. Ind. vol. xii. p. 33 (1879).
[3] *Loc. cit.—Palæopithecus.* [4] *Loc. cit.*

Genus **HYLOBATES** (pt. i. p. 2).

Hylobates, sp.

Hab. Borneo.

M. 1971. Part of the left ramus of the mandible, containing $\overline{\text{pm. 4}}$, $\overline{\text{m. 1}}$, and $\overline{\text{m. 2}}$, and the broken base of $\overline{\text{m. 3}}$; from an alluvial deposit at Sarawak, Borneo.
Presented by P. L. Sclater, Esq., 1884.

Family CERCOPITHECIDÆ (pt. i. p. 2).

Genus **SEMNOPITHECUS** (pt. i. p. 2).

Semnopithecus entellus (Dufresne [1]).

Syn. *Simia entellus,* Dufresne [2].

Hab. India.

M. 2963. Several canines and molars, provisionally referred to this species; from the caves of Billa-Surgam, Karnul district, Madras. Remains belonging to this form are described and figured by the present writer in the 'Palæontologia Indica,' ser. 10, vol. iv. p. 28, pl. vii. *Presented by the Director of the Geological Survey of India*, 1886.

Genus **CYNOCEPHALUS** (pt. i. p. 4).

Cynocephalus falconeri, Lydekker [3].

This name is applied to the specimen No. 15709 described in pt. i. p. 6, which indicates a small species; it is figured in pl. i, fig. 4 of the memoir cited above.

Cynocephalus, sp.

Hab. India.

M. 2959. Cast of the second left lower true molar. The original is from a cave at Billa-Surgam, Karnul district, Madras; and is preserved in the Indian Museum, Calcutta. It is described and figured by the present writer in the 'Palæontologia Indica,' ser. 10, vol. iv. p. 28, pl. vii. figs. 5, 5 *a*. *Presented by the Director of the Geological Survey of India*, 1886.

[1] Bull. Soc. Philom. Paris, 1797, p. 49.—*Simia*. [2] *Loc. cit.*
[3] Palæontologia Indica (Mem. Geol. Surv. Ind.), ser. 10, vol. iv. p. 7 (1886).

Suborder *LEMUROIDEA.*

Family LEMURIDÆ (pt. i. p. 8).

Genus **ADAPIS** (pt. i. p. 8).

Adapis magna (pt. i. p. 262).

M. 3572. The greater part of the left maxilla; from the Hordwell beds of Hampshire. In this specimen all the cheek-teeth except <u>pm. 1</u> are preserved.

Presented by Sir R. Owen, K.C.B., 1884.

Genus **CÆNOPITHECUS,** Rütimeyer [1].

There is considerable doubt if this form is really distinct from *Adapis*, with which it was provisionally identified by Gervais [2]. Gaudry [3] compares the teeth of *Cænopithecus* to those of *Lemur* and *Hapalemur*, while those of *Adapis* are regarded as nearer to *Indris* and *Galago*; but the dentition of *A. magna*, figured by Gaudry [4], closely resembles Rütimeyer's figure of *Cænopithecus.*

Cænopithecus lemuroides, Rütimeyer [5].

This is the type species.

Hab. Switzerland.

M. 2175. Cast of a fragment of the right maxilla, probably belonging to this species, and containing five cheek-teeth. The original is from the Upper Eocene (Oligocene) of Egerkingen, near Soleure, Switzerland; the true molars apparently resemble those of the type specimen, figured by Rütimeyer in the 'Denkschr. schw. Ges. Nat.' vol. xix. art. 3, pl. v. figs. 87–88; the specimen is not sufficiently perfect to admit of exact comparison with *Adapis.*

Presented by Dr. Kowalevsky.

[1] Denkschr. schw. Ges. Nat. vol. xix. art. 3, p. 88 (1862).
[2] Zool. et Pal. Générales, sér. 2, p. 35.
[3] Les Enchaînements, etc.—Mammifères Tertiaires, p. 226.
[4] *Op. cit.* p. 226, fig. 229.
[5] *Loc. cit.*

Order CHIROPTERA.

Suborder *MICROCHIROPTERA.*

Family RHINOLOPHIDÆ (pt. i. p. 11).

Genus **PHYLLORHINA** (pt. i. p. 13).

Phyllorhina diadema (E. Geoffroy [1]).

Syn. *Rhinolophus diadema,* E. Geoffroy [2].

Hab. India and Malay Peninsula.

M. 2965. Fragments of the cranium and mandibular rami; from the caves of Billa-Surgam, Karnul district, Madras. Other specimens are described and figured by the writer in the 'Palæontologia Indica,' ser. 10, vol. iv. p. 34, pl. viii. figs. 10, 10*a*. *Presented by the Director of the Geological Survey of India,* 1886.

Family EMBALLONURIDÆ.

Genus **TAPHOZOUS,** E. Geoffroy [3].

Dentition:—I. $\frac{1}{2}$, C. $\frac{1}{1}$, Pm. $\frac{2}{2}$, M. $\frac{3}{3}$.

Taphozous saccolæmus, Temminck [4].

Hab. India, Malay Peninsula, Burma, Borneo, and Sumatra.

M. 2964. Two imperfect specimens of the cranium; from the caves of Billa-Surgam, Karnul district, Madras. Other specimens are described and figured by the present writer in the 'Palæontologia Indica,' ser. 10, vol. iv. p. 34, pl. viii. figs. 11, 12. *Presented by the Director of the Geological Survey of India,* 1886.

Order INSECTIVORA.

Family TALPIDÆ (pt. i. p. 15).

Genus **TALPA** (pt. i. p. 15).

Talpa tyrrhenaica, Forsyth-Major [5].

Hab. Sardinia.

[1] Ann. du Muséum, vol. xx. p. 263 (1818).—*Rhinolophus.* [2] *Loc. cit.*

[3] Descript. de l'Egypte, vol. ii. p. 126 (1812).

[4] Monographies de Mammalogie, vol. ii. p. 285 (1835–41).

[5] The specimens were sent to the Museum with this unpublished name attached to them.

M. 3489. Several specimens of mandibular rami, of the humerus, radius, ulna, femur, and tibia; from the Pleistocene breccia of Mont San Giovanni, Sardinia. *Purchased*, 1886.

Genus **PROTALPA**, Filhol [1].

Distinguished from *Talpa* by the characters of the humerus.

Protalpa cadurcensis, Filhol [2].

This is the type and only described species.

Hab. France.

The specimens of the humerus from Caylux (No. M. 413), mentioned in part i. p. 15, belong to this species. The type specimen is figured by Filhol in the 'Ann. Soc. Sci. Phys. Nat. Toulouse,' 1884, pl. i. figs. 2, 6, 8 [3].

Family SORICIDÆ (pt. i. p. 16).

Genus **SOREX** (pt. i. p. 16).

Sorex similis, Hensel [4].

This is a small species, slightly larger than *Crossopus fodiens*.

Hab. Sardinia.

M. 3491. An imperfect cranium, two maxillæ, and six mandibular rami; from the Pleistocene breccia of Mont San Giovanni, Sardinia. *Purchased*, 1886.

M. 3491 a. Several specimens of the humerus and femur; from the same deposits. *Purchased*, 1886.

Family ERINACEIDÆ (pt. i. p. 17).

Genus **ERINACEUS** (pt. i. p. 17).

Erinaceus œningensis, Lydekker [5].

In its size and the number of roots to the teeth, this species agrees with the existing African *E. algirus*, but is distinguished by the relative proportions of the upper incisors and true molars.

Hab. Switzerland.

[1] Bull. Soc. Philom. Paris, sér. 7, vol. i. p. 52 (1877).
[2] *Loc. cit.*
[3] There is an error in the description of the plate.
[4] Zeitschr. deutsch. geol. Ges. vol. vii. p. 459 (1855).
[5] Quart. Journ. Geol. Soc. vol. xlii. p. 25 (1886).

42824. The palatal half of the cranium, exhibiting the whole of the
(*Fig.*) dentition; from the Upper Miocene of Œningen, Switzerland. The type specimen; figured by the writer in the 'Quart. Journ. Geol. Soc.' vol. xlii. pl. ii. figs. 3, 4.
Van Breda Collection. Purchased, 1871.

Genus **NEUROGYMNURUS**, Filhol[1].

Dentition:—I. $\frac{3}{3}$, C. $\frac{1}{1}$, Pm. $\frac{4}{4}$, M. $\frac{3}{3}$.

This genus, with which *Cayluxotherium*, Filhol[2], is probably identical, is allied to *Gymnura*, but has no cingulum in the lower true molars, while the premolars differ, and in the cranium (*Cayluxotherium*) palatal vacuities are present.

Neurogymnurus minor, Filhol[3].

This species is somewhat smaller than the typical *N. cayluxi*, the length of the lower true molars being 0,006.

Hab. Europe (France, and (?) England).

M. 2388 g. The nearly entire right ramus of the mandible, showing $\overline{\text{pm. 4}}$ and the alveoli of the other cheek-teeth and canine; from the Phosphorites of Caylux (Tarn-et-Garonne), France. Type mandibles figured by Filhol in the 'Ann. Soc. Sci. Phys. Nat. Toulouse,' 1884, pl. i. figs. 13, 15.
Purchased, 1885.

M. 2813. An upper true molar, provisionally referred to this species; from the Upper Eocene (Lower Oligocene) of Headon Hill, Isle of Wight. This tooth agrees with the upper molars of *Cayluxotherium elegans* figured by Filhol, *op. cit.* pl. i. figs. 9–12, but is of smaller size.
Presented by J. E. Lee, Esq., 1885.

Neurogymnurus major, Lydekker (n. sp.).

Larger than *N. cayluxi*, the length of the lower true molars being 0,0115 (against 0,009), and the depth of the mandible behind $\overline{\text{m. 3}}$ 0,0095 (against 0,004). The molars are relatively shorter than in *Gymnura*.

Hab. Europe (England).

29718. Two imperfect mandibular rami; from the Upper Eocene of Hordwell, Hants. The types; one shows $\overline{\text{pm. 4}}$ and the true molars. *Hastings Collection. Purchased*, 1855.

[1] Bull. Soc. Philom. Paris, sér. 7, vol. i. p. 52 (1877).
[2] Ann. Soc. Sci. Phys. Nat. Toulouse, 1884, p. 1. [3] *Ibid.* p. 10.

36802, 3, 7. Three imperfect mandibular rami; from Hordwell.
Presented by S. Laing, Esq., 1862.

Family MICROCHŒRIDÆ.

This family is apparently allied to the *Erinaceidæ*, with which it agrees in the squared crowns of the true molars; but is distinguished by the arrangement of the tubercles or cusps on the crowns of the hinder cheek-teeth, and by pm. 3 being as complex as pm. 4.

The genus *Hyopsodus* (woodcut, fig. 44), which is regarded by

Fig. 44.

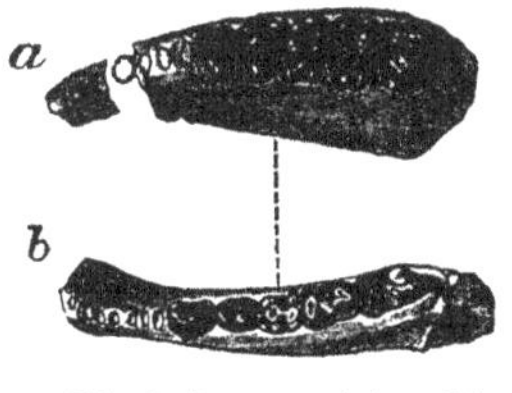

Hyopsodus vicarius, Cope.—The left upper (*a*) and lower (*b*) cheek-dentition; from the (? Bridger) Eocene of Wyoming, U.S.A. $\frac{1}{1}$. (From the 'Amer. Nat.')

some writers as closely allied to the Lemuroidea, should probably be included in the present family[1]; it is stated to have four premolars.

Genus **MICROCHŒRUS,** Wood[2].

Dentition:—I. $\frac{2}{?}$, C. $\frac{1}{1}$, Pm. $\frac{3}{?}$, M. $\frac{3}{3}$, or I. $\frac{3}{?}$, C. $\frac{1}{1}$, Pm. $\frac{2}{?}$, M. $\frac{3}{3}$. The first pair of upper incisors are separated from one another by an interval (as in *Erinaceus*), but there is no other diastema; the first four upper teeth are compressed and trenchant, while the last five are quadrangular and tuberculated; the true molars have four main tubercles with an intervening pair of smaller ones, and a single one on the outer side; pm. 3 and pm. 4 have two tubercles.

Microchœrus erinaceus, Wood[3].

This is the type species and is about equal in size to *Erinaceus europœus*.

Hab. England.

[1] See Quart. Journ. Geol. Soc. vol. xli. p. 529. In that paper the present writer regarded *Hyopsodus* as probably identical with *Microchœrus*; he is, however, informed by Mr. Osborn, of Princeton, New Jersey (who when in England carefully examined the type specimen of *Microchœrus*), that the two appear distinct.

[2] London Geological Journal, no. 1, p. 5 (1846).

[3] *Loc. cit.*

25229. The cranium and the imperfect left ramus of the mandible;
(*Fig.*) from the Upper Eocene (Lower Oligocene) of Hordwell, Hampshire. These are the type specimens, and are figured by Wood in the 'London Geological Journal,' no. 1, pl. ii. figs. 1–3, as belonging to a Perissodactyle; and also by the present writer in the 'Quart. Journ. Geol. Soc.' vol. xli. p. 529 (1885), the latter figure being reproduced in the accompanying woodcut (fig. 45).
Presented by S. V. Wood, Esq., 1847.

Fig. 45.

Microchœrus erinaceus.—The right upper dentition; from the Hordwell beds. $\frac{3}{1}$. (From the 'Quart. Journ. Geol. Soc.')

30346.b. Several fragments of the maxilla and mandible; from Hordwell. In one specimen the upper true molars are unworn, while one mandibular ramus exhibits the last five cheek-teeth. *Hastings Collection. Purchased*, 1855.

Family TUPAIIDÆ (pt. i. p. 19).

Genus **PARASOREX**, Meyer[1].

The identification of this genus by Fraas ('Fauna von Steinheim,' p. 4) with *Plesiosorex*, followed in part i. p. 15, has been found erroneous[2], the latter genus being allied to *Myogale*. The premolars $\left[\frac{4}{4}\right]$ of *Parasorex* are more complex than those of *Tupaia*.

Parasorex socialis, Meyer[3].

Hab. Europe.

The specimens noticed in part i. p. 19 under the name of *Plesiosorex soricinoides* belong to this species.

1 Neues Jahrb. 1865, p. 845.
2 See Introduction.
3 *Loc. cit.*

Order CARNIVORA.

Suborder *CARNIVORA PRIMIGENIA*.

It appears advisable to notice the chief characters of the Suborder [1]. Homology and number of the teeth following the Eutherian type (*i. e.* there are usually four premolars, and neither the true molars nor the incisors exceed three in number); $\underline{\text{pm. 4}}$ and $\overline{\text{m. 1}}$ not differentiated into distinct carnassials; true molars often structurally like those of the Polyprotodont Marsupialia; a complete milk-dentition; no palatal vacuities or inflection of the angle of the mandible; brain-cavity small; a third trochanter frequently present in the femur; scaphoid and lunar generally separate [2]; absence of a distinct groove on the tibial facette of the astragalus [3].

Family HYÆNODONTIDÆ (pt. i. p. 20).

The scaphoid and lunar were apparently united in at least one genus (? *Hyænodon*) [4].

Genus **HYÆNODON** [5] (pt. i. p. 21).

Hyænodon leptorhynchus (pt. i. p. 26).

M. 2346. The greater part of the associated cranium and mandible; (*Fig.*) from the Phosphorites of Caylux (Tarn-et-Garonne), France.

[1] For an amended arangement see Schlosser, 'Morphol. Jahrb.' vol. xii. pp. 287–294 (1886). This writer regards the group as of equivalent value with both Carnivora and Insectivora; but the view adopted in pt. i. is to use the term Carnivora in a wider sense, and to regard its three Suborders as approximately of equal value with the Suborders of the Ungulata. It may be questioned if, on this view, the Insectivora can be distinguished from the Carnivora, although they apparently pass in the other direction into the Lemuroidea.

[2] See note 4.

[3] Grooved in *Mesonyx*.

[4] See pt. i. No. 26752; No. 27583 on the same page also has a scapho-lunar. There are no other Carnivora from the Vaucluse beds to which these specimens could have belonged, and they were evidently associated with the hind feet showing the ungrooved astragalus characteristic of the Suborder. From its specialized dentition it is probable that these bones belonged to *Hyænodon* rather than to *Pterodon*. On account of Gervais's reference of a scapho-lunar to *Hyænodon*, Schlosser (*op. cit.*) refers that genus to the Carnivora Vera, but its dentition is too like that of *Pterodon* to admit of this view. According to Scott (see Cope, 'Amer. Nat.' vol. xx. p. 966 [1886]), the scaphoid and lunar are distinct in the American species of *Hyænodon*, but there is no reason why they should not have united in the European forms; since an analogous instance occurs in the case of the N.-American *Anchitherium bairdi*, where the meso- and entocuneiform remain distinct, although they have united in the French *A. aurelianense*.

[5] In reference to the note in pt. i. p. 21, as to the number of upper cheek-teeth in this genus, the writer finds that Filhol has recorded in the 'Ann. Soc. Sci. Phys. Nat. Toulouse,' 1882, p. 18, the absence of $\underline{\text{m. 3}}$. See also 'Ann. Sci. Géol.' vol. viii. art. 1, p. 16.

This specimen is described and figured (reversed) by Filhol in the 'Ann. Sci. Géol.' vol. vii. art. 7, p. 180, pl. xxix. fig. 143, and pl. xxx. figs. 144–146. *Purchased*, 1885.

M. 2347. Part of the left ramus of the mandible of a smaller specimen, showing the last four cheek-teeth; from Caylux. This specimen agrees with No. M. 1530, and is intermediate in size between the mandible of the last specimen and the larger examples of *H. vulpinus*. *Purchased*, 1885.

M. 2353. Fragment of the left ramus of the mandible of a young individual; from Caylux. In this specimen $\overline{\text{pm. 1}}$, $\overline{\text{mm. 3}}$, $\overline{\text{mm. 4}}$, and $\overline{\text{m. 1}}$ are in use; the alveoli of the milk-canine and $\overline{\text{mm. 2}}$, as well as the germs of the incisors of the permanent canine and of $\overline{\text{pm. 2}}$, are also displayed; the cavities for $\overline{\text{pm. 3}}$ and $\overline{\text{pm. 4}}$ are developed, but the germs of these teeth do not seem to have been calcified. *Purchased*, 1885.

M. 2353 a. Fragment of the left ramus of the mandible of a slightly older individual; from Caylux. The last two milk-molars are in use, and the germs of $\overline{\text{pm. 2}}$ and $\overline{\text{pm. 3}}$ are seen *in alveolo*. *Purchased*, 1885.

Hyænodon vulpinus (pt. i. p. 28).

M. 2352. Three fragments of mandibular rami of young individuals; from the Phosphorites of Caylux (Tarn-et-Garonne), France. The least imperfect of the three specimens shows $\overline{\text{mm. 4}}$ and $\overline{\text{m. 1}}$ in use, and the germs of $\overline{\text{pm. 3}}$ and $\overline{\text{m. 2}}$ *in alveolo*. *Purchased*, 1885.

Specifically undetermined.

M. 2352 a. Small fragment of the right ramus of the mandible of a young individual of a species larger than *H. leptorhynchus*; from the Phosphorites of Caylux (Tarn-et-Garonne), France. The second milk-molar is in use, while the germs of the canine, $\overline{\text{pm. 1}}$, and $\overline{\text{pm. 2}}$ are seen *in alveolo*; $\overline{\text{pm. 1}}$ is unusually late in appearing. *Purchased*, 1885.

Genus **PTERODON**.[1] (pt. i. p. 33).

Pterodon dasyuroides (pt. i. p. 33).

M. 2354. Fragment of the left maxilla, showing $\underline{\text{pm. 4}}$, $\underline{\text{m. 1}}$, and

[1] In reference to the observation in pt. i. p. 35, it should be mentioned that *Oxyæna* differs from *Pterodon* by its elongated mandibular symphysis, which is a good generic character.

$\underline{\text{m. 2}}$; from the Phosphorites of Caylux (Tarn-et Garonne), France. *Purchased*, 1885.

M. 2355. Fragment of a smaller left maxilla, provisionally referred to this species[1]; from Caylux. The specimen shows $\underline{\text{pm. 3}}$ and $\underline{\text{pm. 4}}$. *Purchased*, 1885.

M. 2356. Fragment of the left maxilla of an immature individual, provisionally referred to this species; from Caylux. The *alveoli* of the milk-teeth and the germs of $\underline{\text{pm. 3}}$ and $\underline{\text{pm. 4}}$ are shown. *Purchased*, 1885.

M. 2357. Part of the left ramus of the mandible, containing the last five cheek-teeth; from Caylux. *Purchased*, 1885.

M. 2358. Part of the left ramus of the mandible, showing $\overline{\text{pm. 3}}$, $\overline{\text{pm. 4}}$, and $\overline{\text{m. 1}}$; from Caylux. *Purchased*, 1885.

M. 2359. Hinder portion of the right ramus of a smaller mandible, provisionally referred to this species; from Caylux. The last true molar is shown. *Purchased*, 1885.

M. 2359 a. The imperfect left ramus of the mandible, exhibiting the symphysis; from Caylux. *Purchased*, 1885.

Family PROVIVERRIDÆ[2].

Distinguished from the *Hyænodontidæ* by the greater development of the inner tubercle of the upper true molars, and the presence of an inner cusp to the blade of the lower true molars (fig. 46)[3]. The latter resemble the corresponding teeth of *Dasyurus*, and also $\overline{\text{m. 1}}$ of the *Viverridæ* and *Cynodictis*; and it is thus not improbable that the present group is derived from early marsupials which had four premolars; and that the *Viverridæ* and *Cynodictis* are themselves descended from forms allied to the present. In the one instance the chief dental modification would be the reduction in the number of incisors and true molars; and in the other the differentiation of $\underline{\text{pm. 4}}$ and $\overline{\text{m. 1}}$ into carnassials, and the assumption of a more or less completely tubercular character by the cheek-teeth behind the two mentioned.

[1] The cheek-teeth of smaller individuals of this species cannot apparently be distinguished from those of *P. biïncisivus*, Filhol.

[2] Formed by Schlosser in the 'Morphol. Jahrb.' vol. xii. p. 293 (1886), to supersede the *Leptictidæ* of Cope (Amer. Nat. vol. xviii. p. 347 [1884]), of which the type genus is referred to the Insectivora.

[3] The instances of the *Caninæ* and *Dasyuridæ* suggest that this character need not be of more than generic value.

Genus **PROVIVERRA**, Rütimeyer[1].

Including:—*Stypolophus*, Cope[2].
Cynohyænodon, Filhol[3].

Dentition:—I. $\frac{3}{3}$, C. $\frac{1}{1}$, Pm. $\frac{4}{4}$, M. $\frac{3}{3}$. The upper cheek-teeth are formed on the general plan of those of *Pterodon*, but pm. 4 frequently

Fig. 46.

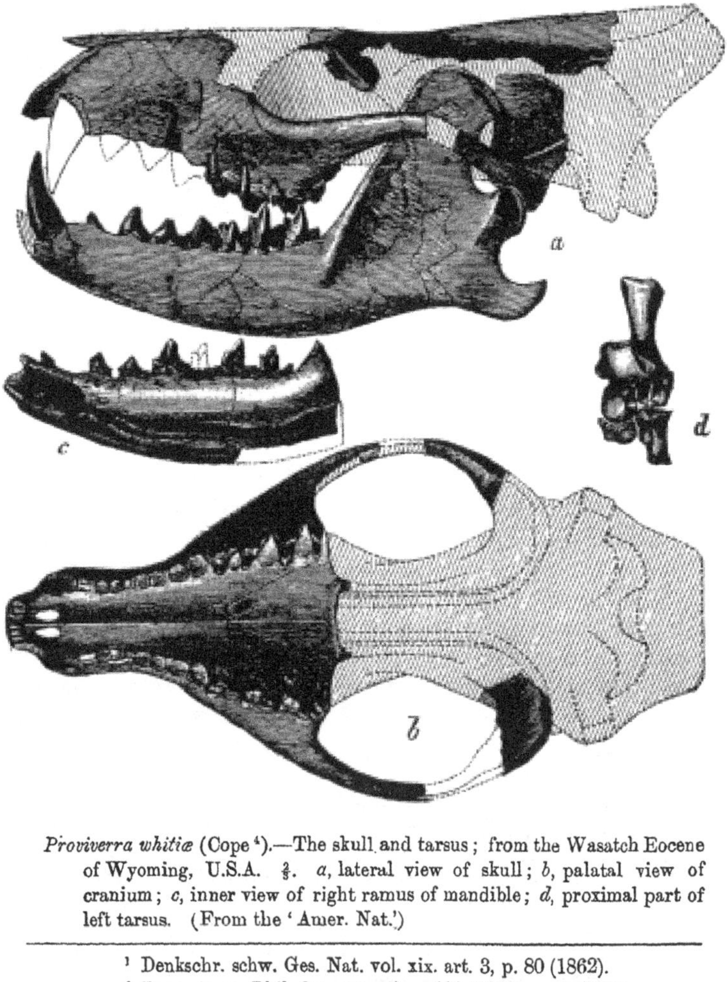

Proviverra whitiæ (Cope[4]).—The skull and tarsus; from the Wasatch Eocene of Wyoming, U.S.A. $\frac{3}{5}$. *a*, lateral view of skull; *b*, palatal view of cranium; *c*, inner view of right ramus of mandible; *d*, proximal part of left tarsus. (From the 'Amer. Nat.')

[1] Denkschr. schw. Ges. Nat. vol. xix. art. 3, p. 80 (1862).
[2] Proc. Amer. Phil. Soc. vol. xii. p. 466 (1873—read 1872).
[3] Bull. Soc. Philom. Paris, sér. 6, vol. x. p. 87 (1873).
[4] Syn. *Stypolophus whitiæ*, Cope.

develops an anterior basal cusp. The cranium (fig. 46) is elongated; and the talon of the lower true molars comparatively small.

Gaudry ('Les Enchaînements, etc.—Mamm. Tert.' p. 20) identifies *Cynohyænodon* with *Proviverra*; while Cope (Amer. Nat. vol. xviii. p. 351), who retains the latter, regards the former as a synonym of *Stypolophus*.

Proviverra cayluxi (Filhol[1]).

Syn. *Cynohyænodon cayluxi*, Filhol[2].
Stypolophus cayluxi, Cope[3].

Hab. Europe (France).

M. 2384. The greater part of the right ramus of the mandible; from the Quercy Phosphorites. Similar to the specimen figured by Gaudry, *op. cit.* fig. 14. *Purchased*, 1885.

Genus **DELTATHERIUM**, Cope[4].

Syn. *Lipodectes*, Cope[5].

Dentition:—I. $\frac{3}{3}$, C. $\frac{1}{1}$, Pm. $\frac{3}{3}$, M. $\frac{3}{3}$. The upper true molars are wider internally, more equal in size, and differ less from $\underline{\text{pm. 3}}$ and $\underline{\text{pm. 4}}$ than in *Proviverra*; the cranium is shorter, and the talon of the lower true molars larger.

Deltatherium fundaminis, Cope[6].

Syn. *Lipodectes penetrans*, Cope[7].

This is the type species.
Hab. North America.

M. 2571. Fragment of the right maxilla, showing portions of three hinder cheek-teeth; from the Puerco Eocene of New Mexico. This and the next specimen agree precisely with the corresponding portions of the skull figured by Cope in the 'Amer. Nat.' vol. xviii. p. 352, fig. 20. *Presented by R. Lydekker, Esq.*, 1885.

M. 2571 a. Part of the left ramus of the mandible, showing $\overline{\text{m. 2}}$; from New Mexico. *Presented by R. Lydekker, Esq.*, 1885.

[1] Bull. Soc. Philom. Paris, sér. 6, vol. x. p. 87 (1873)—*Cynohyænodon.*
[2] *Loc. cit.*
[3] Amer. Nat. vol. xviii. p. 351 (1884).
[4] *Ibid.* vol. xv. p. 337 (1881).
[5] *Ibid.* vol. xv. p. 1019 (1881).
[6] *Ibid.* vol. xv. p. 337 (1881).
[7] *Ibid.* vol. xv. p. 1019 (1881).

Suborder *CARNIVORA VERA.*

Family FELIDÆ (pt. i. p. 41).

Genus **MACHÆRODUS** (pt. i. p. 41).

Machærodus cultridens (pt. i. p. 42).

The upper canines are long and somewhat narrow; and, if the following specimen be correctly referred, $\overline{\text{pm. 3}}$ is absent, and the mandible short, with a comparatively small diastema.

M. 3458. Cast of the right ramus of the mandible of a very old individual, provisionally referred to this species. The original is from the Forest-bed of Kessingland, Suffolk; and is described and figured by Backhouse and the present writer in the 'Quart. Journ. Geol. Soc.' vol. xlii. p. 309, pl. x. *Made in the Museum,* 1886.

Machærodus aphanistus (Kaup [1]).

Syn. *Felis aphanista*, Kaup [2].
Machærodus leoninus, Roth and Wagner [3].
Megantheréon aphanistus, Pomel [4].
Drepanodon aphanistus, Cope [5].

If the preceding specimen be rightly referred, the Pikermi and Eppelsheim *Machærodus* will be distinct from *M. cultridens*. The upper canines are relatively broad, the diastema in the mandible of the male is long, and $\overline{\text{pm. 3}}$ present. Both this species and *M. cultridens* are of large size.

All the specimens from Pikermi and Eppelsheim entered in pt. i. p. 43 belong to this form.

Genus **EUSMILUS**, Gervais [6].

Dentition:—I. $\frac{?}{2}$, C. $\frac{1}{1}$, Pm. $\frac{2}{1}$, M. $\frac{1}{1}$. This genus agrees with *Machærodus* in the general structure of the mandible and the large size of the upper canines; but differs in the excessive depth of the descending portion of the mandibular symphysis, in the presence of

[1] Oss. Foss. d. Darmstadt, pt. ii. p. 18 (1833).—*Felis.*
[2] *Loc cit.*
[3] Abh. math.-phys. Cl. k.-bay. Ak. Wiss. vol. vii. p. 400 (1854).
[4] Catalogue Méthodique, p. 56 (1853).
[5] Amer. Nat. vol. xiv. p. 853 (1880).
[6] Zool. et Pal. Générales, sér. 2, p. 53 (1876).

only two lower incisors, in the small size of the anterior lobe of pm. 4, and the presence of a talon to m. 1; the two latter characters indicating affinity with the less specialized felines. The inner tubercle of pm. 4 is almost wanting.

Eusmilus bidentatus (Filhol [1]).

Syn. *Machærodus bidentatus*, Filhol [2].
Machærodus perarmatus, Gervais [3].
Eusmilus perarmatus, Gervais [4].
Drepanodon bidentatus, Filhol [5].

This is the only species, and is somewhat inferior in size to *Machærodus megantherean* (pt. i. p. 42).

Hab. France.

M. 2369 a. Small fragment of the left maxillary region, showing (*Fig.*) pm. 4 and the alveolus of m. 1; from the Phosphorites of Caylux (Tarn-et-Garonne), France. The length of the carnassial (woodcut, fig. 47) is 0,022, which accords with

Fig. 47.

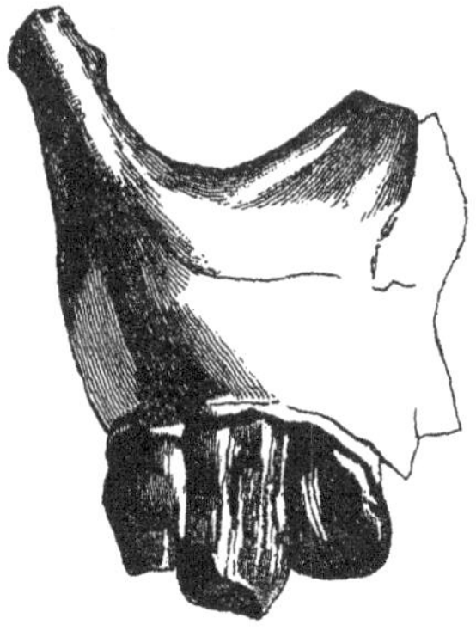

Eusmilus bidentatus.—Outer aspect of the left upper carnassial; from the Phosphorites of Caylux. $\frac{1}{1}$.

that of the corresponding lower tooth figured by Filhol in the 'Ann. Sci. Géol.' vol. vii. pl. xxviii. figs. 140–142.
Purchased, 1885.

[1] Bull. Soc. Sci. Phys. Nat. Toulouse, vol. i. p. 208 (1873).—*Machærodus.*
[2] *Loc. cit.*
[3] Journ. Zool. vol. iv. p. 420 (1875).
[4] Zool. et Pal. Générales, sér. 2, p. 53 (1876).
[5] Ann. Sci. Géol. vol. vii. art. 7, pl. xxviii. (1876)

M. 2369 b. Fragment of a left maxilla with $\underline{\text{m. 1}}$, provisionally referred to this species; from Caylux. *Purchased*, 1885.

Genus **PSEUDÆLURUS** (pt. i. p. 64).

Pseudælurus quadridentatus (Blainville[1])

Syn. *Felis quadridentata*, Blainville[2].
Felis tetraodon, Blainville[3].
Felis hyænoides, Lartet[4].
Megantherеon hyænoides, Pomel[5].

This is the type species, and was nearly as large as *Felis pardus*; three lower premolars are present, $\overline{\text{m. 1}}$ has a very minute talon, and $\overline{\text{m. 2}}$ is wanting in all described specimens.

Hab. Europe (France).

M. 2378. Fragment of the left ramus of the mandible of an immature individual, showing the hinder milk- and permanent cheek-teeth; from the Middle Miocene of Sansan (Gers), France. *Purchased*, 1885.

Pseudælurus edwardsi, Filhol[6].

This species, which is considerably larger than *P. intermedius* (pt. i. p. 64), usually agrees approximately in size with *Felis chaus*, although some specimens are larger; some mandibles[7] show a small $\overline{\text{pm. 1}}$, while in others $\overline{\text{m. 2}}$ is developed; $\overline{\text{m. 1}}$ has a distinct talon.

Hab. Europe (France).

M. 2376. Part of the left ramus of the mandible, showing $\overline{\text{pm. 3}}$, $\overline{\text{pm. 4}}$, and $\overline{\text{m. 1}}$; from the Phosphorites of Caylux (Tarn-et-Garonne), France. This specimen agrees very closely with the one figured by Filhol in the Ann. Sci. Géol. vol. vii. pl. xxvii. fig. 128; $\overline{\text{m. 2}}$ is absent. *Purchased*, 1885.

M. 2377. Fragment of the right ramus of a slightly larger mandible, showing the carnassial; from Caylux. *Purchased*, 1885.

[1] Ostéographie: Genus *Felis*, p. 155 (1843).—*Felis*.
[2] *Loc. cit.* [3] *Ibid.* pl. xvi.
[4] Notice sur la Colline du Sansan, p. 18 (1851).
[5] Catalogue Méthodique, p. 57 (1853).
[6] Comptes Rendus, vol. lxxv. p. 93 (1872).
[7] See Filhol, Ann. Soc. Sci. Phys. Nat. Toulouse, 1882, p. 85.

Genus **PROÆLURUS** (pt. i. p. 65).

Proælurus lemanensis (pt. i. p. 65).

M. 2379. The nearly complete right ramus of the mandible, showing the broken canine, and all the cheek-teeth except $\overline{\text{pm. 1}}$ and $\overline{\text{m. 2}}$; from the Lower Miocene of St. Gérand-le-Puy (Allier), France. *Purchased*, 1885.

Genus **ÆLUROGALE** (pt. i. p. 65).

The American feline termed *Nimravus* (woodcut, fig. 48) by Cope apparently only differs from *Ælurogale*[1] by the absence of $\overline{\text{pm. 2}}$,

Fig. 48.

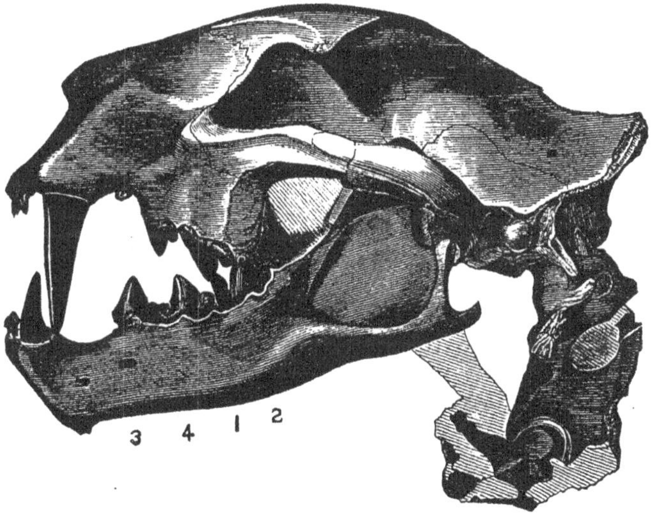

Nimravus gomphodus, Cope.—Skull and vertebræ; from the Miocene of Oregon, U.S.A. $\frac{3}{8}$. (From the Amer. Nat.)

which scarcely seems a generic character, when the variations in the number of these teeth in a single species of many of the primitive Cats is considered.

Ælurogale intermedia, Filhol[2].

This is the type species, and is usually about equal in size to

[1] See 'Palæontologia Indica,' ser. 10, vol. ii. p. 314.
[2] Ann. Sci. Géol. vol. iii. art. 7, p. 14 (1872).

Felis onca, although some specimens are considerably larger; $\overline{\text{pm. 1}}$ and $\overline{\text{m. 2}}$ may be either present or absent [1].

Hab. Europe (France).

M. 2374. The left upper carnassial of a very large individual; from the Phosphorites of Caylux (Tarn-et-Garonne), France. The length of this tooth is 0,032. *Purchased*, 1885.

M. 2369 a. Fragment of the left maxilla of a small individual, containing $\underline{\text{pm. 3}}$, $\underline{\text{pm. 4}}$, and $\underline{\text{m. 1}}$; from Caylux. This specimen agrees very closely with the one figured by Filhol in the 'Ann. Sci. Géol.' vol. viii. pl. iii. figs. 209, 210; the length of $\underline{\text{pm. 4}}$ being 0,0218. *Purchased*, 1885.

M. 2369. Two specimens of the imperfect left maxilla of young individuals, showing $\underline{\text{mm. 3}}$ and the germs of $\underline{\text{pm. 3}}$ and $\underline{\text{pm. 4}}$ *in alveolo*; from Caylux. *Purchased*, 1885.

M. 2370. Fragment of the left ramus of the mandible of an individual agreeing in size with No. M. 2374, and showing $\overline{\text{m. 1}}$ and the alveoli of $\overline{\text{pm. 4}}$ and $\overline{\text{m. 2}}$; from Caylux. This specimen is slightly larger than the one figured by Filhol, *op. cit.* pl. iii. fig. 212; the length of the carnassial being 0,025. *Purchased*, 1885.

M. 2371. The greater part of the left ramus of the mandible of a rather smaller individual, showing the alveoli of the incisors, $\overline{\text{pm. 3}}$ (broken), $\overline{\text{pm. 4}}$, and $\overline{\text{m. 1}}$ (broken); from Caylux. There is no trace of any tooth between the canine and $\overline{\text{pm. 3}}$. *Purchased*, 1885.

M. 2372. The greater part of the left ramus of a mandible of the same size as the last, showing the alveoli of the canine and of three premolars, and the first true molar; from Caylux. The alveolus of $\overline{\text{pm. 2}}$ is single. *Purchased*, 1885.

M. 2373. Part of the right ramus of a smaller mandible, containing the canine, $\overline{\text{pm. 3}}$ (broken), and the alveolus of $\overline{\text{pm. 4}}$, with two small alveoli between the canine and $\overline{\text{pm. 3}}$; from Caylux. This specimen is rather smaller than the one figured by Filhol, *op. cit.* pl. iii. fig. 211, and agrees in relative size with No. M. 2367; the two small alveoli between the canine and $\overline{\text{pm. 3}}$ may have contained the small $\overline{\text{pm. 1}}$ and $\overline{\text{pm. 2}}$; the length of the carnassial is 0,022. *Purchased*, 1885.

[1] Filhol, Ann. Soc. Sci. Phys. Nat. Toulouse, 1882, p. 89.

Family HYÆNIDÆ (pt. i. p. 68).

Genus **HYÆNA** (pt. i. p. 69).

Hyæna crocuta (pt. i. p. 69).

This species is probably descended from the Siwalik *H. colvini* (pt. i. p. 84).

Hab. Africa (recent), Europe and India (Pleistocene).

M. 2960. Cast of the left lower carnassial. The original was obtained from a cavern at Billa-Surgam, Karnul district, Madras, and is preserved in the Indian Museum, Calcutta. It is described and figured by the writer in the 'Palæontologia Indica,' ser. 10, vol. iv. p. 30, pl. vii. figs. 13, 13 *a*.

Presented by the Director of the Geological Survey of India, 1886.

Hyæna felina (pt. i. p. 80).

M. 3563. Cast of the skull figured in pt. i. p. 81.

Presented by the Dublin Museum of Science and Art, 1887.

Hyæna striata (pt. i. p. 87).

Including:—*Hyæna arvernensis*, Croizet and Jobert[1].
Hyæna antiqua (pt. i. p. 87).

M. 3431. Cast of the right upper carnassial. The original (fig. 49) was obtained from the Red Crag at Trimley-St.-Mary,

Fig. 49.

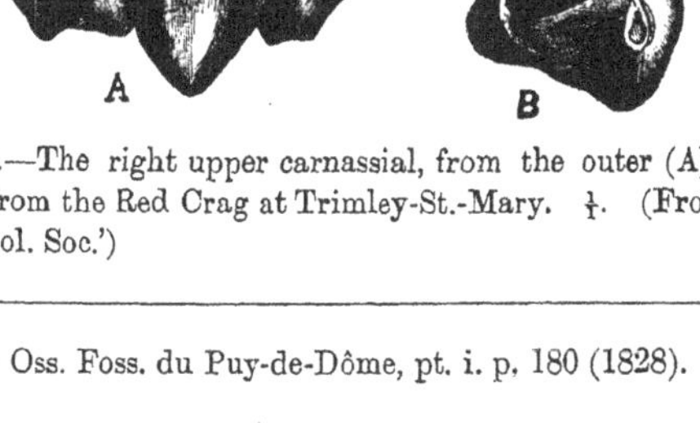

Hyæna striata.—The right upper carnassial, from the outer (A) and oral (B) aspects; from the Red Crag at Trimley-St.-Mary. ½. (From the 'Quart. Journ. Geol. Soc.')

[1] Oss. Foss. du Puy-de-Dôme, pt. i. p. 180 (1828).

Suffolk, and is preserved in the Museum at Ipswich; it is described and figured by the present writer in the 'Quart. Journ. Geol. Soc.' vol. xlii. p. 365, fig. 1. The reasons for the foregoing synonymy are given in that paper.

Made in the Museum, 1886.

Family VIVERRIDÆ (pt. i. p. 93).

Genus **VIVERRA** (pt. i. p. 98).

Viverra karnuliensis, Lydekker [1].

This species is about the same size as *V. zibetha*, from which it is distinguished by the more elongated premolars, in which respects it agrees with the Siwalik *V. bakeri* (pt. i. p. 99) and approaches *Ictitherium*. The species may have been the descendant of *V. bakeri* and the progenitor of *V. zibetha*.

Hab. India.

M. 2961. Cast of a portion of the left ramus of the mandible, showing the carnassial and the alveoli of the last three premolars. The original, which is the type and only specimen, was obtained from a cavern at Billa-Surgam, Karnul district, Madras, and is preserved in the Indian Museum, Calcutta; it is described and figured by the writer in the 'Palæontologia Indica,' ser. 10, vol. iv. p. 31, pl. vii. figs. 6, 6 *a*.

Presented by the Director of the Geological Survey of India, 1886.

Genus **HERPESTES** (pt. i. p. 104).

Herpestes griseus, Desmarest [2].

Hab. India.

M. 2966. Four imperfect specimens of the humerus; from a cave at Billa-Surgam, Karnul district, Madras. Other specimens are described and figured by the writer in the 'Palæontologia Indica,' ser. 10, vol. iv. p. 32, pl. vii. figs. 7–9. *Presented by the Director of the Geological Survey of India*, 1886.

[1] Rec. Geol. Surv. Ind. vol. xix. p. 120 (1886).
[2] Mammalogie, p. 212 (1820).

Herpestes lemanensis, Pomel[1].

Syn. (?) *Herpestes primævus,* Pomel[2].

The specimens referred to this species vary considerably in size, the larger ones indicating an animal fully equal in size to *Viverra zibetha.* The general structure of the mandible and teeth agrees very closely with that of *H. griseus,* but the ramus is somewhat deeper and more convex inferiorly; the cranium is unknown. The species may have been allied to some of the larger African forms.

Hab. Europe.

M. 2380. Hinder portion of the right ramus of the mandible, showing $\overline{m.\ 1}$ and $\overline{m.\ 2}$; from the Lower Miocene (Upper Oligocene) of St. Gerand-le-Puy (Allier), France. Similar specimens are figured by Filhol in the 'Ann. Sci. Géol.' vol. x. pl. xxiii.; $\overline{m.\ 2}$ has one root. *Purchased,* 1885.

M. 2381. The greater part of the right ramus of the mandible of a smaller individual, showing all the cheek-teeth or their alveoli; from St. Gérand-le-Puy. The alveolus of $\overline{m.\ 2}$ is double. *Purchased,* 1885.

Family URSIDÆ (pt. i. p. 106).

Subfamily CANINÆ (pt. i. p. 107).

Genus **CYNODICTIS** (pt. i. p. 107).

Cynodictis leptorhyncha (pt. i. p. 117).

M. 2360. The anterior portion of the cranium together with the mandible in apposition; from the Phosphorites of Caylux. The cranium of this specimen agrees exactly with the one figured by Filhol in the 'Ann. Sci. Géol.' vol. vii. pl. xxiv. fig. 106[3]. *Purchased,* 1885.

Genus **CANIS** (pt. i. p. 123).

Canis dingo, Blumenbach[4].

Hab. Australia.

The following specimens apparently indicate the existence of this species contemporaneously with *Thylacoleo* and other extinct Marsupials.

[1] Catalogue Méthodique, p. 65 (1853).

[2] *Loc. cit.* See Filhol, Ann. Sci. Géol. vol. x. art. 3, p. 171.

[3] For an error in the description of this plate, see pt. i. p. 115, note 2.

[4] Handbuch der Naturgeschichte, 1st French ed. vol. i. p. 103 (1803).

43951 a. An upper canine and four cheek-teeth; from a cave in the Wellington Valley, New South Wales.
Presented by the Trustees of the Australian Museum, 1870.

43951 b. The left astragalus; from the same locality.
Same history.

Genus **AMPHICYON** (pt. i. p. 136).

Amphicyon ambiguus (pt. i. p. 141).

M. 1362. Fragment of the left ramus of the mandible, containing the last three premolars and the alveoli of $\underline{\text{pm. 1}}$ and of the canine; from the Phosphorites of Caylux. The teeth are somewhat smaller than those of the specimen figured by Filhol in the 'Ann. Sci. Géol.' vol. vii. pl. xii. figs. 25, 26, and the anterior basal cusp of $\overline{\text{pm. 4}}$ is less conspicuous.
Purchased, 1884.

Family PROCYONIDÆ.

Genus **NASUA**, Storr[1].

Dentition:—I. $\frac{3}{3}$, C. $\frac{1}{1}$, Pm. $\frac{4}{4}$, M. $\frac{2}{2}$.

Nasua rufa, Desmarest[2].

Syn. *Viverra nasua*, Linn.[3]
Nasua affinis-sociali, Lund[4].

Hab. South America.

18891. The calvarium of an immature individual; from a cave in Minas Geraes, Brazil.
Claussen Collection. Purchased, 1845.

18886, 18931, 18929. Several detached molars, and immature mandibular rami; from the same locality. *Same history.*

[1] Prodromus Meth. Mamm. p. 35 (1780).
[2] Mammalogie, p. 170 (1820).
[3] Syst. Nat. ed. 12, vol. i. p. 64 (1766).
[4] K. Danske Vid. Selskr. Skr. vol. xii. p. 76 (1845). The name *N. socialis* is a synonym of *N. rufa*.

Suborder *CARNIVORA PINNIPEDIA.*

Family PHOCIDÆ (pt. i. p. 205).

The incisive dental formula for the family is I. $\frac{(2-3)}{(1-2)}$; the one given in Part i. being that of the subfamily *Phocinæ*.

Genus **PHOCA** (pt. i. p. 205).

Phoca rugosidens, Owen[1].

The reference of this small undescribed species to the type genus must be regarded as provisional. The teeth are characterized by their rugose enamel.

Hab. Maltese Islands.

37971. (*Fig.*) Three cheek-teeth; from the Miocene of the Island of Gozo, near Malta. These specimens are the types, and are noticed and figured by Leith-Adams in the 'Quart. Journ. Geol. Soc.' vol. xxxv. p. 524, pl. xxv. figs. 2 *a*, 2 *b*, 2 *c*.
Presented by Prof. Leith-Adams, 1863.

M. 3574. A considerable portion of the skeleton of a small Seal, not improbably from the Miocene of the Maltese Islands, and if so perhaps belonging to the present species. The femur agrees very closely with that of *Phoca*. *No history.*

Generically undetermined.

33243. Part of the left ramus of the mandible of a large Seal; from the Miocene of Malta. This specimen is noticed by Leith-Adams in the 'Quart. Journ. Geol. Soc.' vol. xxxv. p. 525, where it is provisionally referred to *Phoca rugosidens*, but it apparently indicates a considerably larger form.
Presented by the Earl of Ducie, 1858.

[1] Quart. Journ. Geol. Soc. vol. xxxv. p. 524 (1879).

Order RODENTIA.

Suborder *RODENTIA SIMPLICIDENTATA.*

Section MYOMORPHA.

Family MURIDÆ (pt. i. p. 225).

Genus **GERBILLUS,** Desmarest [1].

The incisors narrow, and grooved in the upper jaw; molars laminated, the first having three ridges, the second two, and the third one ridge.

Gerbillus indicus (Hardwicke [2]).

Syn. *Dipus indicus,* Hardwicke [3].

Hab. India.

M. 3497. Two specimens of the left ramus of the mandible; from a cave at Billa-Surgam, Karnul district, Madras. Other specimens are described and figured by the writer in the 'Palæontologia Indica.' ser. 10, vol. iv. p. 35, woodcut fig. 1; one of the figures being reproduced in the accompanying woodcut. *Presented by the Director of the Geological Survey of India,* 1886.

Fig. 50.

Gerbillus indicus.—The left ramus of the mandible ($\frac{1}{1}$) and grinding-surfaces of the lower molars ($\frac{5}{2}$); from a cave in the Karnul district. (From the 'Palæontologia Indica.')

[1] Nouv. Dict. d'Hist. Nat. vol. xxiv. p. 22 (1804).
[2] Trans. Linn. Soc. vol. viii. p. 279 (1804–7).—*Dipus.*
[3] *Loc. cit.*

Genus **NESOKIA** (pt. i. p. 226).

Nesokia bandicota (Bechstein[1]).

Syn. *Mus bandicota*, Bechstein[2].
Mus giganteus, Hardwicke[3].
Nesokia gigantea, auct.

Hab. India.

M. 2967. The left ramus of the mandible; from a cave at Billa-Surgam, Karnul district, Madras. Other specimens are described and figured by the writer in the 'Palæontologia Indica,' ser. 10, vol. iv. p. 35, pl. viii. figs. 15, 15 *a*.
Presented by the Director of the Geological Survey of India, 1886.

Nesokia kok, Gray[4].

Syn. *Mus kok*, Gray[5].
Arvicola indica, Gray[6].
Nesokia indica, auct.

Probably a variety of *N. bengalensis*.
Hab. India.

M. 2968. Several mandibular rami; from the caves of Billa-Surgam, Karnul district, Madras. Other specimens are described and figured by the writer in the 'Palæontologia Indica,' ser. 10, vol. iv. p. 35, pl. viii. figs. 13, 14.
Presented by the Director of the Geological Survey of India, 1886.

M. 2970. Specimens of the humerus and femur; from the same locality. *Same history.*

Genus **MUS** (pt. i. p. 226).

Mus orthodon, Hensel[7].

A Rat apparently allied to the group comprising *M. jerdoni* of India and *M. coxinga* of Formosa.
Hab. Sardinia.

[1] Allgem. Uebers. vierfüssige Thiere, vol. ii. p. 713 (1800).
[2] *Loc. cit.*
[3] Trans. Linn. Soc. vol. vii. p. 306 (1804).
[4] Charlesworth's Mag. Nat. Hist. vol. i. p. 585 (1837).—*Mus.*
[5] *Loc. cit.*
[6] Illustrations of Indian Zoology, vol. i. pl. xi. (1830–32).
[7] Not found.

M. 3487. An imperfect cranium and several specimens of the maxilla; from the Pleistocene breccia of Mont San Giovanni, Sardinia. *Purchased*, 1886.

M. 3487 a. Several mandibular rami; from the same locality. *Purchased*, 1886.

M. 3487 b. Specimens of the humerus, radius, ulna, femur, tibia, and astragalus; from the same locality. *Purchased*, 1886.

Genus **EUMYS,** Leidy [1].

This genus, according to Cope [2], is distinguished from *Hesperomys* (pt. i. p. 299) by the upward extension of the orbital fossæ to form an interorbital crest; but some species of the latter show the same feature. Schlosser [3] considers that it is probably identical with *Cricetodon* (pt. i. p. 228), which appears the common ancestral type.

Eumys elegans, Leidy [4].

This is the type and only described species; the series of cheek-teeth is as long as in *Mus decumanus*, but the muzzle is shorter.

Hab. North America.

M. 2563. Two imperfect mandibular rami, showing the molars; from the Miocene of the White River, Colorado, U.S.A. Similar specimens are figured by Cope, *op. cit.* pl. lxv. figs. 9–13. *Presented by R. Lydekker, Esq.*, 1885.

Genus **ARVICOLA** (pt. i. p. 230).

Arvicola henseli, Forsyth-Major [5].

Apparently allied to *A. nivalis* and *A. brecciensis*.

Hab. Sardinia.

M. 3488. Several specimens of the imperfect cranium and mandible; from the Pleistocene breccia of Mont San Giovanni, Sardinia. *Purchased*, 1886.

M. 3488 a. Specimens of the humerus, ulna, femur, and tibia; from the same locality. These specimens were determined by the founder of the species. *Purchased*, 1886.

[1] Proc. Ac. Nat. Sci. Philad. vol. viii. p. 90 (1856, vol. dated 1857).

[2] Vertebrata of the Tertiary Formations of the West (Rep. U.S. Geol. Surv. Terr. vol. iii.), Book i. p. 848 (1884).

[3] Palæontographica, vol. xxxi. p. 137 (1884).

[4] *Loc. cit.*

[5] The specimens were sent to the Museum with this unpublished name attached.

Genus **MYODES,** Pallas [1].

Including *Cuniculus,* Wagler [2].

Molars as in *Arvicola,* but sole of feet hairy.

Myodes torquatus, Pallas [3].

Syn. *Mus torquatus,* Pallas [4].
Arvicola ambiguus (pt. i. p. 231).
Arvicola gulielmi (pt. i. p. 231).
Lemmus torquatus, Sanford [5].
Cuniculus torquatus, auct.

Hab. Europe.

The specimens entered in pt. i. p. 232 under the name of *Arvicola ambiguus* belong to this species; see Alston, 'Proc. Zool. Soc.' 1874, p. 460.

Family GEOMYIDÆ.

Dentition:—I. $\frac{1}{1}$, C. $\frac{0}{0}$, Pm. $\frac{1}{1}$, M. $\frac{3}{3}$. The molars may be either rooted or rootless. All the genera are American.

Genus **ENTOPTYCHUS,** Cope [6].

Crowns of the cheek-teeth prismatic, and in the young presenting a deep enamel-fold, which eventually wears into an islet; no roots to the teeth; the cranium has no vacuities; and the limb-bones resemble those of the existing *Thomomys.* Five species have been recorded.

Entoptychus crassiramis, Cope [7].

The length of the cranium is 0,046. The superciliary ridges are thickened, and the premolar is widened at the base.

Hab. North America.

M. 2567. Middle portion of the cranium, showing the three true molars of the left side; from the Miocene of the White River, Colorado, U.S.A. A similar specimen is figured by Cope in the 'Vertebrata of the Tertiary Formations of the West'

[1] Zoogr. Rosso-Asiatica, vol. i. p. 173 (1811).
[2] Isis, 1832, p. 1220.
[3] Nov. Spec. Glirium, p. 77 (1778).—*Mus.* [4] *Loc. cit.*
[5] Quart. Journ. Geol. Soc. vol. xxvi. p. 125 (1870).
[6] Proc. Amer. Phil. Soc. vol. xviii. p. 64 (1879, vol. dated 1880).
[7] *Ibid.* p. 65.

(Rep. U.S. Geol. Surv. Terr. vol. iii.), Book i. pl. lxiv. fig. 5 (1884). *Presented by R. Lydekker, Esq.*, 1885.

M. 2567 a. Fragment of the mandible with the cheek-teeth; from the White River. *Presented by R. Lydekker, Esq.*, 1885.

Section HYSTRICOMORPHA.

Family HYSTRICIDÆ (pt. i. p. 246).

Genus **HYSTRIX** (pt. i. p. 247).

Hystrix crassidens, Lydekker [1].

Considerably larger than *H. hirsutirostris*, from which it is distinguished by the upper incisors being wider than the lower.

Hab. India.

M. 3448. Several cheek-teeth and imperfect mandibular rami; from the caverns of Billa-Surgam, Karnul district, Madras. Other specimens are described and figured by the writer in the 'Palæontologia Indica,' ser. 10, vol. iv. p. 37, pl. viii. *Presented by the Director of the Geological Survey of India*, 1886.

M. 3448 a. Several imperfect specimens of the humerus and ulna; from the same locality. *Same history.*

Genus **ATHERURA,** Cuvier [2].

Cheek-teeth as in *Hystrix*, but spines flattened, and tail long and scaly.

Atherura karnuliensis, Lydekker [3].

Somewhat larger than *A. fasciculata*, and distinguished by the greater bevelling of the lateral borders of the anterior surface of the upper incisors. The skull is unknown.

Hab. India.

M. 3444. A lower incisor; from a cavern at Billa-Surgam, Karnul district, Madras. Other specimens are described and figured by the writer in the 'Palæontologia Indica,' ser. 10, vol. iv. p. 38, pl. viii. *Presented by the Director of the Geological Survey of India*, 1886.

[1] Palæontologia Indica (Mem. Geol. Surv. Ind.), ser. 10, vol. iv. p. 37 (1886).
[2] Règne Animal, 2nd ed. vol. i. p. 215 (1829).—*Atherure*
[3] Palæontologia Indica (Mem. Geol. Surv. Ind.), ser. 10, vol. iv. p. 38 (1886).

Suborder *RODENTIA DUPLICIDENTATA*.

Family LAGOMYIDÆ (pt. i. p. 255).

Genus **LAGOMYS** (pt. i. p. 255).

Lagomys sardus (pt. i. p. 256).

M. 3459, 3481. Several crania, mostly imperfect; from the Pleistocene breccia of Mont San Giovanni, Sardinia. *Purchased*, 1886.

M. 3460–1. Several imperfect maxillæ; from the same locality. *Purchased*, 1886.

M. 3462, 3468, 3482. Numerous mandibular rami; from the same locality. *Purchased*, 1886.

M. 3466–7. Numerous incisors; from the same locality. *Purchased*, 1886.

M. 3471–86. A large series of limb-bones and vertebræ; from the same locality. *Purchased*, 1886.

Family LEPORIDÆ (pt. i. p. 259).

Genus **LEPUS** (pt. i. p. 258).

Lepus nigricollis, F. Cuvier[1].

Hab. India.

M. 2969. Fragment of a mandibular ramus, and several vertebræ and imperfect limb-bones, provisionally referred to this species; from the caves at Billa-Surgam, Karnul district, Madras. Other specimens are described and figured by the writer in the 'Palæontologia Indica,' ser. 10, vol. iv. p. 38, pl. viii. *Presented by the Director of the Geological Survey of India*, 1886.

Genus **PALÆOLAGUS,** Leidy[2].

This genus is distinguished from *Lepus* by the simpler $\overline{\text{pm. 3}}$ and the absence of a postfrontal process. (*Vide* Cope, 'Vertebrata of the Tertiary Formations of the West,' [Rep. U.S. Geol. Surv. Terr. vol. iii.], Book i. p. 870 [1884].)

[1] Dict. Sci. Nat. vol. xxvi. p. 307 (1823).

[2] Proc. Ac. Nat. Sci. Philad. vol. viii. p. 89 (1856, vol. dated 1857).

Palæolagus haydeni, Leidy [1].

This is the type species; m. 3 is cylindrical, not grooved, and smaller than pm. 2. This species is regarded as the progenitor of the existing *Lepus sylvaticus*.

Hab. N. America.

M. 2565. Fragment of the left maxilla, showing the six cheek-teeth; from the Miocene of the White River, Colorado, U.S.A.
Presented by R. Lydekker, Esq., 1885.

M. 2565 a. Part of the right ramus of the mandible, with the five cheek-teeth; from the White River.
Presented by R. Lydekker, Esq., 1885.

Palæolagus turgidus, Cope [2].

This species is considerably larger than the preceding; and differs in the structure of the upper molars.

Hab. N. America.

M. 2564. Fragment of a mandibular ramus, with three teeth; from the Miocene of the White River, Colorado, U.S.A.
Presented by R. Lydekker, Esq., 1885.

Order UNGULATA.

Suborder *ARTIODACTYLA.*

Family BOVIDÆ (pt. ii. p. 1).

Genus **BOS** (pt. ii. p. 1).

Bos taurus, var. **primigenius** (pt. ii. p. 2).

M. 3686. Cast of the calvarium and horn-cores. The original from the Pleistocene of Italy.
Presented by J. B. Pentland, Esq., 1852.

Genus **COBUS** (pt. ii. p. 53).

Cobus (?) palæindicus (pt. ii. p. 53).

M. 487. The imperfect cranium; from the Siwalik Hills. Noticed by the writer in the 'Palæontologia Indica,' ser. 10, vol. iv. p. 13. *Cautley Collection. Presented*, 1842.

[1] Proc. Ac. Nat. Sci. Philad. vol. viii. p. 89 (1856, vol. dated 1857).

[2] Ann. Rep. U.S. Geol. Surv. Terr. (Hayden) for 1873, p. 479 (1874).

39569 (pt. ii. p. 58). This female cranium may probably be referred either to this species or to *C.* (?) *patulicornis*; it is noticed by the writer, *op. cit.* p. 13. This reference confirms the generic determination of the present species, since the females of the existing species of *Cobus* are hornless.

Cobus (?), sp.

Hab. India.

M. 3683. The calvarium of a small female, either belonging to this or a closely allied genus; from the Siwaliks of Perim Island, Gulf of Cambay, India. The specimen approximates in structure to that of the larger so-called *Eleotragus arundinaceus*, but has the supraorbital foramina much larger. *Presented by Col. J. W. Watson*, 1887.

Genus **TETRACEROS,** Leach [1].

The upper true molars are relatively broad, and have no inner accessory column; the premolars are relatively long, and the inner crescents incomplete; in the lower molars there may be a minute

Fig. 51.

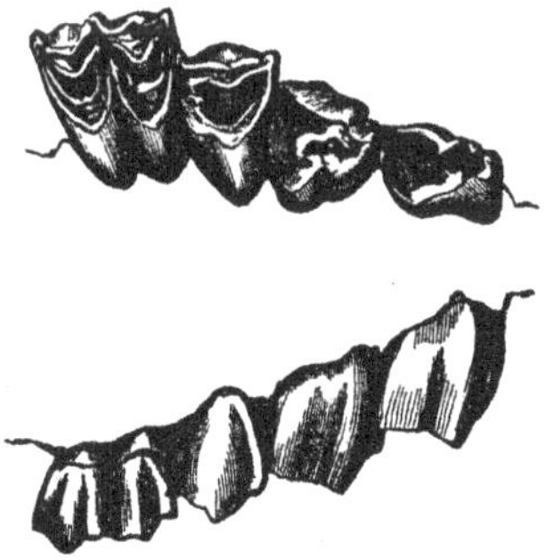

Tetraceros quadricornis (Blainv.).—The first four right upper cheek-teeth of a female, from the grinding and outer aspects. Recent, Madras. ⅔. British Museum (No. 884*a*). (From the 'Palæontologia Indica.')

accessory column. The lachrymal depression is well marked, but there is no vacuity. Horns are absent in the female. Part of the upper dentition of the existing species is represented in woodcut fig. 51.

[1] Trans. Linn. Soc. vol. xiv. p. 524 (1823).

Tetraceros daviesi, Lydekker[1].

This species is somewhat smaller than *T. quadricornis*, and distinguished by the narrower anterior upper premolars.

Hab. India.

M. 3492. (*Fig.*) The middle portion of the cranium; from the Pliocene of the Siwalik Hills. This specimen (fig. 52) is the type, and is described and figured by the writer in the 'Palæontologia Indica,' ser. 10, vol. iv. p. 19, woodcut fig. 3.

Cautley Collection. Presented, 1842.

Fig. 52.

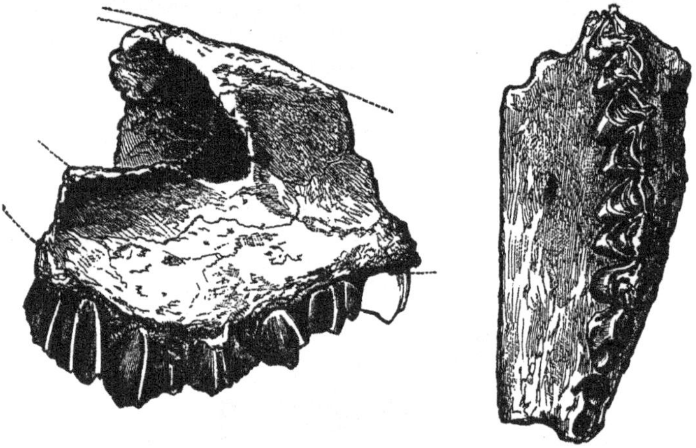

Tetraceros daviesi.—The right half of the middle portion of the cranium and of the palate; from the Siwalik Hills. $\frac{1}{1}$. (From the 'Palæontologia Indica.')

16535. (*Fig.*) Fragment of a right mandibular ramus (fig. 53), provisionally referred to this species; from the Siwalik Hills. Described and figured by the writer, *op. cit.* p. 20, woodcut fig. 4.

Cautley Collection. Presented, 1842.

Fig. 53.

(?) *Tetraceros daviesi.*—Fragment of the right ramus of the mandible; from the Siwalik Hills. $\frac{1}{1}$. (From the 'Palæontologia Indica.')

---

[1] Palæontologia Indica (Mem. Geol. Surv. Ind.), ser. 10, vol. iv. p. 19 (1886.)

GENUS *non det.*

The following specimen indicates a small Antelope, perhaps belonging to *Cephalophus*.

Hab. India.

M. 3493. Fragment of the right maxilla, showing the three true molars; from the Pliocene of the Siwalik Hills. Noticed by the writer in the 'Palæontologia Indica,' ser. 10, vol. iv. p. 21. *Cautley Collection. Presented* 1842.

Family GIRAFFIDÆ (pt. ii. p. 58).

Genus **HYDASPITHERIUM,** Lydekker[1].

The antlers arise from a common base directly above the occiput; and there is a large lachrymal vacuity in the facial region.

Hydaspitherium megacephalum, Lydekker[2].

This is the type species.

Hab. India (Punjab).

M. 3723. Cast of the imperfect cranium. The original, which is the type, was obtained from the Pliocene Siwaliks of Asnot, Punjab, and is preserved in the Indian Museum, Calcutta (No. B. 150); it is described and figured by the present writer in the 'Palæontologia Indica,' ser. 10, vol. i. p. 159, pls. xxvi., xxvii.; a supplemental description being also given in vol. ii. p. 118 of the same work. *Made in the Museum*, 1886.

Family CERVIDÆ (pt. ii. p. 73).

Genus **CARIACUS** (pt. ii. p. 74).

A. *Cariacine group.*

Cariacus paludosus (Desmarest[3]).

Syn. *Cervus paludosus*, Desmarest[4].
Cervus affinis-paludoso, Lund[5].
Cervus magnus, Bravard[6].
Blastoceros paludosus, Gray[7].

Hab. S. America (Brazil and Paraguay).

[1] Rec. Geol. Surv. Ind. vol. ix. p. 154 (1876).—*Hydaspidotherium.*
[2] *Loc. cit.*
[3] Mammalogie, p. 443 (1822).—*Cervus.* [4] *Loc. cit.*
[5] K. Danske Vid. Selsk. Skr. vol. ix. p. 198 (1842).
[6] In Gervais' Zool. et Pal. Générales, sér. 1, p. 133 (1867–69).
[7] Cat. Ruminant Mammalia Brit. Mus. p. 87 (1872).

The following specimens probably belong to this species.

37682. Fragment of the calvarium, showing the bases of the pedicles of the antlers; from the Pleistocene of the Argentine Republic. *Bravard Collection. Purchased*, 1854.

37678. Fragment of the left maxilla of a young individual, containing $\underline{\text{mm. }3}$, $\underline{\text{mm. }4}$, and $\underline{\text{m. }1}$; from the same region. *Same history.*

37679. Fragment of the left ramus of the mandible, belonging to the same individual as the last specimen. *Same history.*

B. *Coassine group.*

The following specimens from the caves of Minas Geraes, Brazil, belong to the group[1] with simple antlers, comprising *C. rufus*, *C. simplicicornis*, *C. rufinus*, and *C. nemorivagus*; it is improbable that they belong to the third species (confined to Ecuador and Guatemala); but it is most likely that they represented at least two of the other three, although it would be almost impossible to distinguish between the first and second by the skull alone. The fossils may be divided into three groups on the ground of size. *All the specimens belong to the Claussen Collection. Purchased*, 1845.

Species a.

The specimens under this head come nearest in point of size to *C. rufus* (F. Cuv.[2]) and *C. simplicicornis* (H. Smith[3]), in which the lengths of the upper series of cheek-teeth[4] in the females are respectively 0,068 and 0,058.

18694. The imperfect calvarium of a male, showing the pedicles the horn-cores.

18959. Part of the frontal region of a male.

18699. The imperfect calvarium of a female[5].

18694 a. The right maxillary region, containing the six cheek-teeth. The length of the dental series is 0,066.

[1] Brooke, Proc. Zool. Soc. 1878, p. 925. [2] Part ii. p. 74.

[3] In Griffith's 'Animal Kingdom,' vol. v. p. 318 (1827).—*Cervus.*

[4] Brooke, *op. cit.*

[5] The calvarium of a female (No. 18701) entered in pt. ii. p. 74 as *C. rufus* is intermediate in size between this specimen and No. 18699 *a* (species *b*).

18695. The left maxillary region of an immature individual, showing the three milk-molars and the first true molar.

18706. Fragment of the left maxilla, containing the three premolars.

18696. The left ramus of the mandible of an immature individual.

18705 a. Part of the left ramus of a similar mandible.

18705 b. Part of the right ramus of a similar mandible.

Species b.

The following smaller specimens agree approximately in size with *C. nemorivagus* (F. Cuv. [1]), in which the length of the upper cheek-series of the female [2] is 0,050.

18705. The occipital segment of the cranium.

18699 a. The calvarium of a female.

18706 d. Two fragments of left maxillæ, one showing the last three and the other the last four cheek-teeth.

18706 a. Fragment of the left maxilla, containing the three true molars, of which the length is 0,0315.

18695 a. Part of the left maxillary region of a young individual, showing the three milk-molars and the first true molar.

18699 b. A similar specimen of the right maxilla.

18705 c. Part of the right ramus of the mandible, containing the three true molars.

18699 c. Part of the left ramus of the mandible of an immature individual, showing the milk-teeth.

18696 a. Part of the right ramus of a similar mandible.

Species c.

The following specimen indicates a still smaller form, which does not appear larger than *Pudua*.

18706 b. Fragment of the right maxilla, containing the three true molars, of which the united length is 0,029.

[1] Dict. Sci. Nat. vol. vii. p. 485 (1817). [2] Brooke, *op. cit.*

Genus **CERVUS** (pt. ii. p. 82).

Cervus aristotelis (pt. ii. p. 113).

M. 3447. Four upper true molars; from the caves of Billa-Surgam, Karnul district, Madras. Similar specimens are figured by the writer in the 'Palæontologia Indica,' ser. 10, vol. iv. pl. xi. figs. 5, 6. *Presented by the Director of the Geological Survey of India*, 1886.

Family POËBROTHERIIDÆ (pt. ii. p. 149).

Poëbrotherium wilsoni (pt. ii. p. 149).

M. 2561 a. Portions of the maxilla and mandible; from the Miocene of the White River, Colorado, U.S.A.
Presented by R. Lydekker, Esq., 1885.

M. 2561 b. An astragalus; from the same formation. *Same history.*

Family TRAGULIDÆ (pt. ii. p. 150).

Genus **LEPTOMERYX,** Leidy [1].

The cheek-dentition is almost or quite indistinguishable from that of *Prodremotherium* (pt. ii. p. 150), on which grounds Cope [2] has proposed to unite the two genera. *Leptomeryx* is, however, distinguised by having four distinct metacarpals, although the two metatarsals are fused into a "cannon-bone," and the navicular and cuboid are united [3].

Schlosser [4], who places this genus in the *Tragulidæ*, refers [5] *Prodremotherium* and *Bachitherium* (pt. ii. p. 155) to the *Cervidæ* [6]; but the resemblance between the dentition of *Leptomeryx* and *Prodremotherium* appears too close to admit of their being classed in distinct families; and the reference of the former to the present family therefore supports the view of Rütimeyer [7], who includes the

[1] Proc. Ac. Nat. Sci. Philad. vol. vi. p. 394 (1853, vol. dated 1854).

[2] Tertiary Vertebrata of the West (Rep. U.S. Geol. Surv. Terr. vol. iii.), Book i. p. 37 (1884).

[3] Schlosser, *inf. cit.*

[4] Morphol. Jahrb. vol. xii. p. 75 (1886).

[5] *Ibid.* p. 65.

[6] The *Cervidæ*, together with the other Pecora, are regarded by this writer as subfamilies of the *Gelocidæ*.

[7] Verh. nat. Ges. Basel, vol. vii. art. 2, p. 43 (1883), and 'Natürliche Geschichte der Hirschfamilie' (Abh. Schw. pal. Ges.), pt. 2, p. 68: this view is adopted in part ii. of the present work.

latter in the same family. The connection shown between the *Dichodontidæ* (*Gelocus*) and the *Cervidæ* (*Palæomeryx*) by means of the three above-mentioned genera is so close as to leave little doubt as to the descent of the one family, through the *Tragulidæ*, from the other; the existing genera of *Tragulidæ* being lateral offshoots from the same early stock.

Leptomeryx evansi, Leidy [1].

Syn. *Trimerodus cedrensis*, Cope [2].

This is the type and only species, and is about the size of *Tragulus meminna*.

Hab. North America.

M. 2562. A fragment of the left maxilla, showing the last two premolars and the first two true molars; from the Miocene of the White River, Colorado, U.S.A. Less imperfect specimens are figured by Leidy in the 'Ancient Fauna of Nebraska,' pl. xiv.

Presented by R. Lydekker, Esq., 1885.

M. 2562 a. Four fragments of mandibular rami with teeth; from the same deposits. *Same history.*

Genus **DORCATHERIUM** (pt. ii. p. 153).

Dorcatherium naui (pt. ii. p. 153).

M. 3714. Cast of the right ramus of the mandible. Original (type) from Eppelsheim; figured by Kaup in 'Oss. Foss. d. Darmstadt,' pt. 5, pl. xxiii. fig. 1.

Egerton Collection. Purchased, 1882.

Family DICHODONTIDÆ (pt. ii. p. 159).

Genus **DICHODON** (pt. ii. p. 164).

Dichodon cuspidatus (pt. ii. p. 164).

M. 3679. Part of the right ramus of the mandible with the three

[1] Proc. Ac. Nat. Sci. Philad. vol. vi. p. 394 (1853, vol. dated 1854).

[2] Palæontological Bulletin, No. 16, p. 8; see Cope, Rep. U.S. Geol. Surv. Terr. p. 503 (1874), 8vo.

true molars; from Hordwell. Described by Wright in the 'Ann. Mag. Nat. Hist.' ser. 2, vol. x. p. 87 (1852).

Wright Collection. Purchased, 1887.

Family ANOPLOTHERIIDÆ (pt. ii. p. 182).

Genus **XIPHODON** (pt. ii. p. 182).

Xiphodon cayluxensis, Lydekker (n. sp.).

This specific name is applied to the small *Xiphodon* of which the immature mandibular ramus is figured in pt. ii. p. 186, under the provisional name of *X. gelyensis*. That species according to Schlosser[1] is not a *Xiphodon* at all, but is allied to *Gelocus*, and is made the type of the new genus *Phaneromeryx*. A very large species of *Xiphodon* from the Quercy Phosphorites has been described by Filhol[2] under the name of *X. magnus*.

Family OREODONTIDÆ (pt. ii. p. 207).

Genus **EPOREODON** (pt. ii. p. 209).

Eporeodon jacksoni (Leidy[3]).

Syn. *Eucrotaphus jacksoni*, Leidy[4].
Oreodon bullatus, Leidy[5].
Oreodon occidentalis, Marsh[6].
Eporeodon occidentalis, Marsh[7].
Eucrotaphus occidentalis, Cope[8].

The synonymy is given on the authority of Cope[9], who includes under this specific name three forms to which he has applied distinct varietal names, with the suggestion that they may be of specific value; the length of the upper series of cheek-teeth varies from 0,099 to 0,086; the largest form being superior in size to *E. major* (pt. ii. p. 209), while the smallest agrees in this respect with *Oreodon culbertsoni* (pt. ii. p. 207). The species is distinguished from *E. trigonocephalus* (Cope) by the palato-narial border being in

[1] Morphol. Jahrb. vol. xii. p. 95 (1886); see also p. 296 of the same volume.
[2] Ann. Soc. Sci. Phys. Nat. Toulouse, 1884, p. 40.
[3] Proc. Ac. Nat. Sci. Philad. vol. v. p. 92 (1850, vol. dated 1852).—*Eucrotaphus*.
[4] *Loc. cit.*
[5] Quoted in Rep. U.S. Geol. Surv. Terr. vol. i. p. 318 (1873).
[6] Amer. Journ. ser. 3, vol. v. p. 409 (1873).
[7] *Ibid.* vol. ix. p. 250 (1875).
[8] Bull. U.S. Geol. Surv. Terr. vol. v. p. 59 (1879).
[9] Proc. Amer. Phil. Soc. vol. xxi. p. 517 (1884).

line with the posterior edges of the maxillæ, and by the infraorbital foramen being situated above the posterior part of pm. 3; and from *E. major* by the paroccipital process being placed behind the auditory bulla. The following specimen belongs to the variety named *pacificus* by Cope [1], which is intermediate between the other two.

Hab. North America.

M. 2558. The cranium, with the dentition in an early stage of wear; from the Miocene of the John Day River, Oregon, U.S.A. The length of the series of cheek-teeth is 0,091.

Presented by R. Lydekker, Esq., 1885.

Family SUIDÆ (pt. ii. p. 250).

Genus **HYOTHERIUM** (pt. ii. p. 254).

The writer has come to the conclusion that this genus may probably be regarded as the ancestor both of *Dicotyles* and *Sus* [2].

Hyotherium perimense, Lydekker [3].

This species may be defined as similar in size to *H. waterhousei* (*suprà*, pt. ii. p. 256), but distinguished by its wider molars, in which respect it approaches *H. sœmmerringi* (*suprà*, pt. ii. p. 257) and the Indian *H. sindiense*, although differing from the former by the greater

Fig. 54.

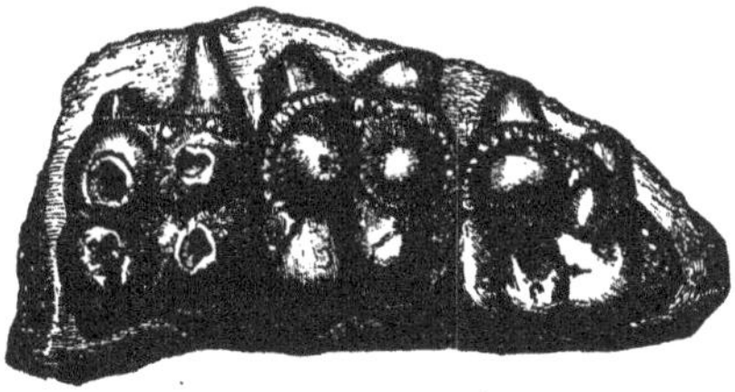

Hyotherium perimense.—Part of the left maxilla; from the Pliocene of Perim Island, India. $\frac{1}{1}$.

width of the lower molars, and from the latter either by its inferior size, or by its wider lower molars, or by both these two characters.

Hab. India (Perim Island).

[1] *Ibid.* p. 519.

[2] Quart. Journ. Geol. Soc. vol. xliii. p. 22 (1887).

[3] *Ibid.* p. 19.

M. 3501. (*Fig.*) Fragment of the left maxilla, showing the three true molars in an early stage of wear; from the Siwaliks of Perim Island, Gulf of Cambay, India. This specimen (fig. 54) is the type, and is described and figured by the writer in the 'Quart. Journ. Geol. Soc.' vol. xliii. p. 19, woodcut. *Présented by Col. J. W. Watson*, 1886.

Genus **SUS** (pt. ii. p. 259).

Sus cristatus, Wagner [1].

This species, in which the talons of the last true molars (woodcut, fig. 55) are usually more complex than in *S. scrofa*, is probably descended from the larger Siwalik *S. falconeri* (pt. ii. p. 263) through a form allied to the following species. The specimens from the

Fig. 55.

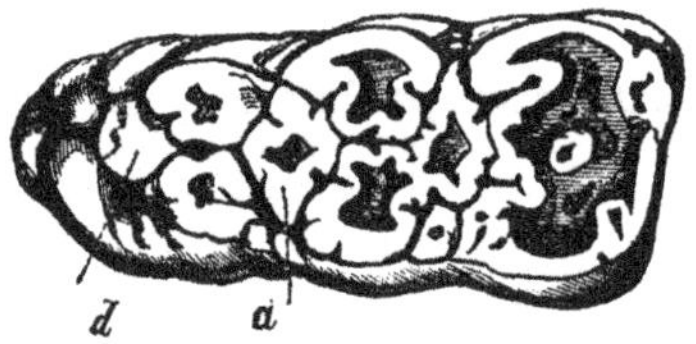

Sus cristatus.—The third right lower true molar of a male. Recent. $\frac{1}{1}$. (From the 'Palæontologia Indica.') *a, d*, the middle columns of the talon.

Narbada Valley (Nos. 36843, 36725), mentioned in pt. ii. p. 266, have been referred by the writer, in the 'Palæontologia Indica,' ser. 10, vol. iv. p. 48, to the present species.

Hab. India.

M. 3445. Ten specimens of premolars and molars and one lower incisor; from the caves of Billa-Surgam, Karnul district, Madras. Other specimens are described and figured by the writer in the 'Palæontologia Indica,' ser. 10, vol. iv. p. 47, pl. ix.; in the last lower molar of the fossil race the complexity of the talon exceeds the average of recent examples (woodcut, fig. 55). *Presented by the Director of the Geological Survey of India*, 1886.

[1] Münch. gelehrt. Anzeiger, vol. ix. p. 435, *errorim* 535 (1839).

Sus karnuliensis, Lydekker[1].

This provisional species agrees in size with the Siwalik *S. falconeri* (pt. ii. p. 263), but in the general structure of the molars with *S. cristatus*, although some specimens of these teeth approach the more complicated structure obtaining in the molars of the former species. It may be the descendant of the hypothetical form connecting the Siwalik species with *S. cristatus*.

Hab. India.

M. 3503. Cast of the almost unworn third left upper true molar. The original, which is one of the types, was obtained from a cave at Billa-Surgam, Karnul district, Madras, and is preserved in the Indian Museum, Calcutta; it is described and figured by the writer in the 'Palæontologia Indica,' ser. 10, vol. iv. p. 49, pl. ix. fig. 9.

Made in the Museum, 1886.

M. 3502. Cast of the much-worn third left lower true molar. The history of the original (figured *op. cit.* pl. ix. fig. 3) is the same as that of the preceding specimen.

Same history.

Sus, sp. *c* (pt. ii. p. 268).

M. 3430. Cast of an unworn but rolled third left upper true molar. The original was obtained from the Red Crag of Suffolk, and is preserved in the Museum at Ipswich; it is noticed by the present writer in the 'Quart. Journ. Geol. Soc.' vol. xlii. p. 366, and considered to belong probably either to *S. erymanthius* or *S. antiquus*.

Made in the Museum, 1886.

Sus hysudricus (pt. ii. p. 271).

M. 3438. The hinder part of the palate, showing the last four teeth of either side; from the Siwaliks of Perim Island.

Presented by Col. J. W. Watson, 1886.

M. 3681. The middle portion of the cranium of a nearly adult individual, showing $\underline{\text{pm. 3}}$, $\underline{\text{pm. 4}}$, $\underline{\text{m. 1}}$, and $\underline{\text{m. 2}}$ in use and $\underline{\text{m. 3}}$ *in alveolo*; from Perim Island.

Presented by Col. J. W. Watson, 1887.

[1] Palæontologia Indica (Mem. Geol. Surv. Ind.), ser. 10, vol. iv. p. 49 (1886).

Sus palæochœrus (pt. ii. p. 273).

M. 3711. Cast of part of right ramus of the mandible, with last five cheek-teeth. Original (type) from Eppelsheim; figured by Kaup in 'Oss. Foss. d. Darmstadt,' pt. 2, pl. ix. fig. 1.
Egerton Collection. Purchased, 1882.

Suborder *PERISSODACTYLA.*

Family EQUIDÆ (pt. iii. p. 49).

Genus **EQUUS** (pt. iii. p. 65).

Equus, sp.

The following specimen indicates a Horse closely allied to, if not identical with, *Equus sivalensis* (pt. iii. p. 66). It may be observed that in the upper true molars of that species the elongation of the grinding surface of the antero-internal pillar is somewhat greater than in *E. stenonis* (pt. iii. p. 69).

Hab. Nubia.

M. 3595. (*Fig.*) A right upper cheek-tooth (? $\underline{\text{m. 1}}$), in a comparatively early stage of wear; from the Upper Pliocene (?) of Wady Halfa, near the second cataract of the Nile, Nubia. This specimen is described and figured by the writer in the 'Quart. Journ. Geol. Soc.' vol. xliii. p. 161 (woodcut).
Presented by Brigade-Surgeon Archer, A.M.D., 1886.

Family RHINOCEROTIDÆ (pt. iii. p. 90).

Genus **RHINOCEROS** (pt. iii. p. 91).

Rhinoceros perimensis (pt. iii. p. 155).

M. 3682. Fragment of the right ramus of the mandible, containing an imperfect (? third) true molar; from the Siwaliks of Perim Island. The tooth is remarkable for the extreme flatness of its outer surface, and has a length of 0,079.
Presented by Col. J. W. Watson, 1887.

Suborder *PROBOSCIDEA*.

Family ELEPHANTIDÆ (pt. iv. p. 13).

Genus **MASTODON** (pt. iv. p. 14).

Mastodon pandionis (pt. iv. p. 37).

M. 3680. The imperfect crown of the third right upper true molar; from the Siwaliks of Perim Island. This specimen agrees very closely with No. M. 2490.
Presented by Col. J. W. Watson, 1887.

Genus **ELEPHAS** (pt. iv. p. 78).

Elephas antiquus (pt. iv. p. 122).

M. 3598. The four ultimate molars of a single individual about one-third worn; from the Pleistocene of Barnstaple, Devonshire. *Wright Collection. Purchased*, 1887.

M. 3599. The two fourth lower milk-molars of one individual; from the same locality. *Same history.*

Order EDENTATA.

Family MEGATHERIIDÆ (pt. v. p. 85).

Genus **SCELIDOTHERIUM** (pt. v. p. 91).

The following specimens not improbably belong to *S. leptocephalum* (pt. v. p. 92).

37583. The pelvis and sacrum; from Buenos Ayres.
Bravard Collection.

37561–3. Three trunk vertebræ; from Buenos Ayres.
Same history.

Family GLYPTODONTIDÆ (pt. v. p. 114).

Genus **HOPLOPHORUS** (pt. v. p. 126).

Hoplophorus (?), sp.

Hab. South America.

M. 3684. Cast of the terminal tube of the caudal sheath. The original was obtained from the Pleistocene of Rio Carcarana, near Villa Jesu Maria, Santa Fe; it differs from the specimens figured on pp. 129, 132.
Made by permission of Mr. E. Hunter, the owner of the original, 1887.

ALPHABETICAL INDEX

OF

GENERA AND SPECIES, INCLUDING SYNONYMS.

PRINTED BY TAYLOR AND FRANCIS,
RED LION COURT, FLEET STREET.

Made in the USA
Monee, IL
04 November 2026

67391561R00213